T0137427

Advanced Structured Materials

Volume 111

Common engineering materials reach in many applications their limits and new developments are required to fulfil increasing demands on engineering materials. The performance of materials can be increased by combining different materials to achieve better properties than a single constituent or by shaping the material or constituents in a specific structure. The interaction between material and structure may arise on different length scales, such as micro-, meso- or macroscale, and offers possible applications in quite diverse fields.

This book series addresses the fundamental relationship between materials and their structure on the overall properties (e.g. mechanical, thermal, chemical or magnetic etc) and applications.

The topics of *Advanced Structured Materials* include but are not limited to

- classical fibre-reinforced composites (e.g. glass, carbon or Aramid reinforced plastics)
- metal matrix composites (MMCs)
- micro porous composites
- micro channel materials
- multilayered materials
- cellular materials (e.g., metallic or polymer foams, sponges, hollow sphere structures)
- porous materials
- truss structures
- nanocomposite materials
- biomaterials
- nanoporous metals
- concrete
- coated materials
- smart materials

Advanced Structured Materials is indexed in Google Scholar and Scopus.

More information about this series at http://www.springer.com/series/8611

Engin Burgaz

Polyurethane Insulation Foams for Energy and Sustainability

Engin Burgaz
Department of Metallurgical and Materials
Engineering, Faculty of Engineering
Ondokuz Mayis University
Atakum, Samsun, Turkey

ISSN 1869-8433 ISSN 1869-8441 (electronic)
Advanced Structured Materials
ISBN 978-3-030-19560-1 ISBN 978-3-030-19558-8 (eBook)
https://doi.org/10.1007/978-3-030-19558-8

This Springer imprint is published by the registered company Springer Nature Switzerland AG
The registered company address is: Gewerbestrasse 11, 6330 Cham, Switzerland

Preface

For more than three decades, polyurethane rigid foams consisting of micron- and nano-sized additives have been fabricated and characterized for the development of new and highly advanced insulation materials that can function in wide range of environmental, chemical, mechanical, and thermomechanical conditions while taking aim at generating high energy efficiency and sustainability. In the current book, I have tried to present a recent survey and review of previously published significant works in this important application area in which the priority should be given to energy saving and sustainability for fulfilling the needs of modern world. Particularly, this book focuses on structure–property relationships of polyurethane rigid nanocomposite foams consisting of plate-like nanofillers, cylindrical nanofillers, and spherical nanofillers in comparison with those of conventional polyurethane rigid composite foams containing micron-sized fillers. Thermal insulation property of polyurethane rigid composite foams is discussed along with their significant characteristics such as closed-cellular morphology, thermal, mechanical, and thermomechanical properties, thermal degradation and flammability, energy absorption and saving capability, recycling and recovery behavior, modeling and simulation results. Potential applications of polyurethane rigid composite foams are discussed, and the main problems that are still not resolved and the future work related to this important topic are addressed. This book is systematically arranged in accordance with following main topics. In Chap. 1, a brief introduction to the fascinating area of polyurethane foams is given by focusing on subtopics such as types of polyurethane foams, types of polyurethane rigid composite foams, polyurethane rigid composite foams, including micron-sized fillers, polyurethane rigid composite foams, including nano-sized fillers and experimental conditions for polyurethane rigid foam fabrication at the introductory level. In Chap. 2, polyurethane rigid composite foams containing micron-sized fillers are presented in detail by highlighting subtopics such as morphology, mechanical, thermal and thermomechanical properties, thermal degradation and flammability, and recycling and recovery behavior. In Chaps. 3–5, closed-cellular morphology, mechanical, thermal, and thermomechanical properties, thermal degradation, and flammability of polyurethane rigid nanocomposite foams containing plate-like, cylindrical, and spherical nanofillers are discussed, respectively. It is believed that the information that is covered in this book can be useful to colleagues and researchers in both academic and industrial laboratories and students at both undergraduate and graduate levels. Moreover, it is expected that systematically organized and reviewed data that is given about polyurethane insulation foams in this book can help and guide researchers in their distinct research projects for the development of advanced polyurethane rigid foams with improved properties by particularly focusing on energy saving and sustainability issues.

Samsun, Turkey
July 2019

Engin Burgaz

Contents

1 Introduction 1
1.1 Introduction 1
1.2 Types of Polyurethane (PU) Foams 3
1.3 Types of PU Rigid Composite Foams 7
1.3.1 PU Rigid Composite Foams Including Micron-Sized Fillers 8
1.3.2 PU Rigid Composite Foams Including Nano-sized Fillers 10
1.4 Experimental Conditions for PU Rigid Foam Fabrication 14
References 19

2 PU Rigid Composite Foams Containing Micron-Sized Fillers 27
2.1 Introduction 27
2.2 Morphology 32
2.3 Mechanical Properties 52
2.4 Thermal and Thermomechanical Properties 65
2.5 Thermal Degradation and Flammability 75
2.6 Recycling and Recovery Behavior 96
References 98

3 PU Rigid Nanocomposite Foams Containing Plate-Like Nanofillers 103
3.1 Introduction 103
3.2 Morphology 107
3.3 Mechanical Properties 131
3.4 Thermal and Thermomechanical Properties 142
3.5 Thermal Degradation and Flammability 153
References 162

4 PU Rigid Nanocomposite Foams Containing Cylindrical Nanofillers 165
4.1 Introduction 165
4.2 Morphology 169
4.3 Mechanical Properties 194
4.4 Thermal and Thermomechanical Properties 206
4.5 Thermal Degradation and Flammability 214
References 228

5 PU Rigid Nanocomposite Foams Containing Spherical Nanofillers 233
5.1 Introduction 233
5.2 Morphology 236
5.3 Mechanical Properties 255
5.4 Thermal and Thermomechanical Properties 263
5.5 Thermal Degradation and Flammability 274
References 287

Chapter 1
Introduction

1.1 Introduction

The fabrication and characterization of new and highly advanced insulation materials that can function in extraordinary circumstances such as very high and low temperatures, extensive pressures and loadings, highly corrosive and chemically reactive environments with high energy efficiency and sustainability should be realized in order to find solutions for insulation problems in today's modern world. Thermal insulation materials have been used in various applications including thermal protection of houses and buildings, conveyors and containers, food storage and transportation, deep-freezing containers for space liftoff and transportation systems, appliances such as refrigerators at homes and thermal insulation and efficient delivery of natural gas in liquid form [1–4]. Among these application areas, thermal insulation of buildings in the construction industry is the most important issue to handle since more than 10% of the world's total energy consumption equals to the energy needed to provide a comfortable living environment in buildings [5]. Thus, for this reason, researchers in academic and industrial laboratories have used their expertise to enhance energy saving capability and thermal insulation of systems in which preservation of heat and temperature is a critical issue. To satisfy the needs of modern world, new insulation materials with improvements in stiffness, reduced compressive modulus and strength, thermal insulation characteristics, fracture toughness, impact energy absorption and vibration damping, thermal degradation, flammability and thermomechanical properties have been tried to develop in terms of reaching the essential thermal insulation resistance.

Thermal insulation has been recognized as one of the most efficient methods of decreasing the energy consumption in new and old buildings [6–10]. In addition, previous studies [11] on thermal insulation of buildings showed that energy efficiency actions such as insulation retrofitting of old buildings are proven to be more cost-effective ones in comparison with new alternative methods such as solar

E. Burgaz, *Polyurethane Insulation Foams for Energy and Sustainability*,
Advanced Structured Materials 111,
https://doi.org/10.1007/978-3-030-19558-8_1

photovoltaics and wind energy. Thus, for this reason, thermal insulation is the main action that has to be taken in overseeing the energy efficiency of buildings. Furthermore, thermal conductivity values of actually consumed thermal protection materials such as polyurethane (PU), expanded polystyrene (EPS), fibre-glass and mineral wool must be significantly reduced to meet the demands coming from the construction industry [11, 12]. The detailed information about the thermal degradation behavior, flammability and fire resistance of thermal insulation materials should be exactly known to correctly estimate the conditions for heating and cooling systems with suitable sizes [13]. Thus, based on this experimental data, the most desirable insulation conditions can be calculated for specific building applications, and this procedure can provide the essential energy efficiency for thermal insulation in buildings. As mentioned earlier, the most typical thermal insulation materials that have been used in buildings are rigid thermoset polymer foams (polyurethane and EPS), fibre-glass and mineral wool. However, among these materials, the most widely studied thermal insulation material both in industry and academia is the rigid polymer foam due to its broad range of properties such as superior thermal insulation, favorable reduced strength, and durability in various thermal insulation applications [10, 14].

Polymer foams are composed of a cellular morphology in which voids or cells with certain sizes are distributed within the polymer matrix. The voids are usually filled with a gas which is surrounded by the polymer matrix [15, 16]. This type of cellular morphology is very abundant in nature. Natural versions of cellular materials can be very easily found in wood, palm, plant leaves and stems, cancellous bone, pit, cork, sponges and honeycombs [17]. Moreover, advanced materials with cellular morphologies such as metal foams and polymer, ceramic and composite foams with scaffold structures have been also used in medical applications including orthopedic implants and tissue regenerations [17]. Thus, most of the cellular materials have the common cellular morphology-property relationship, and the knowledge that has been gathered from natural cellular materials has been successfully used in the fabrication of man-made cellular materials. However, careful analysis of the close connection between natural and man-made cellular materials can guide researchers both in academia and industry to find out outstanding synthesis methods and to a greater extent direct the formulation and fabrication of fresh and sophisticated cellular foams with advanced characteristics.

This book focuses on the structure-property relationships of polyurethane (PU) rigid nanocomposite foams consisting of plate-like nanofillers, cylindrical nanofillers and spherical nanofillers in comparison with those of conventional PU rigid composite foams containing micron-sized fillers. Thermal insulation property of PU rigid composite foams will be discussed along with their significant characteristics such as closed-cellular morphology, thermal, mechanical and thermomechanical properties, thermal degradation and flammability, energy absorption and saving capability, recycling and recovery behavior, modeling and simulation results. Potential applications of PU rigid composite foams will be discussed, and the main problems that are still not resolved and the future work related to this important topic will be addressed.

A list of previous books and reviews on PU foams and composites is given in the references. Previously, several books were published on the topics of polyurethane composite foams, science and technology of polyurethanes and polyurethane nanocomposite foams [18–20]. All of these previously published books were written on the topics of synthesis, processing and characterization techniques of polyurethane flexible and rigid foams, coatings, adhesives, elastomers, composites and nanocomposites. Moreover, previously published books cover all of the polyurethane industry such as polyurethane adhesives, coatings, elastomers and foams. Thus, these books did not specifically focus on structure-property relationships of polyurethane (PU) rigid nanocomposite foams consisting of plate-like nanofillers, cylindrical nanofillers and spherical nanofillers in comparison with those of conventional PU rigid composite foams containing micron-sized fillers.

1.2 Types of Polyurethane (PU) Foams

Polymer foams should be categorized based on the size of their cells since they can be found in various sizes from nano to micron scale. Thus, based on this classification, according to their cell sizes, polymer foams are usually named as nano-sized foams which have cell sizes in the range of 0.1–100 nm, ultramicron-sized foams which have cell sizes in the range of 0.1–1 μm, micron-sized foams which have cell sizes in the range of 1–100 μm and macro-sized foams which have cell sizes larger than 100 μm [15, 21]. Polymer foams exhibit different degrees of mechanical strength and physical properties based on the difference in their cellular morphology. Thus, in addition to the classification based on cell size, polymer foams are commonly classified under two main categories such as rigid and flexible foams. These two main categories are well-accepted by polymer foam community for systematic establishment and innovation of both theoretical and applied information about these two types of polymer foams with diverse properties. In general, rigid foams are composed of closed-cellular morphology whereas flexible foams have cellular morphology based on open cells. In closed-cellular morphology, material gaps are totally isolated by cell walls. On the other hand, in the open cell morphology, the cell walls are generally disconnected or fractured, and this broken type of cellular structure is dominated by struts and ribs [15, 21]. Thus, this difference of cellular structure leads to extreme diversion in terms of mechanical and physical properties between two main groups of polymer foams. Usually, because of their lower permeability characteristics, closed or isolated cellular morphology in polymer rigid foams leads to better thermal insulation properties compared to flexible foams. On the other hand, flexible foams consisting of open or connected cell morphology have better absorption properties in comparison with rigid foams [15, 16, 21]. In this regard, two of the most important applications of flexible foams is sound absorption and vibration dampening. In addition to these applications, flexible foams containing interrelated voids were utilized as temporary frameworks or platforms for binding and development of cells in tissue engineering [22]. In more general sense, the main application areas of rigid polymer foams are thermal insulation, structural and buoyancy. On the other hand, the main application ares of flexible polymer foams are sound insulation, cushioning and vibration dampening [15, 16, 21]. However, in the future, polymer foam structures can be potentially used in important applications such as hydrogen production and storage, diversified catalysis, optics, photonics, photocatalysis, carbon dioxide split-up, selective filtration and energy conversion [23, 24].

Polymer foams which are famous for their lightweight properties have been extensively used in so many applications such as thermal and sound insulations, cushion, absorbents and weight-bearing structures due to their high reduced strength and energy absorption, outstanding sound and thermal insulations, etc. [15, 16, 21, 25–27]. In addition to these beneficial properties, polymer foams do not meet the criteria for most of their applications due to the presence of voids or holes within their structures. Generally, voids with polydisperse diameters are formed and not very homogeneously distributed within the foam matrix during the free rise of foams during their processing. Thus, these voids present within the foam structure decrease its important properties including surface quality, dimensional and thermal stability and reduced strength [15, 26]. Furthermore, during chemical or physical fabrication of polymer foams, the morphology and exact structure of cells and their sizes cannot be wholly controlled due to the fast reaction rates and complicated gas diffusion mechanism [15, 28].

Many thermoset and thermoplastic polymers such as polyurethane (PU), polystyrene (PS), poly lactic acid (PLA), ethylene propylene diene monomer (EPDM), poly (methyl methacrylate) (PMMA) and its copolymers, polyetherimide (PEI), poly(caprolactone) (PCL), polyethylene (PE), polypropylene (PP), poly(vinyl chloride) (PVC), polycarbonate (PC) have been used as matrix materials for the fabrication of polymer foams since 1930s [21, 29]. Among these polymers, PU is the most consumed polymer whereas PS is the second most consumed polymer in the application area of polymer foams [21, 29]. Polyurethane is one of widely used polymer in the world since it is produced in large quantities such as 18 millions tons per year. This large amount of production corresponds to fabrication of specific PU products with more than a million of cubic meters on a daily basis [30].

Literature data shows that some polymer insulation foams display improvements in stiffness, reduced compressive modulus and strength, thermal insulation characteristics, fracture toughness, impact energy absorption and vibration damping, thermal degradation, flammability and thermomechanical properties. These enhanced properties are very important in so many diverse applications such as thermal protection, energy preservation, padding, floatability, damping, packaging, formulation of lower weight and rigid material parts in aerospace and automotive industry, structural geometries in construction and boat production industries and fabrication of composite and nanocomposite materials.

Rigid polymer foams (RPFs) have been primarily used as thermal insulation materials due to their beneficial properties in applications such as construction industry, domestic appliances and refrigerators, transportation of liquefied natural gas, steel pipe insulations in low-energy district heating and offshore oil and gas production and insulation of cryogenic space launchers [4, 31–36]. Specifically, RPFs have been widely used in thermal insulation of cryogenic tanks for space vehicles due to its mechanical strength, light weight and thermal insulation performance [3, 37, 38].

Thus, in space vehicle applications, RPFs provide thermal protection for a temperature difference of 280 K by keeping up the mass and density of the propellant at the desired levels that are required for a successful flight [3, 38] There is a very good amount of knowledge about RPFs that has been published in the literature [15, 35, 39–42]. Thus, the data available in the literature can be used to fabricate new and advanced RPFs with improved properties in thermal insulation applications.

Rigid polymer foams have a unique advantage in terms of their fabrication in addition to their important properties such as superior thermal insulation, favorable reduced strength, and durability. More precisely, the fabrication technology of RPFs is very practical and easy since it is based on the mixing medium conditions of reactive foaming procedure and strong adhesion of fabricated RPFs to surfaces or materials that are already used in industry [4, 14]. This ease of fabrication of RPFs into specific products with various geometries has also contributed to the widespread use of these materials in thermal insulation applications. Among RPFs, polyurethane rigid foams (PURFs) are the most widely used thermal insulation foams due to their superior thermal protection, improved chemical resistance and toughness values in combination with convenient bendability at low temperatures [4]. Specifically, typical thermal conductivity values of PURF are in the range of 20–30 mW/mK which are way too smaller compared to that of polystyrene, and substantially lower than those of molten glass or stone wool insulation and cellulose based materials [4, 11, 43]. Isolated or closed-cellular structure, smaller cell sizes and types and properties of blowing agents that occupy the interior volume of cells are all responsible for the occurrence of low conductivity in PURFs [44, 45]. Due to these important properties, PURFs have been also used quite a lot in thermal insulation applications such as frozen or refrigerated food storage and transportation, and appliances such as household and industrial refrigerators. In addition, rigid polyurethane foams have been also extensively used for thermal protection of buildings and steel pipelines and hot-water containers, and also thermal insulation for transportation of natural gas in liquid form and cryogenic space launchers [14, 31].

Polyurethanes (PUs) have numerous types of applications due to their diverse physical and chemical properties [35, 46–50]. According to their use in different types of applications, polyurethanes consisting of enormous range of compositions have been mainly used as binders, elastomers, coatings, adhesives, rigid and flexible foams, sealants, synthetic leathers, membranes and many biomedical products [35, 40, 47, 51, 52]. Due to their durability, comfort, cost and environmental benefits and energy savings, PUs are among the mostly used and consumed commercial materials in the polymer market [40, 47, 51]. The durability of PU products that are used in the daily life and industry predominantly has a very positive effect in extending the shelf times, and directly contributes to the less consumption of polyurethane raw materials and energy [53–55]. Besides durability, good recyclability of PUs is also another positive factor in terms of choosing PUs as the raw materials for the production of daily consumer products. PUs including foams and elastomers can be recycled from various daily products such as appliances, automobiles, bedding, cushion and furniture [56–58].

Numerous recycling techniques such as energy retrieval, mechanical, advanced chemical, thermochemical and product recycling methods have been used in the transformation of PU products to initially starting raw materials. Among the mechanical and chemical recycling techniques, regrinding and glycolysis have been used effectively and economically for recycling polyurethane foams and composites. Although glycolysis of PUs is economically feasible based on the current technology, more scientific work should be required to enhance the characteristics of final product at the final stage of this recycling process [51, 57]. Thus, using less amounts of energy and raw materials and effective and economic recycling of PU based consumer products really bring in lots of benefits for the protection of environment. In addition to durability, recyclability and conservation of energy and environment, PUs have a great degree of functionality in terms of their chemical structure and composition. The morphology, thermal, mechanical, thermomechanical properties and also applications of PUs can be adjusted systematically by proper selection of raw materials for the synthesis from a huge list of macrodiols, di-/tri-isocyanates and chain extenders [47, 51, 59].

Polyurethanes are typically synthesized by the chemical reactions occurring between oligomeric polyols and diisocyanates or polyisocyanates [11, 40, 51, 60]. In their chemical structure, besides having other important functional groups, PUs have a large amount of urethane functional groups which are generally responsible for the characteristic properties of PUs [40, 51, 60]. Polyols that are going through standard PU chemical reactions are usually chosen from polyols with multiple hydroxyl end groups [11, 40]. The mostly used polyols that are chosen for the synthesis of PUs are polyfunctional polyethers such as polyethylene glycols, polypropylene glycols, acrylic polyols and polyester polyols [47, 59]. The closed-cellular morphology, thermal, physical and mechanical characteristics of PUs are significantly related to molecular weights and chemical structures of isocyanates and polyols which are the main ingredients for the formation of PUs. Especially, the molecular weight of polyols clearly dictates the final properties of PUs [47, 59, 61]. Based on the molecular weight, the polyols that are used for the synthesis of PUs are categorized into two main groups such as polyols having higher and lower molecular weights. For simple and mostly used low molecular weight polyols in fabrication of PUs, neopentyl glycol, glycerol, ethylene glycol, propylene glycol, etc. can be given as examples [59, 61]. On the other hand, the high molecular weight polyols are generally defined as the oligomers with a maximum average molecular weight of 10,000 Da [60, 62].

Rigid PUs are usually formed via the reaction of isocyanates with polyols having lower molecular weights which are usually smaller than 1000 Da. Thus, higher numbers of urethane groups per unit volume in rigid PUs are generated via reactions of polyisocyanates with polyols having lower molecular weights. The molecular weights of polyols for rigid foams are generally below 1000 Da and their functionality and equivalent weight are usually in the ranges of 3–6 and 100–200, respectively [39]. However, on the contrary, using high molecular weight polyols in PU reactions usually results in the formation of flexible PUs since smaller number of urethane groups and more mobile alkyl chains are formed due to the longer chain length and higher molecular weight of polyols [59–61]. Generally, the functionality of most flexible polyols is around 3 [39]. Usually, semi-rigid PUs are formed by using polyols that have molecular weights between those polyols that are used for the fabrication of rigid and flexible PUs. In addition to the importance of molecular weights of polyols, secondary inter or intramolecular forces such as hydrogen bonding, van der Waals, dipole-dipole types of interactions, etc. have direct and substantial influences on closed-cellular morphology, physical and mechanical properties of PUs [47, 59, 61]. Thus, among these secondary forces, hydrogen bonding is the major type of interaction due to higher number of polyether, urea and urethane functional groups that exist in PUs [63].

All of the polyols and polyisocyanates that have been used in the fabrication of PUs are petroleum-based products. For about more than two decades, researchers both in academia and industry have focused on the production of these raw materials from renewable bioresources since raw products that have been derived from petroleum are forecasted to be run out of production in large quantities [40, 64–68]. Specifically, the amount of academic and industrial research on using vegetable oils in the fabrication of PUs has been increased in a great extent in recent years [69–72]. Among polyols, typically, polyether and polyester polyols are the mostly consumed polyol types in the fabrication of PUs. In numbers, 75% of polyols which have been used for PU synthesis is polyether-based whereas the rest of the polyols is polyester-based [40, 62]. There are several ways to obtain biobased polyols from their starting bio-based raw materials such as saccharides, glycerin and vegetable or animal oils [4, 39, 40]. However, among these routes, the most cheapest method is to use vegetable oils which can be found in large quantities in nature [39, 40, 73]. In addition, among these vegetable oils, specifically, the unsaturated oils have a great potential to be used in the fabrication of PUs since many reactions can be carried out from their unsaturated sites in the synthesis of biobased polyols [4, 39, 40].

1.3 Types of PU Rigid Composite Foams

PU rigid foams (PURFs) have unique properties such as low thermal conductivity due to their smaller cell sizes, closed cell structures that are filled with blowing agents and easier processability. Most importantly, as mentioned earlier in the above paragraphs, different combinations of polyols and polyisocyanates with various chemical structures enable PU rigid foams to find use in diverse applications. In most of the previously mentioned PURF applications, good amount of mechanical and thermal stability is the most important criterion to further improve the efficient use of PURF. However, PURFs do not really fulfill this criterion since they have low thermal stability and mechanical strength due to their fragile structure and morphology [35, 74]. In addition to their low thermal stability and mechanical strength, PURFs also have poor performance of thermal insulation which reveals itself over an extended period of time. Specifically, the decay of thermal insulation over time is the result of diffusion of blowing agents from the interior volume of closed cell morphology to the surroundings [75]. Thus, for durability, sustainability, recyclability and conservation of energy and environment, a lot of research has been done to improve the thermal insulation property of PURFs. One of the most effective methods to improve thermal insulation performance is to fabricate PURFS with smaller cell sizes [35, 76]. Consequently, for the development of PURF technologies with smaller sizes and good mechanical strength and thermal stability, researchers both in academia and industry have fabricated PU rigid composite foams (PURCFs) consisting of micron-sized or nano-sized fillers [35, 45, 76]. Here, in this book, PURCFs consisting of micron-sized fillers are named as PU rigid microcomposite foams (PURMCFs). Also, PURCFs consisting of nano-sized fillers are named as PU rigid nanocomposite foams (PURNCFs).

1.3.1 PU Rigid Composite Foams Including Micron-Sized Fillers

Various micron-sized fillers with different geometries have been used to improve the impact behaviour, thermal insulation and stability, compression and tensile strengths and flame retardant behavior of PURFs [27]. The micron-sized fillers can be classified based on the shape of fillers such as cylindrical, plate-like and spherical geometries [77]. For cylindrical micron-sized fillers, fibers can be given as the mostly used filler example in PURMCFs [27, 78–80]. For more than two decades, in addition to synthetic fibers, researchers in academic and industrial laboratories have used their expertise and knowledge to produce bio-based fibers for durability, sustainability and protection of the environment. Thus, for this reason, vegetable and cellulose-based fibers have been used in the fabrication of PURMCFs due to their low cost, easier accessibility, high tensile strength and modulus and good amount of biodegradability and recyclability [79, 81]. In recent years, fire resistance, thermal decomposition behaviour and mechanical properties of PURMCFs were improved by incorporation of aramid fibers which are seen as economically feasible and environment friendly alternatives to halogenated flame retardants [82].

Among plate-like micron-sized fillers, talc [83, 84] and expandable graphite [85–88] can be given as examples that have been used in PURMCFs. However, among previously published PURF/plate-like micron-sized filler studies in literature, the most predominantly investigated filler is expandable graphite (EG) due to its outstanding properties such as high flame retardancy, low price, environmentally friendliness, high porosity, excellent expansibility, high temperature resistance, thermal insulation and exchange surface [4, 87, 89]. On the other hand, a few studies about PURMCFs consisting of talc are available in the literature.

The flame retardant materials that have been used in PURFs typically consist of halogenated, phosphate compounds and antimony trioxide or their mixtures [88, 90]. However, for more than two decades, the use of halogen-containing flame-retardant materials has been restricted since corrosive and toxic smoke is released by these materials to the environment in the presence of fire [88, 90]. Thus, because of the safety and prevention of pollution in environment, many regulations have been implemented on the use of halogen-free flame retardant materials instead of halogenated ones. For this reason, many studies have been performed both in academia and industry to develop highly efficient and environmentally friendly flame retardants for a very long time. A short time ago, phosphorus and intumescent based materials have been focused in a great extent since they form char on the surface of polymers and this property induces flame protection as well as prevention of any toxic smoke during fire [88]. Although intumescent flame-retardant materials have been used in the field of coatings for more than 50 years, these materials have been successfully applied to the area of PU rigid foams for about 20 years [88, 90].

EG is a typical halogen-free intumescent flame retardant material for PURFs which is synthesized in the medium of sulfuric and nitric acid mixture to form the intercalated structure from graphite layers. At higher temperatures, EG expands to higher volumes, forms an intumescent carbonaceous layer on its surface. When EG is used in PURFs, also at higher temperatures, this insulative worm-like layer covers the surface of PURF matrix. Thus, the transfer of heat and oxygen throughout the polymer matrix is obstructed by this thermally insulated worm-like layer [89, 91]. For this reason, researchers both in academia and industry have concentrated on the effects of EG addition on flame retardancy of PURFs since the solution of this topic has become very important due to the beneficial properties of expandable graphite for the PURF system.

Among spherical micron-sized fillers, silica [92–94], calcium carbonate [92, 93], aluminum powder, glass powder and microspheres [89, 92], carbon black [95] can be given as examples that have been used in PURMCFs. Previously, PURFs consisting of diverse micron-sized fillers such as precipitated silica and calcium carbonate and glass powder were prepared via reactive foaming event in which distilled water was used as the blowing agent, and effects of filler incorporation on closed-cellular morphology, water absorption and thermal conductivity behaviour, mechanical and thermal properties of PURFs were systematically studied [92]. Moreover, compressive stress and modulus values and hardness data of PURMCFs containing these micron-sized fillers such as precipitated silica and calcium carbonate and glass powder diminished compared to those of pure PURF. In PURMCFs consisting of silica or calcium carbonate, the insulation properties were found to be decreasing with the increase of micron-sized filler content. On the contrary, in the case of glass powder, it was shown that thermal conductivity initially decreased, but subsequently increased with further addition of filler [92]. In addition, the micron-sized silica with an median size of 1.5 μm was utilized in fabrication of PURMCFs [94], and the results showed that the reduced density of foams does not change that much with the addition of silica. Furthermore, it was also revealed that the strengths of PURMCFs consisting of micron-sized silica particles with concentrations of 5 and 10% are slightly increased, but at higher concentrations of micro-silica, the strength of foams is reduced due to the irregularity of the closed-cell structure caused by the higher contents of micron-sized silica particles [94].

1.3.2 PU Rigid Composite Foams Including Nano-sized Fillers

PU rigid foams which have closed-cell structures usually have a cell wall thickness of about 1 μm, and cell sizes of at least 100 μm [35, 45, 96, 97]. Thus, for this reason, these size ranges of the actual foam structure should be taken into consideration in the selection of the most suitable filler type for the design of advanced PURFs with improved properties. The most appropriate filler for this selection process should be the filler type in which at least one of its dimensions is smaller than 1 μm [96]. Thus, for the satisfaction of this criterion, fillers or additives having median sizes on the order of 1–100 nm are better options compared to fillers having average sizes greater than 1 μm. Due to their smaller sizes, higher surface areas and higher degrees of interfacial adhesion with the polymer matrix compared to micron-sized fillers, the addition of very low amounts of various nano-sized fillers with different geometries is enough to improve thermal insulation and stability, mechanical and dimensional stability, reduced gas permeability, impact behaviour, compression and tensile strengths and flame retardant behavior of PURFs [35, 98, 99]. The physical and mechanical properties of PURNCFs strongly depend on the chemical composition, size and shape of nanofillers [96]. Nano-sized fillers can be classified based on their shapes such as plate-like, cylindrical, and spherical geometries [77]. For plate-like nano-sized fillers, clays, graphene oxide and graphene can be given as the nano-filler examples that have been used in PURNCFs. However, among previously published PURF/plate-like nano-sized filler studies in literature, the most extensively researched nanoadditive is nanoclay due to its outstanding properties such as reduced gas permeability, flame retardancy, low price, environmentally friendliness, excellent expansibility, high temperature resistance, thermal insulation [35, 45, 76, 97, 100–104].

Among PURF/clay published works, the most extensively researched clays are montmorillonite [35, 45, 76, 97, 101, 103] and vermiculite [102, 104] clays which established an extraordinarily significant place both in academia and industry because of their outstanding and improved properties in terms of higher aspect ratio, heat deflection temperature, dimensional stability, reduced gas permeability, impact behaviour, compression and tensile strengths and flame retardant behavior. Nanoclay is a special and important nanomaterial which can be used to enhance mechanical properties of PURFs due to its very large lateral dimension to thickness ratio in the range of 100–1000, a maximum lateral dimension of 10 μm and a minimum thickness of 1 nm [65]. The effects of clay addition to PURFs have been investigated by several researchers in terms of morphology, density, thermal stability, thermal conductivity, flame retardant, mechanical and thermal properties [35, 45, 76, 97, 100–104]. Previously, several research groups prepared PURNCFs consisting of montmorillonite nanoclay by using unmodified original nanoclay and nanolayered silicates modified with organic functional groups via in situ polymerization method. Moreover, the effects of nanoclay addition to PURFs were systematically investigated [35, 45, 105]. In order to generate a well-dispersion of nanoclay in PURFs, usually clay was ultrasonicated either in polyol or isocyanate component [45]. In addition, based on gas chromatography results, it was shown that the blowing agent diffusion throughout the closed cell structure was suppressed significantly with incorporation of 1 wt% clay to PURFs. The reason behind permeability reduction of blowing agent was explained such that nanoclay behaves like a blockade against spreading of blowing agent due to its smaller particle size and homogenous dispersion within the PURNCF matrix [45].

PU foams are fabricated via a very complicated reaction mechanism in which reactive polymerization, foaming and blowing occur at the same time [35]. During the reactive foaming process, nanoclays can also serve as nucleation centers for the creation of many bubbles with smaller sizes. In addition to their major role on the formation of more bubbles, nanoclays can also obstruct the growth of extra more bubbles by increasing the viscosity of reaction medium [45]. Throughout the time of reactive foaming event, nanoclay particles perform as additional seeding cores in PURNCFs for the generation of uniform distribution of cells with smaller sizes and higher cell densities. Hence, this positive improvement in closed-cellular morphology paves the way for inducing better gas barrier properties in PURNCFs compared to the PURF system without any clay [35]. In accordance with improved gas barrier properties of clay containing PURNCFs, the use of nanoclay in PURFs also decreases the thermal aging behavior of rigid foams. Thermal aging which is defined as the increase of thermal conductivity over time usually occurs in PURFs since CO_2 with low thermal conductivity leaves the closed cell structure and is exchanged with N_2 and O_2 which have higher thermal conductivities [106, 107]. It was previously unveiled that generation of smaller cell sizes in PURFs paved the way for obtaining lower thermal conductivity values, and smaller cell sizes in PURFs can be obtained by using different surfactants, catalysts, and nanoclay particles [108, 109]. Thus, in order to reduce the thermal conductivity increase over time, in addition to different combinations of surfactants and catalysts, nanoclays should be used in PURFs since their nanoplate-like morphology is very suitable for acting as an obstacle against gas diffusion. Previously, it was shown that dispersion of vermiculite clay in the isocyanate component leads to the formation of PURNCFs with better barrier properties [76].

There are a few studies about PURNCFs consisting of graphene oxide (GO) and graphene [98, 99, 110–112] in the literature. Graphene is a different form of carbon and it has planar sheet geometry with a thickness of one carbon atom. Each graphene plate consists of carbon atoms with sp^2 bonds which are densely loaded in a mass of hexagonal crystal lattice [113]. Graphene has been considered to be the thinnest nanomaterial with substantial thermal, mechanical and electronic properties [110, 112, 113]. Graphene oxide is generally synthesized by using Hummers method through oxidation of graphite. Graphene oxide is a cheaper, more functional and effective nanofiller compared to graphene since it has oxygen-containing functional groups on its surface. Due to these functional groups, the solubility of GO in solvents such as water and other liquids is pretty high. As a consequence, graphene oxide was utilized as an efficient plate-like nanoadditive in fabrication of polymer nanocomposites with enhanced properties and diverse applications [98, 110, 111].

For cylindrical nano-sized fillers, carbon nanotubes, carbon nanofibers and bio-based cellulosic nanofibers can be given as the nanocylinder examples that have been used in PURNCFs. However, among previously published PURF/cylindrical nano-sized additive researches in literature, the most extensively investigated nano-cylindrical fillers are carbon nanotubes (CNTs) due to their outstanding properties in PURFs such as thermal and electrical conductivity, thermal stability, mechanical, thermal and thermomechanical properties [114–118].

The effects of CNT addition to PURFs have been investigated by several researchers in terms of morphology, density, thermal and electrical conductivity, thermal stability, mechanical, thermal and thermomechanical properties [114–118]. However, the most important problems that should be solved in the preparation of PURNCFs consisting of CNTs are clustering and uniform distribution problems of carbon nanotubes within PURF matrix. The deficient distribution of CNTs in PURNCFs is mainly caused by their large aspect ratio and strong π–π interactions [114, 119]. Consequently, too much aggregation in PURF/CNT systems leads to the decrease of interfacial area and thus thermal and mechanical properties. For this reason, many studies have been performed in order to solve the aggregation problem of CNTs [114, 119]. For this purpose, the compatibility between CNTs and polymer matrix was improved by performing surface modification of CNTs such as silane modification [114, 119], amination [120], hydroxylation [119, 121], carboxylation [114], fluorination [122] and isocyanate treatment [123].

Although surface modified CNTs has been used in PURFs, the physicochemical compatibility between the matrix and CNTs is still not satisfactory because of poor intermolecular secondary forces between surface organic functionalities on CNTs and PU molecules in PURNCFs [114]. However, new methods for surface modification of CNTs have been developed and these methods are more superior compared to previous ones in connection with elevating suitable adhesion forces in interfaces between PU molecules and CNTs and consequently the dispersion degree of CNTs within the matrix. Only very recently, PURNCFs filled with PEO-*graft*-MWCNTs that were successfully synthesized by a ring-opening polymerization of ethylene oxide were fabricated to comprehend impacts of anchored PEO molecules on closed-cellular morphology, mechanical properties and thermal decomposition behavior in PURNCFs [114]. It was shown that the cell size diminished but both glass transition temperature and mechanical properties of PURNCFs elevated with the addition of PEO-*graft*-MWCNTs [114]. Thus, the improvement in mechanical properties and reduction of cell size are clear indications for enhanced physicochemical compatibility and strong interfacial adhesion between surface-modified MWCNTs and PURF matrix. Previously, hydroxylated MWCNTs were surface modified by silanization with 3-aminopropyltriethoxysilane (APTS) and dipodal silane (DSi), and these modified MWCNTs were used in the fabrication of PURNCFs. Based on morphological and mechanical results, the values of cell density, tensile strength and elastic modulus increased in PURNCFs consisting of 1.5 wt% of DSi-MWCNT more than those containing the same amount of APTS-MWCNT [119]. Therefore, this result also proves the strong interfacial interactions between DSi-MWCNTs and PURF matrix.

For spherical nano-sized fillers, nanosilica, polyhedral oligomeric silsesquioxanes (POSS), TiO_2 and ZnO can be given as the nanofiller examples that have been used in PURNCFs [124, 125]. However, among previously reported PURF/spherical nano-sized filler studies in literature, the most extensively researched spherical nanoadditive is nanosilica due to its outstanding properties in PURFs such as improved thermal stability, mechanical, thermal and thermomechanical properties [94, 125].

Many different types of nanosilica particles such as nanosilica surface-treated with oligomers [126], hydrophilic and hydrophobic nanosilica and nanosilica containing pores with diameters of 2–50 nm [127, 128] have been used as efficient nano-sized spherical additives in polymer nanocomposites. Moreover, nanosilica particles prepared via generation of colloidal solution and integrated network of separate particles [129] have been extensively exploited as nanoadditives consisting of both distinct physical and chemical surface characteristics in polymer nanocomposites. All of these diverse nanosilica particles were utilized in polymer nanocomposites for enhancement of structure, morphology, intermolecular interactions, mechanical and thermomechanical properties and thermal decomposition behaviour [128, 129]. Under the category of nanosilica particles, pyrogenic silica which is also called as fumed silica has been extensively utilized in academic and industrial research projects as a filler which serves efficiently as a medium in fluids and powders and as a nanofiller for the property improvement of elastomers [130]. Silica (especially fumed silica) is an important and effective nanofiller in terms of enhancing the viscosity of organic solutions [130, 131]. Especially, hydrophilic silica is a very effective and multipurpose type of silica which generates steady and thin solutions consisting of lower values of viscosity. Furthermore, another important characteristic of hydrophilic silica is that it can establish powerful hydrogen bonding interactions if it is dispersed in liquid media such as low molecular weight poly (ethylene glycol) and alcohols containing short chain structures [130]. In particular, in terms of geometry, fumed silica is composed of amorphous silicon dioxide in which stable aggregates of approximately 100–250 nm are formed due to the fusion of primary particles with sizes of 1–3 nm [131]. Due to their large surface area (50–400 m^2/g), the formation of particle-particle interactions leads to higher degrees of aggregation, and consequently can form a three-dimensional network in polymer solutions, molten polymers and casted polymer nanocomposite films [130, 131].

Most of the studies that were published about PURNCFs in the literature mainly focuses on the fabrication and characterization by including the degree of nanofiller dispersion or aggregation in the PURF matrix [96]. On the other hand, most of the final properties of PURNCFs are determined during the complicated reactive foaming process. Thus, there is a need to understand the effects of nanofillers on the foaming reaction kinetics since there is very little amount of published work on this topic [96, 124]. During the reactive foaming process, diisocyanates react with hydroxyl groups of polyether or polyester type of polyols in order to form the urethane functional groups. Thus, the presence of nanofillers in the reaction medium changes the kinetics of these chemical reactions. As a consequence, the presence of nanofillers paves the way for altering closed-cellular morphology with regard to cell size and density and amount of closed cells and also thermal and mechanical properties in connection with variations of crosslinking density.

For the development of advanced PURNCFs with improved thermal, mechanical and thermomechanical properties, it is very crucial to realize influences of nanofillers with diverse types, shapes, concentrations and surface areas on closed-cellular morphology and reaction rates during reactive foaming event. For this purpose, studies on PURNCFs consisting of spherical nanofillers such as TiO_2 [124], SiO_2 [132], ZnO [124] displayed that the cell size of foams was reduced in comparison with that of pure PURFs due to the formation of extra nucleation centers by spherical nanofillers during the reactive foaming process [35, 124]. Previously, impacts of using diverse sorts and geometries of TiO_2, Fe_3O_4, and ZnO nanoadditives on deformation and flow characteristics, reaction rates during reactive foaming event and closed-cellular morphology of PURFs were studied [124]. In accordance with experimental data that was obtained from reaction rate analysis during reactive foaming event, it was shown that reactive foaming speed increased with elevation of surface area, and specific parameters (geometry, size, surface property, etc.) of nanoadditives did not have any direct influence on the reactive foaming speed. In addition, foam cell sizes in PURNCFs diminished with incorporation of nanofillers whose surface areas are below the critical surface area. However, cell size distribution was observed to be broader above the critical surface area [124].

1.4 Experimental Conditions for PU Rigid Foam Fabrication

PURFs are fabricated via characteristic signatures of rigid foams such as physical foam expansion with the help of blowing agents and PU polymerization reaction in which polyols typically react with polyisocyanates. During synthesis process of PURFs, both polymerization reaction and physical foaming event happens simultaneously in the reaction medium. PURFs are primarily synthesized based on the polyaddition reactions between polyfunctional isocyanates and polyols. Basically, Otto Bayer invented the reactions that are responsible for the formation of polyurethanes. The functional group of urethane basically shares the same chemical structure of ethyl carbamate (–NH–CO–O–), and this chemical structure is typically the signature of polyurethanes. PURFs consist of various types of PUs such as polyurea-modified PUs and polyisocyanurate (PIR)-modified PUs based on the chemical structure and types of chemical reactions. Generally, the cyclo-trimerization chemical reaction of excessive amount of polyisocyanates produces PIR-modified PUs. In addition, polyurea-modified PUs are obtained from the reactions of isocyanates with amines [14, 133].

During the reactive foaming process, not only the fragile structure of synthesized foam should be protected, but also the reaction rate should not be that high to maintain the growth of bubbles in a continuous manner within the reaction medium [35].

Researchers both in industry and academia have developed a common terminology (cream time, gel time and rise time) to distinguish important phases for the events that take place during the reactive foaming process. The reactivity of components in the reaction medium is simply evaluated by cream time which is the amount of time that should be spent for appearance of reacting mixture to become cream-like due to formation of gas bubbles [14]. Furthermore, gel time is identified as time value at which solid foam sample can be selected and pulled from rising foam. Lastly, rise time is identified as time value at which foam does not grow or expand any more [14].

The physical, thermal and mechanical properties of PURFs can be adjusted to a greater extent by changing the types of components that are used during the reactive foaming process. These components are typically identified as polyols, polyisocyanates, catalysts, blowing agents, surfactants and additives. Changing the types or physicochemical properties of these components definitely alters the final properties of PURFs since each component in the reaction medium has a major role in terms of changing the closed-cellular morphology of PURFs. The formation of closed-cellular morphology during the reactive foaming process primarily determines the ultimate properties of PURFs. Thus, based on the types of specific applications, researchers both in academia and industry should adjust the parameters of each component, accordingly. However, among these components, polyols and polyisocyanates are the most important ones since these two raw materials are the major compounds that are responsible for the poly-addition reactions, and they also have the highest weight percentages in the reaction medium.

For the fabrication of PURFs, polyols are usually selected from aromatic or aliphatic polyesters as well as polyethers. In addition, most of the times, various polyol blends are used in the preparation of PURFs. Molecular weights of polyols that are used in the fabrication of PURFs are usually below 1000 g/mol, and their functionality typically varies in the range of 3–6. In addition, the equivalent weight of polyols for PURFs is typically within the range of 100–200. Instead of synthetic polyols, vegetable oil-based polyols can be used in the fabrication of PURFs. In addition, polyols that are synthesized from vegetable oils have a high chance to compete with synthetic polyols [39]. In terms of isocyanates, generally polymeric diphenylmethane diisocyanate (PMDI) and sometimes toluene diisocyanate (TDI) can be utilized as typical PU starting components during reactive foaming of PURFs. However, in production of PURFs, the most widely consumed isocyanate is PMDI since it has much lower health risks to humans in comparison with TDI [35, 39, 51, 134].

During the fabrication of foams, usually organic molecules with low molecular weights are used as chemical blowing agents. Generally, the presence of these molecules in the reaction medium leads to the formation of gas and bubbles during the foaming reaction. The final properties of PURFs are substantially affected by the type and content of the blowing agents that are used during the reactive foaming process. During reactive foaming process, the viscosity of reaction medium increases, and polyurethane linkages are formed between isocyanates and polyols because of exothermic essence of reactions. In reaction containers of PURFs, blowing agents are usually mixed with polyol components, and they are also responsible from the simultaneous "blowing" or "foaming" of the liquid mixture in the reaction medium [76]. The blowing agents which are low-boiling liquids (such as pentane, hexane, hydrochlorofluorocarbons, etc.) vaporize due to the generation of heat in the reaction medium. The low thermal conductivity property of PURFs is obtained due to accumulation of the blowing agent gas in closed-cellular structure of foams. This reactive foaming process is called physical blowing. On the other hand, if there is water present in the reaction medium, urea functional groups and carbon dioxide molecules can be generated as products via reaction of isocyanates with water. This reactive foaming process is called chemical blowing [76, 133].

However, the use of chemical blowing agents is not very practical and effective in terms of controlling the final properties of foam such as morphology (irregular size and shape of closed-cellular structure), physical and mechanical properties [42]. On the other hand, physical blowing agents can be used instead of chemical blowing agents to form porosity and closed-cellular morphology in the final foam product. Conventionally, chlorofluorocarbons (CFCs), hydrochlorofluorocarbons (HCFCs), hydrocarbons (HCs) and hydrofluorocarbons (HFCs) were utilized as commonly preferred physical blowing chemicals. However, using these compounds in the reactive foaming process has been strictly controlled in many countries since CFCs and HCFCs lead to the depletion of ozone layer and global warming, HFCs worsen the greenhouse effect and HCs cause risks due to their high flammability [27, 135–137].

The use of CFCs as the physical blowing agent in the fabrication of PURFs has a positive effect in terms of reducing thermal conductivity. Typically, PURFs consisting of CFCs have the lowest thermal conductivity values among non-vacuum insulation panel products [138]. As mentioned earlier, due to their bad environmental effects, the production of CFCs was discontinued in 1990s [139]. Previously, HCFC-141b was utilized as a blowing chemical in refrigerators due to its small degree of ozone depletion potential. In addition, although physical blowing chemicals such as HFC-245fa and cyclopentane have some drawbacks in terms of reducing the thermal resistance of foams, they have been used as non-chlorine containing blowing agents in the fabrication of PURFs quite successfully [137, 139]. In addition, these traditional physical blowing agents necessitate the use of large amounts of organic solvents which remain inside the porous structure and eventually lead to the disruption of the closed-cellular morphology over time [136]. However, due to several disadvantages of these previously mentioned physical blowing agents, researchers both in academia and industry searched for alternative blowing agents to use these compounds in the fabrication of PURFs with improved physical, thermal and mechanical properties. One of these alternatives is using water as the blowing agent in the reactive foaming process to produce PURFs for vacuum insulation panels [140, 141]. However, the presence of large amounts of water in the reaction medium leads to pressure fluctuations and consequently causes deformation in the closed-cellular morphology of PURFs [135, 136]. Furthermore, it is very well-known that PURFs consisting of CO_2 which is basically formed by the reaction of water and isocyanate has a higher thermal conductivity compared to PURFs that are prepared from CFCs [139].

Thus, for these reasons, nowadays, instead of traditional physical blowing agents, CO_2 has been used as a physical blowing agent since it is easy to use and it has good amount of compatibility with polymer matrices. Although using supercritical carbon dioxide in terms of physical blowing chemical in PURFs is still at its infancy, a few studies about this subject have been performed to fabricate PURFs consisting of microcellular morphology [133, 136, 142]. Furthermore, the use of CO_2 in PURFs really protects the environment since there is no need to use toxic solvents and no extra solvent is left inside the final foam product in this procedure. Previously, the fabrication of PURFs in the presence of supercritical carbon dioxide was performed, and it was shown that microcellular PURFs with a cell size of 30–180 μm, a cell density of 1.26×10^7 per cm^3 and a bulk of 0.2–0.4 g/cm^3 were obtained [133, 136, 142].

In addition to traditional physical blowing agents, other vaporous substances such as cyclopentane and hexane have been utilized both in academic and industrial research laboratories for reactive foaming event of PURFs. In contrast, compared to traditional physical blowing agents, these new agents have considerably higher thermal conductivity and flammability. So, for this reason, lots of optimizations need to be done to use these new physical blowing agents more effectively in the fabrication of PURFs [143].

During the reactive foaming process, nucleation and growth of bubbles strongly depend on the amount of dispersed gas in reaction medium. If the impregnation pressure is too high, then this can lead to very high rates in the seeding and development of bubbles [143]. Usually, fabrication of PURFs with improved physical, thermal and mechanical properties is closely related to the production of uniform closed-cellular morphology consisting of small cell sizes. Thus, for generation of homogenous closed-cellular morphology with smaller cell sizes and higher cell densities, a higher speed of seeding and a slower expansion of cells should be maintained during reactive foaming event [143]. Recently, PURFs with a cellular structure were produced with a new reactive foaming process in which ultrasonic excitation was applied to a supersaturated solution of nitrogen gas. In that new foaming process, ultrasonication energy was used to provide seeding of bubbles, and also development of bubbles was suppressed by using a low saturation pressure [143, 144].

Besides selecting a proper type of blowing agent for steps such as polymerization reaction and physical foaming process, understanding effects of hydroxyl functionality values of polyols, isocyanate index, surfactant type on the final properties of PURFs is also very important [27]. During the reactive foaming process, foam expands approximately to 30 or 40-fold of its original volume. Thus, during the foam expansion, it is very important to eliminate any bubble collapse and coalescence and stabilize the reacting liquid-gas interface to obtain PURFs with uniform cell sizes. For this purpose, suitable surfactants such as polydimethylsiloxane (PDMS)-co-polyethers can be used to solve effectively the issues during foam expansion. Surfactants which are essential components of the foaming reaction medium are typically responsible for lowering the surface tension and providing sufficient surface elasticity to expanding liquid films [14]. Previously, PURFs were prepared from PMDI consisting of median functionality of 2.9 and polyether polyol having average functionality of three, and the influences of silicone surfactant on characteristics of PURFs were systematically studied [137]. It was reported that the cell size of PURFs diminished from with increasing amount of surfactant. In contrast, when surfactant amount exceeded the optimum value, no noticeable change was detected in connection with cell size. Also, based on results of mechanical properties, it was shown that the mechanical strength of PURFs elevated with increasing amount of surfactant. The increase in mechanical strength of PURFs was explained due to the reduction of cell size when the surfactant amount reached to optimum value. On the other hand, mechanical strength of PURFs decreased when the surfactant amount increased more than the optimum value [137].

Among the vital components for the fabrication of PURFs, additives in the form of micron or nano-sized fillers have also important positive effects in terms of improving the final physical, thermal and mechanical properties of PURFs [76]. However, incorporation of these additives, especially nanofillers into PURFs brings many problems such as increase of viscosity and aggregation of nanofillers because of their enormous surface area values during blending of these components with PURF matrix. The increase of viscosity and aggregation of nanofillers are unwanted events since reaction rates and morphology are negatively affected by the increase of viscosity. Also, aggregation of nanofillers directly decreases the final PURF properties due to less amount of interaction between nanofillers and PURF matrix. Consequently, closed-cellular morphology, thermal and mechanical properties of PURFs that are fabricated from a reaction medium consisting of high viscosity and aggregation of nanofillers would be inferior compared to pure PURF. Thus, in order to solve these problems during the reactive foaming process, researchers both in academia and industry have tried to find various types of fabrication methods for the uniform dispersion of nanofillers and decrease of aggregation within the PURF matrix.

Different synthesis methods such as ultrasonication [145], solution mixing [146], in situ polymerization [147], melt mixing [148], surface modification of nanofillers via coupling agents [97, 149] were used to enhance homogeneous distribution of nanofillers throughout PURF matrix. Previously, uniform dispersion of organoclay plate-like nanofillers was accomplished by using approach of ultrasonic cavitation during reactive foaming event in PURFs. In addition to ultrasonic cavitation method, researchers also surface-modified the organoclay surface with silane coupling agents, and thus provided better compatibility between nanoclay and PURF matrix [97]. Consequently, the use of both surface modification and ultrasonication leads to improvements in terms of physical, thermal and mechanical properties by breaking up nanoclay agglomerates within the PURNCF matrix.

Generally, during the reactive foaming process of PURNCFs consisting of nano-sized fillers, firstly, dispersions of nanofillers were prepared either in polyol or isocyanate component. Then, these oligomer-nanofiller dispersions are usually passed through ultrasonication step in order to make homogeneously well-dispersed solutions, and then these dispersions are mixed with the other remaining component of PURF system for the foaming reactions in the presence of catalysts and foaming agents [76]. Previously, researchers used the microwave processing method to homogenously disperse nanoclay within PURNCFs. Their experimental findings showed that the microwave processing method is a pretty fast and successful method in terms of breaking up nanoclay agglomerations in PURNCFs [101].

Besides optimization of experimental conditions for the fabrication of PURFs, the proper methods for recycling and recovery of these materials should be systematically determined in order to recover various types of waste PURFs, reduce fossil fuel consumption, and consequently increase the energy value and sustainability of PURFs. However, the recycling and recovery of PURFs into useful dailylife consumer products should be executed without giving any damage to environment in order to provide both energy conservation and sustainability. The first aim of recycling methods is to decrease the size of PURF waste products such that they can be reused for the fabrication of new PURFs. Based on lots of research data both in academia and industry, researchers have found a number of recycling and recovery methods for PURFs which are beneficial in terms of economically and environmentally [30, 51].

Some of the recycling and recovery methods that were published as patents or articles by researchers both in industry and academia in the past have been used in the establishment of new factories where waste PURF products are efficiently recycled and used as raw materials in the fabrication of new useful PURF products. For example; a company called Aprithane which was located at Abtsgmünd, Germany conducted glycolysis of PURFs and successfully used this final single-phase product in the synthesis of new PURFs [30, 150]. However, this factory does not operate any more since the glycolysis process consisting of expensive glycolysis agents is not profitable [151]. On the other hand, another factory called Troy Polymers initiated glycolysis of PU foams based on a single-phase operation and successfully achieved in commercializing polyols for synthesis of PURFs. In addition, this recovered polyol has been used in production of PURFs for automotive industry since 2002 [30, 152]. These examples showed that there are still some economical and environmental feasibility problems associated with glycolysis procedure to use this recycling method effectively for recycling and recovery of PURFs on the industrial scale. Thus, for this reason, researchers both in academic and industrial research laboratories should try to find new alternatives or modified methods of glycolysis to conduct industrial scale recycling methods more efficiently, economically and environmental-friendly for producing high standard recovered PURF components that will be used in the fabrication of fresh PURF products.

References

1. Yu, Y.H., Kim, B.G., Lee, D.G.: Cryogenic reliability of composite insulation panels for liquefied natural gas (LNG) ships. Compos. Struct. **94**(2), 462–468 (2012)
2. Vladimirov, V.S., Lukin, E.S., Popova, N.A., Ilyukhin, M.A., Moizis, S.E., Moizis, E.S., Artamonov, M.A.: New types of light-weight refractory and heat-insulation materials for long-term use at extremely high temperatures. Glass Ceram. **68**(3–4), 116–122 (2011)

3. Fesmire, J.E., Coffman, B.E., Sass, J.P., Williams, M.K., Smith, T.M., Meneghelli, B.J.: Cryogenic moisture uptake in foam insulation for space launch vehicles. J. Spacecraft Rockets **49**(2), 220–230 (2012)
4. Kirpluks, M., Cabulis, U., Zeltins, V., Stiebra, L., Avots, A.: Rigid polyurethane foam thermal insulation protected with mineral intumescent mat. Autex Res. J. **14**(4), 259–269 (2014)
5. International Energy Agency. Technology Roadmap: Energy Efficient Building Envelopes (2013)
6. Cabeza, L.F., Castell, A., Medrano, M., Martorell, I., Perez, G., Fernandez, I.: Experimental study on the performance of insulation materials in Mediterranean construction. Energy Build. **42**(5), 630–636 (2010)
7. Daouas, N.: A study on optimum insulation thickness in walls and energy savings in Tunisian buildings based on analytical calculation of cooling and heating transmission loads. Appl. Energy **88**(1), 156–164 (2011)
8. Mahlia, T.M.I., Iqbal, A.: Cost benefits analysis and emission reductions of optimum thickness and air gaps for selected insulation materials for building walls in Maldives. Energy **35**(5), 2242–2250 (2010)
9. Tambach, M., Hasselaar, E., Itard, L.: Assessment of current Dutch energy transition policy instruments for the existing housing stock. Energy Policy **38**(2), 981–996 (2010)
10. Kallaos, J., Bohne, R.A., Hovde, P.J.: Long-term performance of rigid plastic foam building insulation. J. Mater. Civ. Eng. **26**(2), 374–378 (2014)
11. Jelle, B.P.: Traditional, state-of-the-art and future thermal building insulation materials and solutions—properties, requirements and possibilities. Energy Build. **43**(10), 2549–2563 (2011)
12. Wicklein, B., Kocjan, A., Salazar-Alvarez, G., Carosio, F., Camino, G., Antonietti, M., Bergstrom, L.: Thermally insulating and fire-retardant lightweight anisotropic foams based on nanocellulose and graphene oxide. Nat. Nanotechnol. **10**(3), 277–283 (2015)
13. Stovall, T.K., Fabian, B.A., Nelson, G.E., Beatty, D.R.: A comparison of accelerated aging test protocols for cellular foam insulation. Am. Soc. Test. Mater. **1426**, 379–391 (2002)
14. Fangareggi, A., Bertucellli, L.: Thermoset insulation materials in appliances, buildings and other applications. In: Thermosets. Woodhead Publ Mater, pp. 254–288 (2012)
15. Brun, N., Ungureanu, S., Deleuze, H., Backov, R.: Hybrid foams, colloids and beyond: from design to applications. Chem. Soc. Rev. **40**(2), 771–788 (2011)
16. Klempner, D., Frisch, K.C. (eds.): Handbook of Polymeric Foams and Foam Technology. Oxford University Press, New York (1991)
17. Gibson, L.J., Ashby, M.F., Harley, B.A.: Cellular Materials in Nature and Medicine. Cambridge University Press, Cambridge (2010)
18. Mittal, V. (ed.): Polymer Nanocomposite Foams. CRC Press, Boca Raton (2014)
19. Sonnenschein, M.F.: Polyurethanes: Science, Technology, Markets, and Trends, Vol. 11. Wiley, Hoboken (2014)
20. Thomas, S., Datta, J., Haponiuk, J., Reghunadhan, A. (Eds.): Polyurethane Polymers: Composites and Nanocomposites. Elsevier, Amsterdam (2017)
21. Lee, L.J., Zeng, C.C., Cao, X., Han, X.M., Shen, J., Xu, G.J.: Polymer nanocomposite foams. Compos. Sci. Technol. **65**(15–16), 2344–2363 (2005)
22. Mikos, A.G., Temenoff, J.S.: Formation of highly porous biodegradable scaffolds for tissue engineering. J. Biotechnol. **3**(2), 1 (2000)
23. Rosa, M.E.: An introduction to solid foams. Philos. Mag. Lett. **88**(9–10), 637–645 (2008)
24. Roucher, A., Depardieu, M., Pekin, D., Morvan, M., Backov, R.: Inorganic, hybridized and living macrocellular foams: "Out of the Box" heterogeneous catalysis. Chem. Rec. **18**(7–8), 776–787 (2018)
25. Klemper, D., Frisch, K.C. (Eds.): Handbook of Polymeric Foams and Foam Technology. Oxford University Press, NY (1991)
26. Yeh, J.M., Chang, K.C., Peng, C.W., Chand, B.G., Chiou, S.C., Huang, H.H., Lin, C.Y., Yang, J.C., Lin, H.R., Chen, C.L.: Preparation and insulation property studies of thermoplastic PMMA-silica nanocomposite foams. Polym. Compos. **30**(6), 715–722 (2009)

27. Kim, S.H., Park, H.C., Jeong, H.M., Kim, B.K.: Glass fiber reinforced rigid polyurethane foams. J. Mater. Sci. **45**(10), 2675–2680 (2010)
28. Lee, S.-T., Ramesh, N.S.: Polymeric Foams: Mechanisms and Materials, 2nd edn. CRC Press, Boca Raton, FL (2009)
29. Okolieocha, C., Raps, D., Subramaniam, K., Altstadt, V.: Microcellular to nanocellular polymer foams: progress (2004-2015) and future directions—a review. Eur. Polym. J. **73**, 500–519 (2015)
30. Simon, D., Borreguero, A.M., de Lucas, A., Rodriguez, J.F.: Recycling of polyurethanes from laboratory to industry, a journey towards the sustainability. Waste Manage. **76**, 147–171 (2018)
31. Stirna, U., Beverte, I., Yakushin, V., Cabulis, U.: Mechanical properties of rigid polyurethane foams at room and cryogenic temperatures. J. Cell. Plast. **47**(4), 337–355 (2011)
32. Kim, S.H., Kim, B.K., Lim, H.: Effect of isocyanate index on the properties of rigid polyurethane foams blown by HFC 365mfc. Macromol. Res. **16**(5), 467–472 (2008)
33. Bird, R.B., Stewart, W.E., Lightfoot, E.N.: Transport Phenomena. Wiley, New York (2006)
34. Song, B., Lu, W.Y., Syn, C.J., Chen, W.N.: The effects of strain rate, density, and temperature on the mechanical properties of polymethylene diisocyanate (PMDI)-based rigid polyurethane foams during compression. J. Mater. Sci. **44**(2), 351–357 (2009)
35. Cao, X., Lee, L.J., Widya, T., Macosko, C.: Polyurethane/clay nanocomposites foams: processing, structure and properties. Polymer **46**(3), 775–783 (2005)
36. Mondal, P., Khakhar, D.V.: Regulation of cell structure in water blown rigid polyurethane foam. Macromol. Symp. **216**, 241–254 (2004)
37. Zhang, X.B., Yao, L., Qiu, L.M., Gan, Z.H., Yang, R.P., Ma, X.J., Liu, Z.H.: Experimental study on cryogenic moisture uptake in polyurethane foam insulation material. Cryogenics **52**(12), 810–815 (2012)
38. Lee, J.R., Dhital, D.: Review of flaws and damages in space launch vehicle: structures. J. Intell. Mater. Syst. Struct. **24**(1), 4–20 (2013)
39. Petrovic, Z.S.: Polyurethanes from vegetable oils. Polym. Rev. **48**(1), 109–155 (2008)
40. Desroches, M., Escouvois, M., Auvergne, R., Caillol, S., Boutevin, B.: From vegetable oils to polyurethanes: synthetic routes to polyols and main industrial products. Polym. Rev. **52**(1), 38–79 (2012)
41. Kornev, K.G., Neimark, A.V., Rozhkov, A.N.: Foam in porous media: thermodynamic and hydrodynamic peculiarities. Adv. Colloid Interface Sci. **82**(1–3), 127–187 (1999)
42. Sauceau, M., Fages, J., Common, A., Nikitine, C., Rodier, E.: New challenges in polymer foaming: a review of extrusion processes assisted by supercritical carbon dioxide. Prog. Polym. Sci. **36**(6), 749–766 (2011)
43. Zatorski, W., Brzozowski, Z.K., Kolbrecki, A.: New developments in chemical modification of fire-safe rigid polyurethane foams. Polym. Degrad. Stab. **93**(11), 2071–2076 (2008)
44. Grunbauer, H.J.M., Bicerano, J., Clavel, P., Daussin, R.D., de Vos, H.A., Elwell, M.J., Kawabata, H., Kramer, H., Latham, D.D., Martin, C.A., Moore, S.E., Obi, B.C., Parenti, V., Schrock, A.K., van den Bosch, R.: Rigid polyurethane foams. In: Lee, S.T., Ramesh, N.S. (Eds.) Polymeric Foams: Mechanism and Materials. CRC Press, Boca Raton, FL (2004)
45. Widya, T., Macosko, C.W.: Nanoclay-modified rigid polyurethane foam. J. Macromol. Sci. Phys. **B44**(6), 897–908 (2005)
46. Kim, H., Miura, Y., Macosko, C.W.: Graphene/polyurethane nanocomposites for improved gas barrier and electrical conductivity. Chem. Mater. **22**(11), 3441–3450 (2010)
47. Chattopadhyay, D.K., Webster, D.C.: Thermal stability and flame retardancy of polyurethanes. Prog. Polym. Sci. **34**(10), 1068–1133 (2009)
48. Krol, P.: Synthesis methods, chemical structures and phase structures of linear polyurethanes. Properties and applications of linear polyurethanes in polyurethane elastomers, copolymers and ionomers. Prog. Mater. Sci. **52**(6), 915–1015 (2007)
49. Ghosh, B., Urban, M.W.: Self-repairing oxetane-substituted chitosan polyurethane networks. Science **323**(5920), 1458–1460 (2009)

50. Engels, H.W., Pirkl, H.G., Albers, R., Albach, R.W., Krause, J., Hoffmann, A., Casselmann, H., Dormish, J.: Polyurethanes: versatile materials and sustainable problem solvers for today's challenges. Angew. Chem. Int. Ed. **52**(36), 9422–9441 (2013)
51. Zia, K.M., Bhatti, H.N., Bhatti, I.A.: Methods for polyurethane and polyurethane composites, recycling and recovery: a review. React. Funct. Polym. **67**(8), 675–692 (2007)
52. Dweib, M.A., Vahlund, C.F., Bradaigh, M.C.O.: Fibre structure and anisotropy of glass reinforced thermoplastics. Compos. Part A: Appl. Sci. Manuf. **31**(3), 235–244 (2000)
53. Wirpsza, Z.: Polyurethane, Chemistry, Technology and Applications. Ellis Harwood, England (1993)
54. Dodge, J.: Polyurethane Chemistry, 2nd edn. Bayer Corp., Pittsburgh, PA (1999)
55. Bayer, A.G.: Polyurethane Application Research Department. "Bayer - Polyurethanes," Leverkusen, Germany. Edition January 1979
56. DeGaspari, J.: Mechanical Engineering Magazine (ASME), June 1999
57. Modesti, M., Simioni, F., Munari, R., Baldoin, N.: Recycling of flexible polyurethane foams with a low aromatic amine content. React. Funct. Polym. **26**(1–3), 157–165 (1995)
58. Zhang, J.H., Kong, Q.H., Yang, L.W., Wang, D.Y.: Few layered $Co(OH)_2$ ultrathin nanosheet-based polyurethane nanocomposites with reduced fire hazard: from eco-friendly flame retardance to sustainable recycling. Green Chem. **18**(10), 3066–3074 (2016)
59. Chattopadhyay, D.K., Raju, K.V.S.N.: Structural engineering of polyurethane coatings for high performance applications. Prog. Polym. Sci. **32**(3), 352–418 (2007)
60. Szycher, M.: Szycher's Handbook of Polyurethanes. CRC Press, Boca Raton, FL (1999)
61. Sawpan, M.A.: Polyurethanes from vegetable oils and applications: a review. J. Polym. Res. **25**(8), 184 (2018)
62. Ionescu, M.: Chemistry and Technology of Polyols for Polyurethanes. Rapra Technology Limited (2005)
63. Yilgor, E., Burgaz, E., Yurtsever, E., Yilgor, I.: Comparison of hydrogen bonding in polydimethylsiloxane and polyether based urethane and urea copolymers. Polymer **41**(3), 849–857 (2000)
64. Hill, K.: Fats and oils as oleochemical raw materials. Pure Appl. Chem. **72**(7), 1255–1264 (2000)
65. de Espinosa, L.M., Meier, M.A.R.: Plant oils: the perfect renewable resource for polymer science. Eur. Polym. J. **47**(5), 837–852 (2011)
66. Williams, C.K., Hillmyer, M.A.: Polymers from renewable resources: a perspective for a special issue of polymer reviews. Polym. Rev. **48**(1), 1–10 (2008)
67. Gandini, A.: Polymers from renewable resources: a challenge for the future of macromolecular materials. Macromolecules **41**(24), 9491–9504 (2008)
68. van Haveren, J., Scott, E.L., Sanders, J.: Bulk chemicals from biomass. Biofuels Bioprod. Biorefin. **2**(1), 41–57 (2008)
69. Campanella, A., Bonnaillie, L.M., Wool, R.P.: Polyurethane foams from soyoil-based polyols. J. Appl. Polym. Sci. **112**(4), 2567–2578 (2009)
70. Zlatanic, A., Lava, C., Zhang, W., Petrovic, Z.S.: Effect of structure on properties of polyols and polyurethanes based on different vegetable oils. J. Polym. Sci. Part B: Polym. Phys. **42**(5), 809–819 (2004)
71. Miao, S.D., Zhang, S.P., Su, Z.G., Wang, P.: A novel vegetable oil-lactate hybrid monomer for synthesis of high-T_g polyurethanes. J. Polym. Sci. Part A: Polym. Chem. **48**(1), 243–250 (2010)
72. Cabulis, U., Kirpluks, M., Stirna, U., Lopez, M.J., Vargas-Garcia, M.D., Suarez-Estrella, F., Moreno, J.: Rigid polyurethane foams obtained from tall oil and filled with natural fibers: application as a support for immobilization of lignin-degrading microorganisms. J. Cell. Plast. **48**(6), 500–515 (2012)
73. Lligadas, G., Ronda, J.C., Galia, M., Cadiz, V.: Plant oils as platform chemicals for polyurethane synthesis: current state-of-the-art. Biomacromol **11**(11), 2825–2835 (2010)
74. Kim, Y.H., Kang, M.J., Park, G.P., Park, S.D., Kim, S.B., Kim, W.N.: Effects of liquid-type silane additives and organoclay on the morphology and thermal conductivity of rigid polyisocyanurate-polyurethane foams. J. Appl. Polym. Sci. **124**(4), 3117–3123 (2012)

75. Hoogendorn, C.J.: Thermal ageing. In: Cunningham, A., Hilyard, N.C. (Eds.) Low Density Cellular Plastics: Physical Basis of Behavior. Chapman and Hall, London (1994)
76. Harikrishnan, G., Lindsay, C.I., Arunagirinathan, M.A., Macosko, C.W.: Probing nanodispersions of clays for reactive foaming. ACS Appl. Mater. Interfaces **1**(9), 1913–1918 (2009)
77. Burgaz, E.: Thermomechanical analysis of polymer nanocomposites. In: Huang, X., Zhi, C. (eds.) Polymer Nanocomposites. Springer, Cham (2016)
78. Yu, Y.H., Nam, S., Lee, D., Lee, D.G.: Cryogenic impact resistance of chopped fiber reinforced polyurethane foam. Compos. Struct. **132**, 12–19 (2015)
79. Aranguren, M.I., Racz, I., Marcovich, N.E.: Microfoams based on castor oil polyurethanes and vegetable fibers. J. Appl. Polym. Sci. **105**(5), 2791–2800 (2007)
80. Park, S.B., Choi, S.W., Kim, J.H., Bang, C.S., Lee, J.M.: Effect of the blowing agent on the low-temperature mechanical properties of CO_2- and HFC-245fa-blown glass-fiber-reinforced polyurethane foams. Compos. Part B: Eng. **93**, 317–327 (2016)
81. Xue, B.L., Wen, J.L., Sun, R.C.: Lignin-based rigid polyurethane foam reinforced with pulp fiber: synthesis and characterization. ACS Sustain. Chem. Eng. **2**(6), 1474–1480 (2014)
82. Yu, Y.H., Choi, I., Nam, S., Lee, D.G.: Cryogenic characteristics of chopped glass fiber reinforced polyurethane foam. Compos. Struct. **107**, 476–481 (2014)
83. Czuprynski, B., Paciorek-Sadowska, J., Liszkowska, J.: Properties of rigid polyurethane-polyisocyanurate foams modified with the selected fillers. J. Appl. Polym. Sci. **115**(4), 2460–2469 (2010)
84. Nikje, M.M.A., Garmarudi, A.B., Haghshenas, M.: Effect of talc filler on physical properties of polyurethane rigid foams. Polym. Plast. Technol. **45**(11), 1213–1217 (2006)
85. Li, J., Mo, X.H., Li, Y., Zou, H.W., Liang, M., Chen, Y.: Influence of expandable graphite particle size on the synergy flame retardant property between expandable graphite and ammonium polyphosphate in semi-rigid polyurethane foam. Polym. Bull. **75**(11), 5287–5304 (2018)
86. Liu, D.Y., Zhao, B., Wang, J.S., Liu, P.W., Liu, Y.Q.: Flame retardation and thermal stability of novel phosphoramide/expandable graphite in rigid polyurethane foam. J. Appl. Polym. Sci. **135**(27), 46434 (2018)
87. Shi, X.X., Jiang, S.H., Zhu, J.Y., Li, G.H., Peng, X.F.: Establishment of a highly efficient flame-retardant system for rigid polyurethane foams based on bi-phase flame-retardant actions. RSC Adv. **8**(18), 9985–9995 (2018)
88. Thirumal, M., Khastgir, D., Singha, N.K., Manjunath, B.S., Naik, Y.P.: Effect of expandable graphite on the properties of intumescent flame-retardant polyurethane foam. J. Appl. Polym. Sci. **110**(5), 2586–2594 (2008)
89. Cheng, J.J., Shi, B.B., Zhou, F.B., Chen, X.Y.: Effects of inorganic fillers on the flame-retardant and mechanical properties of rigid polyurethane foams. J. Appl. Polym. Sci. **131**(10), 40253 (2014)
90. Ye, L., Meng, X.Y., Ji, X., Li, Z.M., Tang, J.H.: Synthesis and characterization of expandable graphite-poly(methyl methacrylate) composite particles and their application to flame retardation of rigid polyurethane foams. Polym. Degrad. Stab. **94**(6), 971–979 (2009)
91. Yuan, Y., Yang, H.Y., Yu, B., Shi, Y.Q., Wang, W., Song, L., Hu, Y., Zhang, Y.M.: Phosphorus and nitrogen-containing polyols: synergistic effect on the thermal property and flame retardancy of rigid polyurethane foam composites. Ind. Eng. Chem. Res. **55**(41), 10813–10822 (2016)
92. Thirumal, M., Khastgir, D., Singha, N.K., Manjunath, B.S., Naik, Y.P.: Mechanical, morphological and thermal properties of rigid polyurethane foam: effect of the fillers. Cell. Polym. **26**(4), 245–259 (2007)
93. Saint-Michel, F., Chazeau, L., Cavaille, J.Y.: Mechanical properties of high density polyurethane foams: II effect of the filler size. Compos. Sci. Technol. **66**(15), 2709–2718 (2006)
94. Javni, I., Zhang, W., Karajkov, V., Petrovic, Z.S., Divjakovic, V.: Effect of nano- and micro-silica fillers on polyurethane foam properties. J. Cell. Plast. **38**(3), 229–239 (2002)
95. Chatterjee, A., Mishra, S.: Nano-calcium carbonate ($CaCO_3$)/polystyrene (PS) core-shell nanoparticle: it's effect on physical and mechanical properties of high impact polystyrene (HIPS). J. Polym. Res. **20**(10) (2013)

96. Santiago-Calvo, M., Tirado-Mediavilla, J., Ruiz-Herrero, J.L., Rodriguez-Perez, M.A., Villafane, F.: The effects of functional nanofillers on the reaction kinetics, microstructure, thermal and mechanical properties of water blown rigid polyurethane foams. Polymer **150**, 138–149 (2018)
97. Han, M.S., Kim, Y.H., Han, S.J., Choi, S.J., Kim, S.B., Kim, W.N.: Effects of a silane coupling agent on the exfoliation of organoclay layers in polyurethane/organoclay nanocomposite foams? J. Appl. Polym. Sci. **110**(1), 376–386 (2008)
98. Wang, Z.Z., Li, X.Y.: Mechanical properties and flame retardancy of rigid polyurethane foams containing SiO_2 nanospheres/graphene oxide hybrid and dimethyl methylphosphonate. Polym. Plast. Technol. **57**(9), 884–892 (2018)
99. Lorenzetti, A., Roso, M., Bruschetta, A., Boaretti, C., Modesti, M.: Polyurethane-graphene nanocomposite foams with enhanced thermal insulating properties. Polym. Adv. Technol. **27**(3), 303–307 (2016)
100. Palanisamy, A.: Water-blown polyurethane-clay nanocomposite foams from biopolyol—effect of nanoclay on the properties. Polym. Compos. **34**(8), 1306–1312 (2013)
101. Semenzato, S., Lorenzetti, A., Modesti, M., Ugel, E., Hrelja, D., Besco, S., Michelin, R.A., Sassi, A., Facchin, G., Zorzi, F., Bertani, R.: A novel phosphorus polyurethane FOAM/montmorillonite nanocomposite: preparation, characterization and thermal behaviour. Appl. Clay Sci. **44**(1–2), 35–42 (2009)
102. Park, Y.T., Qian, Y.Q., Lindsay, C.I., Nijs, C., Camargo, R.E., Stein, A., Macosko, C.W.: Polyol-assisted vermiculite dispersion in polyurethane nanocomposites. ACS Appl. Mater. Interfaces **5**(8), 3054–3062 (2013)
103. Kim, S.H., Lee, M.C., Kim, H.D., Park, H.C., Jeong, H.M., Yoon, K.S., Kim, B.K.: Nanoclay reinforced rigid polyurethane foams. J. Appl. Polym. Sci. **117**(4), 1992–1997 (2010)
104. Patro, T.U., Harikrishnan, G., Misra, A., Khakhar, D.V.: Formation and characterization of polyurethane-vermiculite clay nanocomposite foams. Polym. Eng. Sci. **48**(9), 1778–1784 (2008)
105. Mondal, P., Khakhar, D.V.: Rigid polyurethane-clay nanocomposite foams: preparation and properties. J. Appl. Polym. Sci. **103**(5), 2802–2809 (2007)
106. Tan, S.Q., Abraham, T., Ference, D., Macosko, C.W.: Rigid polyurethane foams from a soybean oil-based polyol. Polymer **52**(13), 2840–2846 (2011)
107. Modesti, M., Lorenzetti, A., Dall'Acqua, C.: Long-term performance of environmentally-friendly blown polyurethane foams. Polym. Eng. Sci. **45**(3), 260–270 (2005)
108. Kim, Y.H., Choi, S.J., Kim, A.M., Han, M.S., Kim, W.N., Bang, K.T.: Effects of organoclay on the thermal insulating properties of rigid polyurethane foams blown by environmentally friendly blowing agents. Macromol. Res. **15**(7), 676–681 (2007)
109. Zhang, X.D., Macosko, C.W., Davis, H.T., Nikolov, A.D., Wasan, D.T.: Role of silicone surfactant in flexible polyurethane foam. J. Colloid Interface Sci. **215**(2), 270–279 (1999)
110. Santiago-Calvo, M., Blasco, V., Ruiz, C., Paris, R., Villafane, F., Rodriguez-Perez, M.A.: Synthesis, characterization and physical properties of rigid polyurethane foams prepared with poly(propylene oxide) polyols containing graphene oxide. Eur. Polym. J. **97**, 230–240 (2017)
111. Kim, J.M., Kim, J.H., Ahn, J.H., Kim, J.D., Park, S., Park, K.H., Lee, J.M.: Synthesis of nanoparticle-enhanced polyurethane foams and evaluation of mechanical characteristics. Compos. Part B: Eng. **136**, 28–38 (2018)
112. Hoseinabadi, M., Naderi, M., Najafi, M., Motahari, S., Shokri, M.: A study of rigid polyurethane foams: the effect of synthesized polyols and nanoporous graphene. J. Appl. Polym. Sci. **134**(26) (2017)
113. Novoselov, K.S., Geim, A.K., Morozov, S.V., Jiang, D., Zhang, Y., Dubonos, S.V., Grigorieva, I.V., Firsov, A.A.: Electric field effect in atomically thin carbon films. Science **306**(5696), 666–669 (2004)
114. Chen, K.P., Cao, F., Liang, S.E., Wang, J.H., Tian, C.R.: Preparation of poly(ethylene oxide) brush-grafted multiwall carbon nanotubes and their effect on morphology and mechanical properties of rigid polyurethane foam. Polym. Int. **67**(11), 1545–1554 (2018)

115. Wang, X., Kalali, E.N., Wan, J.T., Wang, D.Y.: Carbon-family materials for flame retardant polymeric materials. Prog. Polym. Sci. **69**, 22–46 (2017)
116. Madaleno, L., Pyrz, R., Crosky, A., Jensen, L.R., Rauhe, J.C.M., Dolomanova, V., Timmons, A.M.M.V.D., Pinto, J.J.C., Norman, J.: Processing and characterization of polyurethane nanocomposite foam reinforced with montmorillonite-carbon nanotube hybrids. Compos. Part A: Appl. Sci. Manuf. **44**, 1–7 (2013)
117. Caglayan, C., Gurkan, I., Gungor, S., Cebeci, H.: The effect of CNT-reinforced polyurethane foam cores to flexural properties of sandwich composites. Compos. Part A: Appl. Sci. Manuf. **115**, 187–195 (2018)
118. Yan, D.X., Dai, K., Xiang, Z.D., Li, Z.M., Ji, X., Zhang, W.Q.: Electrical conductivity and major mechanical and thermal properties of carbon nanotube-filled polyurethane foams. J. Appl. Polym. Sci. **120**(5), 3014–3019 (2011)
119. Yaghoubi, A., Nikje, M.M.A.: Silanization of multi-walled carbon nanotubes and the study of its effects on the properties of polyurethane rigid foam nanocomposites. Compos. Part A: Appl. Sci. Manuf. **109**, 338–344 (2018)
120. Raja, M., Ryu, S.H., Shanmugharaj, A.M.: Influence of surface modified multiwalled carbon nanotubes on the mechanical and electroactive shape memory properties of polyurethane (PU)/poly(vinylidene diflouride) (PVDF) composites. Colloids Surf. A **450**, 59–66 (2014)
121. Zhang, L.F., Yilmaz, E.D., Schjodt-Thomsen, J., Rauhe, J.C., Pyrz, R.: MWNT reinforced polyurethane foam: processing, characterization and modelling of mechanical properties. Compos. Sci. Technol. **71**(6), 877–884 (2011)
122. Ganesan, Y., Salahshoor, H., Peng, C., Khabashesku, V., Zhang, J.N., Cate, A., Rahbar, N., Lou, J.: Fracture toughness of the sidewall fluorinated carbon nanotube-epoxy interface. J. Appl. Phys. **115**(22) (2014)
123. Dolomanova, V., Rauhe, J.C.M., Jensen, L.R., Pyrz, R., Timmons, A.B.: Mechanical properties and morphology of nano-reinforced rigid PU foam. J. Cell. Plast. **47**(1), 81–93 (2011)
124. Akkoyun, M., Suvaci, E.: Effects of TiO_2, ZnO, and Fe_3O_4 nanofillers on rheological behavior, microstructure, and reaction kinetics of rigid polyurethane foams. J. Appl. Polym. Sci. **133**(28) (2016)
125. Nikje, M.M.A., Tehrani, Z.M.: Thermal and mechanical properties of polyurethane rigid foam/modified nanosilica composite. Polym. Eng. Sci. **50**(3), 468–473 (2010)
126. Liu, W.S., Kong, J.H., Toh, W.E., Zhou, R., Ding, G.Q., Huang, S., Dong, Y.L., Lu, X.H.: Toughening of epoxies by covalently anchoring triazole-functionalized stacked-cup carbon nanofibers. Compos. Sci. Technol. **85**, 1–9 (2013)
127. Ding, J., Maitra, P., Wunder, S.L.: Characterization of the interaction of poly(ethylene oxide) with nanosize fumed silica: surface effects on crystallization. J. Polym. Sci. Part B: Polym. Phys. **41**(17), 1978–1993 (2003)
128. Kweon, J.O., Noh, S.T.: Thermal, thermomechanical, and electrochemical characterization of the organic-inorganic hybrids poly(ethylene oxide) (PEO)-silica and PEO-silica-$LiClO_4$. J. Appl. Polym. Sci. **81**(10), 2471–2479 (2001)
129. Sun, Y.Y., Zhang, Z.Q., Moon, K.S., Wong, C.P.: Glass transition and relaxation behavior of epoxy nanocomposites. J. Polym. Sci. Part B: Polym. Phys. **42**(21), 3849–3858 (2004)
130. Raghavan, S.R., Walls, H.J., Khan, S.A.: Rheology of silica dispersions in organic liquids: new evidence for solvation forces dictated by hydrogen bonding. Langmuir **16**(21), 7920–7930 (2000)
131. Cassagnau, P.: Melt rheology of organoclay and fumed silica nanocomposites. Polymer **49**(9), 2183–2196 (2008)
132. Javni, I., Zhang, W., Petrovic, Z.S.: Effect of different isocyanates on properties of soy-based polyurethanes. Abstr. Pap. Am. Chem. Soc. **223**, D96–D96 (2002)
133. Mahmood, N., Yuan, Z.S., Schmidt, J., Xu, C.: Depolymerization of lignins and their applications for the preparation of polyols and rigid polyurethane foams: a review. Renew. Sustain. Energy Rev. **60**, 317–329 (2016)
134. Demharter, A.: Polyurethane rigid foam, a proven thermal insulating material for applications between $+130$ °C and -196 °C. Cryogenics **38**(1), 113–117 (1998)

135. Kim, S.H., Lim, H., Song, J.C., Kim, B.K.: Effect of blowing agent type in rigid polyurethane foam. J. Macromol. Sci. Part A **45**(4), 323–327 (2008)
136. Cooper, A.I.: Polymer synthesis and processing using supercritical carbon dioxide. J. Mater. Chem. **10**(2), 207–234 (2000)
137. Seo, W.J., Jung, H.C., Hyun, J.C., Kim, W.N., Lee, Y.B., Choe, K.H., Kim, S.B.: Mechanical, morphological, and thermal properties of rigid polyurethane foams blown by distilled water. J. Appl. Polym. Sci. **90**(1), 12–21 (2003)
138. Page, M.C., Glicksman, L.R.: Measurements of diffusion coefficient of alternate blowing agents in closed cell foam insulation. J. Cell. Plast. **28**(3), 268–283 (1992)
139. Tao, W.H., Hsu, H.C., Chang, C.C., Hsu, C.L., Lin, Y.S.: Measurement and prediction of thermal conductivity of open cell rigid polyurethane foam. J. Cell. Plast. **37**(4), 310–332 (2001)
140. Tao, W.H., Sung, W.F., Lin, J.Y.: Development of vacuum insulation panel systems. J. Cell. Plast. **33**(6), 545–556 (1997)
141. Tao, W.H., Chang, C.C., Lin, J.Y.: An energy-efficiency performance study of vacuum insulation panels. J. Cell. Plast. **36**(6), 441–450 (2000). (Reprinted)
142. Parks, K.L., Beckman, E.J.: Generation of microcellular polyurethane foams via polymerization in carbon dioxide. 2. Foam formation and characterization. Polym. Eng. Sci. **36**(19), 2417–2431 (1996)
143. Kim, C., Youn, J.R.: Environmentally friendly processing of polyurethane foam for thermal insulation. Polym. Plast. Technol. **39**(1), 163–185 (2000)
144. Park, H., Youn, J.R.: Study on reaction injection molding of polyurethane microcellular foam. Polym. Eng. Sci. **35**(23), 1899–1906 (1995)
145. Seo, W.J., Sung, Y.T., Kim, S.B., Lee, Y.B., Choe, K.H., Choe, S.H., Sung, J.Y., Kim, W.N.: Effects of ultrasound on the synthesis and properties of polyurethane foam/clay nanocomposites. J. Appl. Polym. Sci. **102**(4), 3764–3773 (2006)
146. Choi, W.J., Kim, S.H., Kim, Y.J., Kim, S.C.: Synthesis of chain-extended organifier and properties of polyurethane/clay nanocomposites. Polymer **45**(17), 6045–6057 (2004)
147. Choe, K.H., Lee, D.S., Seo, W.J., Kim, W.N.: Properties of rigid polyurethane foams with blowing agents and catalysts. Polym. J. **36**(5), 368–373 (2004)
148. Pattanayak, A., Jana, S.C.: Properties of bulk-polymerized thermoplastic polyurethane nanocomposites. Polymer **46**(10), 3394–3406 (2005)
149. Kuan, H.C., Ma, C.C.M., Chuang, W.P., Su, H.Y.: Hydrogen bonding, mechanical properties, and surface morphology of clay/waterborne polyurethane nanocomposites. J. Polym. Sci. Part B: Polym. Phys. **43**(1), 1–12 (2005)
150. Raßhofer, W., Weigand, E.: Advances in Plastics Recycling. Vol. 2: Automotive Polyurethanes. CRC Press, Boca Raton (2001)
151. Behrendt, G., Naber, B.W.: The recycling of polyurethanes (review). J. Univ. Chem. Technol. Metall. **44**(1), 3–23 (2009)
152. Sendijarevic, I., Harris, J.G., Hoffmann, S., Heric, S., Lathia, N., Sendijarevic, V.: Polyether polyols from scrap polyurethanes—for use in rigid and flexible foams. Polyurethanes Mag. Int. **8**(1), 55–60 (2010)

Chapter 2
PU Rigid Composite Foams Containing Micron-Sized Fillers

2.1 Introduction

The impact behaviour, thermal insulation and stability, compression and tensile strengths and flame retardant behavior of PURFs have been tried to enhance by using various micron-sized fillers with different geometries [1]. The micron-sized fillers can be classified based on the shape of fillers such as cylindrical, plate-like and spherical geometries [2]. For cylindrical micron-sized fillers, fibers can be given as the mostly used filler example in PURMCFs [1, 3–5]. For more than two decades, in addition to synthetic fibers, researchers have concentrated their efforts and energy on the production of bio-based fibers for durability, sustainability and protection of the environment. Thus, for this reason, vegetable and cellulose-based fibers have been used in the fabrication of PURMCFs due to their cheap price, easier accessibility and improved aspects of biodegradability, tensile strength, modulus and recyclability [4, 6]. Previously, thermal properties such as thermal stability and fire resistance and mechanical properties of PURMCFs were enhanced with the incorporation of aramid fibers which are seen as economically feasible and environment friendly alternatives to halogenated flame retardants [7].

Halogenated phosphate compounds and antimony trioxide or combination of their mixtures have been used as typical flame retardant materials in PURFs [8, 9]. However, for more than two decades, using flame-retardant materials consisting of halogen elements has been kept under control since corrosive and toxic smoke is released by these materials to the environment in the presence of fire [8, 9]. Thus, because of the safety and prevention of pollution in environment, many regulations have been implemented on the use of flame-retardant materials consisting of halogen free elements instead of halogenated ones. For this reason, many studies have been performed both in academia and industry to develop highly efficient and environmentally friendly flame retardants for a very long time. A short time ago, phosphorus and intumescent based materials have been focused in a great extent since they form char on the

E. Burgaz, *Polyurethane Insulation Foams for Energy and Sustainability*,
Advanced Structured Materials 111,
https://doi.org/10.1007/978-3-030-19558-8_2

surface of polymers and this property induces flame protection as well as prevention of any toxic smoke during fire [8]. Although flame-retardant materials with swelling and charring abilities have been used in the field of coatings for more than 50 years, they have been only used with success in the area of PU rigid foams for about 20 years [8, 9]. EG is a typical halogen-free intumescent flame retardant material for PURFs which is synthesized in the medium of sulfuric and nitric acid mixture to form the intercalated structure from graphite layers. At higher temperatures, EG expands to higher volumes, forms an intumescent carbonaceous layer on its surface. When EG is used in PURFs, also at higher temperatures, this insulative protective layer with charring and swelling properties covers the surface of PURFs. Thus, the oxygen and heat transport throughout PURFs is obstructed by this thermally insulated worm-like layer [10, 11]. For this reason, researchers both in academia and industry have concentrated on the effects of EG addition on flame retardancy of PURFs since the solution of this topic has become very important due to the beneficial properties of expandable graphite for the PURF system.

Among publications which were written on the topic of PURF/fiber in the literature, the most extensively investigated fibers are glass [1, 5, 7, 12–17], carbon [15] vegetable [4, 6, 18–25], chopped [3, 5], aramid [26] and cellulose [27–29] fibers which have attracted a great deal of attention from researchers due to their outstanding load bearing capacities, higher thermal insulation, thermal degradation, thermomechanical and fire-resistance properties. The effects of using glass fibers in PURFs have been investigated by several researchers in terms of closed cellular morphology, mechanical and thermal properties [1, 5, 7, 12–17]. Previously, it was reported that 3.5 wt% addition of carbon fibers to PURFs leads to the optimum reinforcement effect in terms of mechanical properties [15].

The addition of different maple pulp fiber contents to soy and lignin-based rigid PU foams was specifically studied from the perspective of thermal and mechanical properties to PURF composites with improved environmental properties [6, 22]. Previously, the impact resistance of PURMCFs consisting of chopped E-glass fibers was investigated at very low temperatures, and it was found out that the addition of chopped fiber to pure PURF considerably increases the impact resistance of PURMCFs [3]. In addition, various blowing agents such as CO_2 and HFC-245fa were used in the reactive foaming process of PURCFs consisting of different amounts of glass fibers at various temperatures and strain rates. Based on the results of mechanical properties, it was observed that the compressive stress values of PURCFS filled with different amounts of glass fibers were reduced with increasing amounts of strain rates at temperatures much lower than the freezing point of blowing agent [5].

Recently, PURMCFs were prepared by flame retardant materials such as a specific type of phosphate compounds and aramid fiber which was surface modified with special type of surfactant molecules. Based on thermal decomposition results, it was clearly shown that using both of these flame retardant materials with optimum contents in PURFs displays lower amount of fume and poisonous gas discharge and increases the amount of burned material in comparison with PURF consisting of only phosphate compound [26]. In recent years, lignin-based cellulose fibers consisting of many hydroxyl groups have been used as an economical and green alternative reinforcement in PURMCFs due to its easier decomposition through fungal strike in nature [6, 21, 22, 27]. Previously, PURMCFs consisting of woven flax and jute fabrics were prepared, and it was reported that mechanical strengths of PURFs consisting of flax fibers were improved much more than those foams containing jute fibers [28]. In addition to using only fibers in PURFs, ternary composite foams consisting of both fiber and hydrophilic nanosilica were fabricated by varying compositions of glass fiber and nanosilica. The addition of nanosilica to PURFs consisting of 2% fiber content increased the values of foam density, reactive foaming duration and compression strength whereas the glass transition and thermal decomposition temperatures reached to their maximum values at 3% silica loading. Furthermore, it was also pointed out that with the addition of 3% nanosilica, the value of thermal conductivity in PURCFs consisting of glass fibers exhibited a systematic decrease way below than that of pure PURF [17]. Thus, this study clearly showed that using simultaneously a micron-sized fiber and a nano-sized particle in PURFs provide improvements in terms of thermal insulation, reaction kinetics, thermal degradation, mechanical and thermomechanical properties compared to pure PURF and PURFs consisting of only fibers as the fillers.

Among plate-like micron-sized fillers, talc [30, 31] and expandable graphite [8, 9, 11, 32–41] can be given as examples that have been used in PURMCFs. However, among previously published PURF/plate-like micron-sized filler studies, one of the extensively used filler material has been expandable graphite (EG) because of its excellent properties such as high flame retardancy, low price, environmentally friendliness, high porosity, excellent expansibility, high temperature resistance, thermal insulation and exchange surface [32, 33, 41, 42]. On the other hand, a few studies about PURMCFs consisting of talc are available in the literature. Nikje et al. used talc as a micron-sized filler in PURFs to understand the effects of using talc on physical and mechanical properties of PURMCFs [31]. Based on physical and mechanical tests, it was shown that elastic modulus increased in both directions while tensile strength and elongation at break values did not exhibit big changes compared to pure PURF. Therefore, it was concluded that talc with a content of 20% can be used in PURFs as a cheaper micron-sized plate-like filler which does not deteriorate the physical properties of PURFs [31]. In another study, using both talc and aluminum hydroxide with an amount of 2.5–20 wt% in PURFs improved physical and mechanical properties in comparison with pure PURF. It was reported that the compression properties, foam density, amount of closed cells and thermal decomposition temperatures increase, and the brittleness and flammability of PURMCFs decrease with the addition of talc to PURFs [30].

Recently, in the literature, many experimental works were reported about the decrease of flammability properties and improvement of thermal stability, fire resistance and mechanical properties of PURFs. Previously, influences of EG particle size on flame retardancy behavior and closed-cellular morphology of PURFs were examined, and it was found out that suitable particle size could increase flame retardancy [32, 43]. Furthermore, the effects of different particle sizes and contents on thermal, mechanical, and flame-retardant properties and water absorption behavior of water-blown, low-density PURFs have been investigated, and it was reported that mechanical and insulation properties of PURFs decrease, and the flame retardant properties increase with increasing EG loading. In addition, PURMCFs consisting of EG with a larger particle size displayed improved flame retardant behavior and mechanical properties compared to microcomposite foams containing EG with a smaller particle size [8]. In addition, effects of using different amounts of EG on flame retardant and mechanical properties of PURFs were studied by several researchers, and they reported that the flame retardancy behavior of PURMCFs could be enhanced with increasing amounts of EG, however their mechanical properties were worsened in comparison with pure PURF [43–49]. The main reason for the decrease of mechanical properties in PURMCFs filled with EG particles is the disintegration of the proper closed cell morphology of rigid foams due to the larger sizes of EG particles. In addition, the presence of poor interactions between PU matrix and EG particles leads to heterogeneously dispersion of EG particles within PURF matrix, and thus decreases the mechanical properties [9]. Consequently, homogenous dispersion of EG particles, the favorable interaction between PURF matrix and EG particles and the well-preservation the closed cell structure should be achieved in the design of advanced PURMCFs with improved mechanical properties. In order to increase both mechanical properties and flame retardancy behavior of PURFs, researchers have used the synergistic method of using binary mixtures of EG with other micron-sized fillers in the fabrication of PURMCFs.

Previously, rigid PU microcomposites consisting of glass fibers (GFs), expandable graphite and hollow glass microspheres (HGMs) were produced, and it was found out that the addition of HGMs and GFs to PURFs improves the mechanical properties, while the addition of EG increases the flame retardancy behavior of PURF system [32, 50, 51]. Thus, the proper selection of binary mixtures of micron-sized fillers for PURMCFs is very important in order to obtain improved properties in terms of flammability, thermal degradation and mechanical properties. In recent years, researchers have practised the method of binary mixtures of EG with other flame retardants in order to effectively and systematically elevate the flame-retardancy behavior of PURFs. Various binary flame retardant material combinations such as EG and ammonium polyphosphate (APP) [45, 46], EG/hexaphenoxy-cyclotriphosphazene (HPCP) [37], EG and hollow glass microsphere [52], EG and whisker silicon oxide [53], and EG and dimethyl methylphosphonate (DMMP) [54, 55] have been studied and it was clearly shown that choosing the proper composition and combination of flame retardants might systematically improve the flame retardancy behavior of PURFs [37–39].

Among spherical micron-sized fillers, silica [56–58], calcium carbonate [56, 57], aluminum powder [59], glass powder and microspheres [32, 50, 51, 56], carbon black [60] can be given as examples that have been used in PURMCFs. Previously, PURFs consisting of different fillers such as precipitated types of silica and calcium carbonate and glass powder were prepared by using distilled water as the blowing agent, and the effects of increasing amount of fillers on thermal, mechanical, closed-cellular morphology and water absorption behavior of PURFs were systematically examined [56]. Moreover, it was also reported that the compressive stress and modulus and hardness values of PURMCFs consisting of fillers such as precipitated types of silica and calcium carbonate and glass powder decrease compared to pure PURF. In PURMCFs consisting of silica or calcium carbonate, the insulation properties decreased with increasing amounts of micron-sized fillers. On the contrary, in the case of glass powder, it was shown that thermal conductivity initially decreases, but subsequently increases with further addition of filler [56]. In addition, the micron-sized silica with a particle size of 1.5 μm was utilized in the fabrication of PURMCFs [58], and it was shown that the reduced density of PURMCFs does not change that much with the addition of silica. Furthermore, it was also revealed that the strengths of PURMCFs consisting of micron-sized silica particles with concentrations of 5 and 10% are slightly increased, but at higher concentrations of micro-silica, the strength of foams is reduced due to the irregularity of the closed-cell structure caused by the higher contents of micron-sized silica particles [58].

The effects of spherical micron-sized particle size on morphology and mechanical properties of PURMCFs were explored by using calcium carbonate with two different sizes and crystallized silica particles [57]. The mechanical results were compared with modelling and simulation results, and it was shown that the reinforcement of PURFs is not satisfactory if the size of filler is much larger than the size of cells. Moreover, it was also revealed that the addition of certain amount of fillers enhances the crosslink density of the PURMCF matrix above the glass transition temperature [57].

Like glass fibers, hollow glass microspheres (HGM) have been generally used as reinforcement fillers in PURFs [32]. Previously, several research groups concentrated on using HGM as micron-sized fillers in PURCFs and they basically examined influences of using these specific type of fillers on mechanical properties of low density PURMCFs [32, 50, 51]. Their results showed that compressive property of PURFs is enhanced with the increase of HGM content [32, 50]. PURMCFs consisting of micron-sized particles such as EG and whisker silicon oxide were fabricated in order to induce positive improvements in both dynamic mechanical properties and flame retardancy behavior. It was shown that the PURMCF sample consisting of 10 wt% WSi is the optimum foam in terms of enjoying the highest values with regard to compressive strength and modulus [53]. It was also revealed that using 10 wt% WSi in PURFs filled with EG particles provided much more synergetical enhancement concerning mechanical properties and flame retardancy behaviour in comparison with PURMCFs consisting of only EG particles. Thus, these results clearly showed that using binary mixtures of fillers such as EG and WSi in PURMCFs is more beneficial in terms of obtaining improved mechanical, dynamic mechanical and flame retardant properties and thermal degradation behavior compared to the case of using only one type of filler in PURMCFs [53].

2.2 Morphology

Park et al. fabricated PURCFs consisting of glass fibers that were foamed by using the blowing agents such as CO_2^- and HFC-245fa. Authors studied mechanical properties of PURCFs with respect to changing temperature and diverse strain rate values [5]. In accordance with data obtained from mechanical properties and closed-cellular morphology, it was revealed that closed cellular morphology and mechanical properties of PURCFs were substantially influenced by the type of blowing agent that was used during the reactive foaming process of PURCFs [5]. SEM micrographs of specimen sections which were cut perpendicular and parallel to foam growth direction are given in Fig. 2.1.

Morphological results in Fig. 2.1 clearly verify the closed-cellular morphology of PURFs. In addition, it was pointed out that all of the PURCF samples displayed a closed-cellular morphology consisting of an anisotropic cell geometry which is a direct influence of the reactive foaming process. Furthermore, based on the analysis of SEM images, the cell size was measured in specimen planes that are positioned both parallel and perpendicular to foam growth direction in PURCFs containing glass fibers [5]. Based on these results, it was found out that PURCFs that were blown with CO_2 has the larger-sized cells. This result was explained by the authors such that

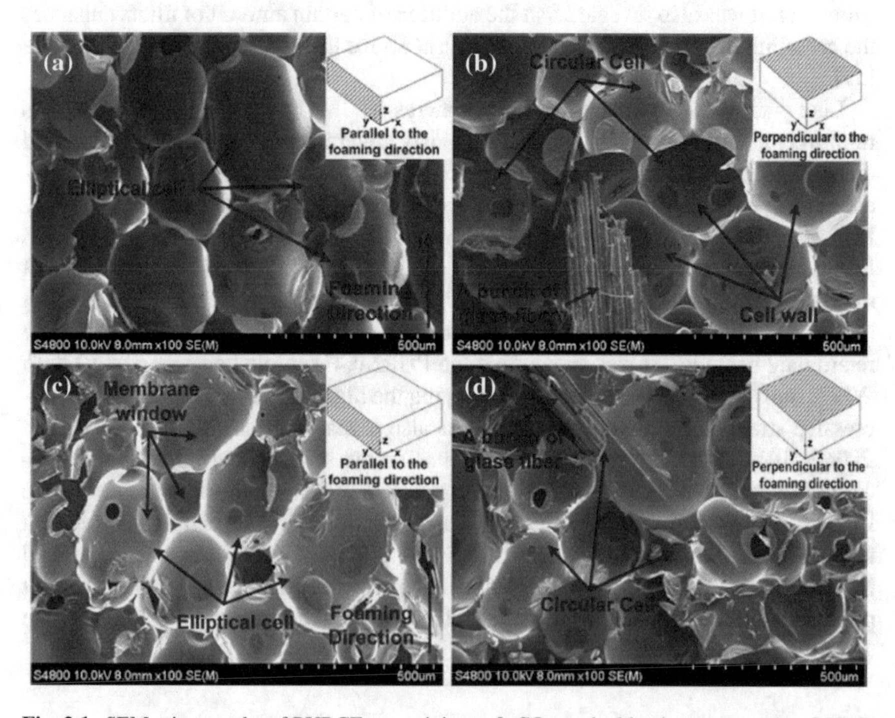

Fig. 2.1 SEM micrographs of PURCFs containing **a**, **b** CO_2 as the blowing agent and **c**, **d** HFC-245fa as the blowing agent. Reproduced with permission from Ref. [5]. Copyright 2016 Elsevier Ltd.

chemical reactions taking place between blowing agent and isocyanate functionalities mainly determines reactive foaming conditions and thus properties in PURCFs that were blown with CO_2. On the contrary, smaller cells were formed in PURCFs that were blown with HFC-245fa since the fast reaction conditions are typically responsible for the formation of smaller cell sizes in these foams [5]. So, the difference between the cell sizes in PURCFs that were prepared by using two different types of blowing agents clearly shows that using two different blowing agents with different chemical and physical properties dramatically affects the foaming conditions and finally changes the morphology and physical properties of PURCFs. Furthermore, authors reported that PURCFs that were blown with CO_2 are considerably longer and larger along the specimen section that is parallel to the foaming direction in comparison with HFC-245fa-blown PURCFs [5]. Also, this result also shows that using CO_2 as a blowing agent in PURCFs induces a much more anisotropic cell geometry compared to using HFC-245fa during the reactive foaming process. Thus, these results clearly point out that choosing the most suitable blowing agent with proper physical and chemical properties is very important in the design of advanced PURCFs with improved physical properties.

In Fig. 2.2, SEM results of PURCFs after compression tests at 20 and −163 °C are shown. It can be easily seen that at higher temperatures, most of the cell walls in PURFs are folded. However, at lower temperatures around −163 °C, the cell structures in closed cellular morphology are more easily fractured due to harsh effects of compression forces on cell morphology at lower temperatures. Additionally, the breaking of the cell structures at −163 °C was explained by the authors such that this brittle destruction of cell morphology happened because of the decrease of pressure and sublimation of blowing agent inside the cells [5]. Thus, in accordance with experimental data, deformation in closed-cellular morphology of PURFs at lower temperatures was found out to be very different than that at room temperature, and also the deformation mechanisms of PURFs that were produced by using different blowing agents did not share the same characteristics [5].

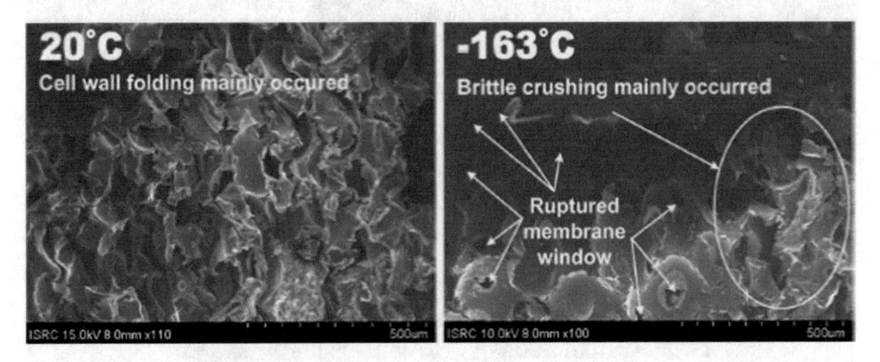

Fig. 2.2 SEM micrographs of PURFs after compression tests. Reproduced with permission from Ref. [5]. Copyright 2016 Elsevier Ltd.

Kim et al. fabricated glass fiber and silica reinforced PURCFs and also only glass fiber filled PURCFs from polymeric diphenylmethane diisocyanate (PMDI) and polyol by using HFC 365mfc as the blowing agent [17]. Authors systematically investigated the closed-cellular morphology, mechanical properties, foam reaction kinetics, glass transition temperature, thermal conductivity and thermal degradation temperatures of PURCFs. Based on the results, it was found out that using both glass fibers and silica in PURCFs induced synergistic improvements in terms of mechanical properties, thermal conductivity, thermomechanical and thermal decomposition properties [17]. Figure 2.3 shows the SEM images of closed-cellular morphology and structure of cells consisting of various sizes due to the addition of different amounts of fillers. Based on these SEM results, it was found out that the cells in closed cellular morphology consist of spherical and polyhedral shapes. In addition, authors also reported that content of the closed cells slightly decreased from 92.5% (G2) to around 91% (GS2-GS5) with the addition of silica particles [17]. Also, it was found

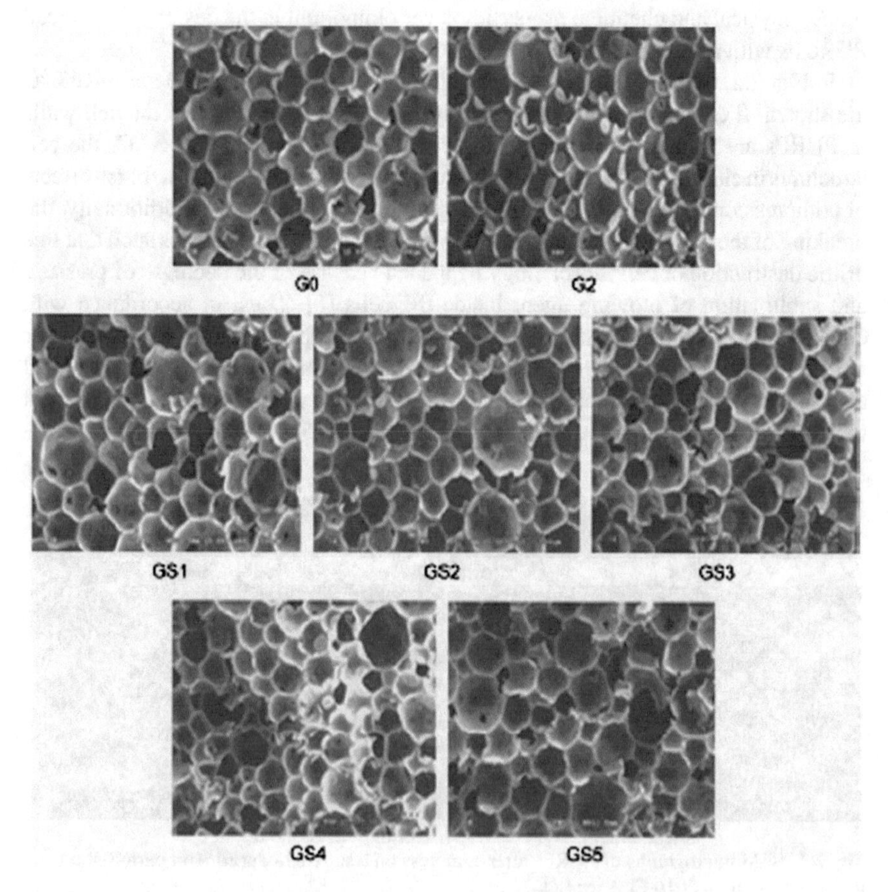

Fig. 2.3 SEM morphology of ternary composites. Reproduced with permission from Ref. [17]. Copyright 2017 BME-PT

that the size of cells decreased from 208 (G2) to a minimum of 169 μm (G3) with addition of silica. The cell size reduction was explained such that silica functioned as a seeding factor at very small concentrations. Thus, the decreased of cell size also increases the cell density and consequently decreases thermal conductivity of PURFs [17].

Xue et al. prepared rigid polyurethane foams consisting of lignin in which some of the synthetic polyol was exchanged with different amounts of lignin [6]. In their study, PURCFs consisting of 37.19% lignin was also mixed with 5 wt% pulp fiber as the maximum amount. According to experimental data obtained from lignin containing PURCFs, authors systematically investigated the closed cellular morphology, chemical structure, mechanical and thermal properties to comprehend influences of partial replacement of polyol with lignin on performance of PURCFs [6].

SEM results of PURCFs containing 37.19% lignin and pulp fiber consisting of various amounts are given in Fig. 2.4. SEM results revealed that the closed cellular morphology was considerably changed with incorporation of lignin and pulp fiber which resulted in heterogenous, non-uniform distribution of cell geometries with larger sizes. Based on the SEM results in Fig. 2.4, the surface morphology of pure PURF was confirmed to be mostly well-ordered and even [6]. Based on these surface morphology results, it is clearly seen that using both lignin and pulp fiber in PURFs transforms the closed cellular morphology into a non-uniform cellular morphology with larger cell sizes. Thus, the inclusion of both lignin and pulp fiber in PURFs does not lead to the improvement of closed cellular morphology which is the main criterion that should be full-filled in terms of thermal insulation properties. Furthermore, the difference in closed cellular morphology between sample LP5 and samples LP1 and LP3 was reasoned by the authors such that the lignin did not disperse well in PURFs when the concentration of lignin was set as 37.19%. In Fig. 2.4, it was also shown that with increasing amounts of pulp fiber in PURCFs consisting of lignin, the overall cell shapes become more non-uniform, and the cells become much larger in samples LP5 consisting of 1 and 5% pulp fiber [6].

Silva et al. fabricated polyurethane rigid composite foams (PURCFs) consisting of industrial residue of blanched cellulose fibers as micron-sized cylindrical fillers [27]. Authors investigated the effects of using different amounts of cellulose fibers on the closed-cellular morphology, thermal and mechanical properties of PURCFs. Moreover, influences of using cellulose fibers on capacity of foams to hinder parasitic assault of organisms were systematically examined [27]. For this specific purpose, authors devised a particular experiment in which the morphological evolution of the surface of foam that was in contact with fungus suspension was examined within a certain period of time. Based on SEM experiments, it was reported that the addition of cellulose filler in the amount of 16 wt% to PURFs based on the amount of polyol diminished the cell sizes of PURFs [27]. The SEM images of PURCFs consisting of cellulose fibers are shown in Fig. 2.5. It was reported that the addition of cellulose fibers into PURCFs did not modify the directionally dependent property of foams, however the cells of closed-cellular morphology were observed to be much more non-uniform and deformed in comparison with those of pure PURF containing no cellulose fiber [27]. In agreement with other works [6], it is clearly noted that using

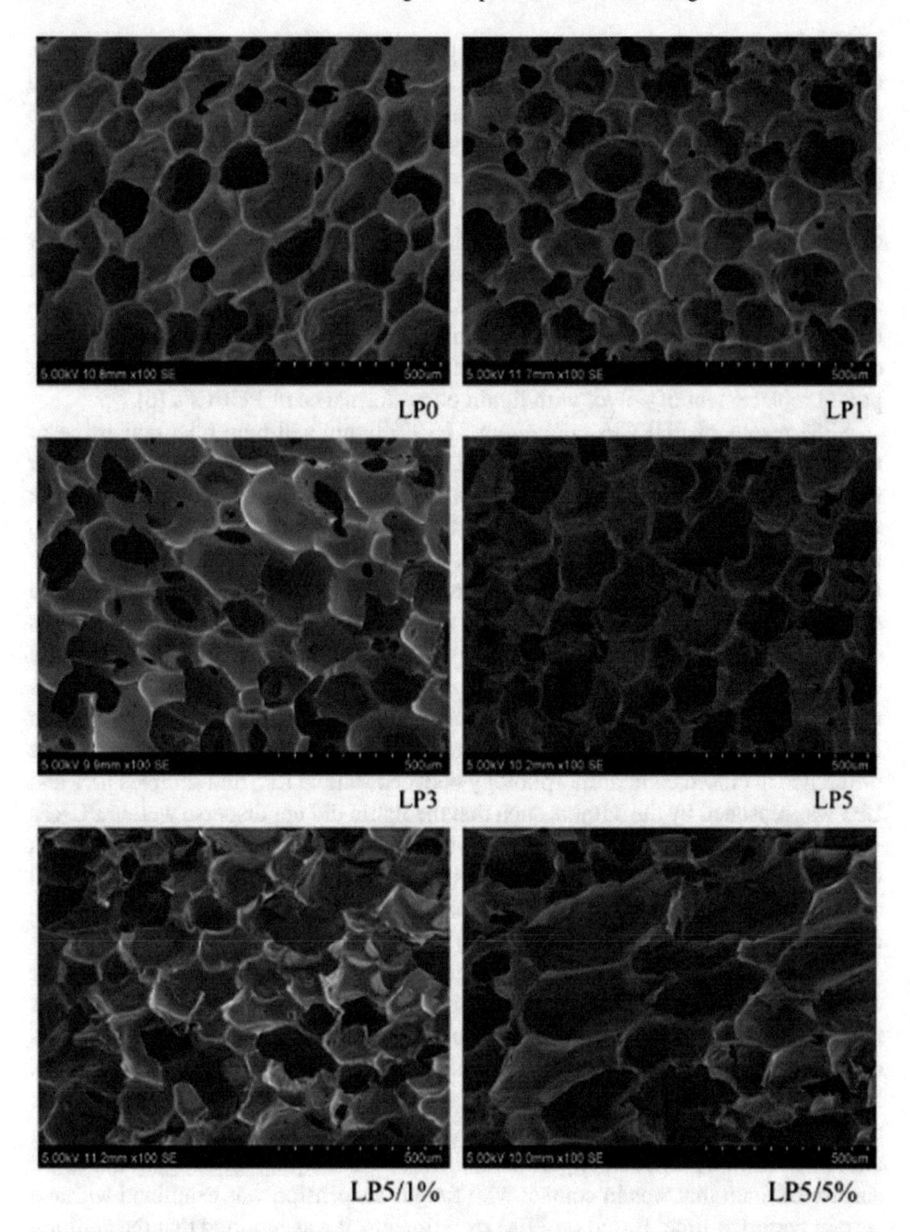

Fig. 2.4 SEM images of pure PURF and different PURFs containing different amounts of lignin and pulp fiber. Reproduced with permission from Ref. [6]. Copyright 2014 American Chemical Society

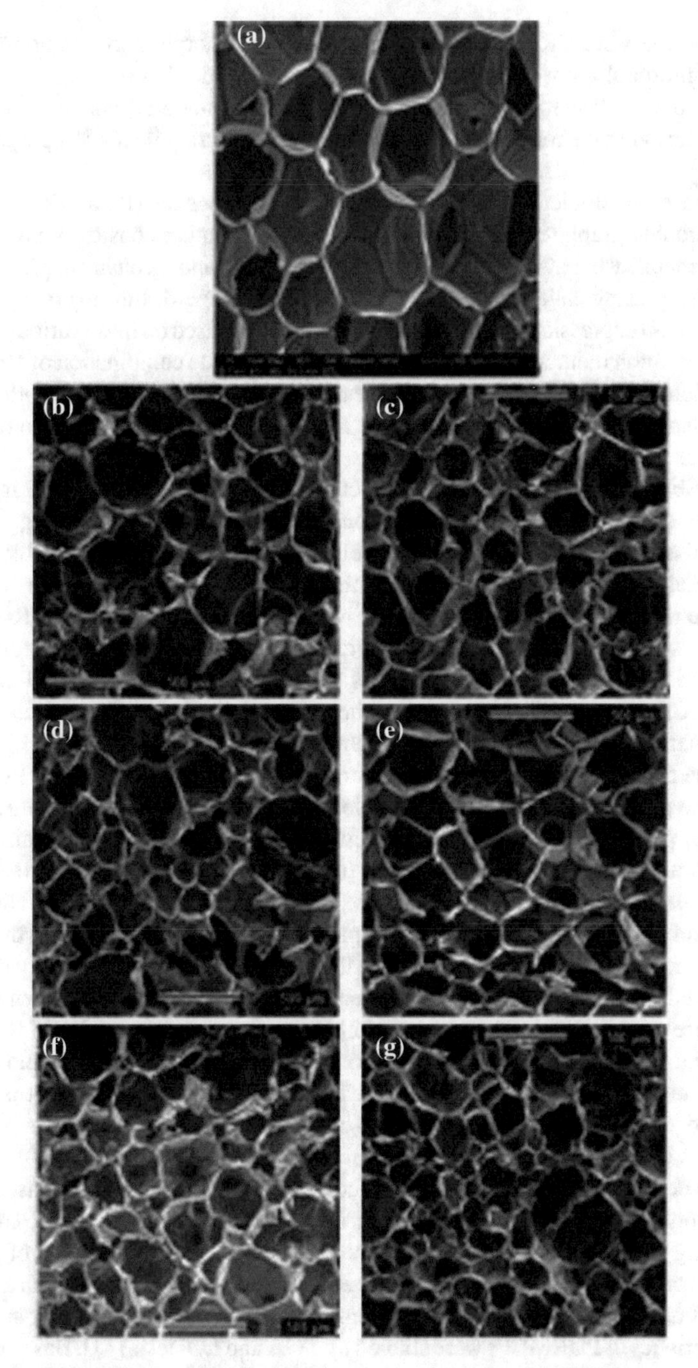

Fig. 2.5 SEM micrographs of **a** pure PURF and PURMCFs consisting of different amounts of cellulose fibers such as **b** 1%, **c** 3%, **d** 4%, **e** 8%, **f** 12%, and **g** 16% (the scale bar is 500 μm). Reproduced with permission from Ref. [27]. Copyright 2010 Wiley Periodicals, Inc.

cellulose fibers in PURFs changes the closed cellular morphology to a non-uniform cellular morphology with larger cell sizes. Thus, the addition of cellulose fibers to PURFs does not lead to the enhancement of closed cellular morphology which is the main criterion that should be full-filled in terms of thermal insulation properties in PURFs.

Cheng et al. studied polyurethane rigid foam composites (PURCFs) consisting of expandable graphite (EG), HGMs and GFs [32]. Authors basically investigated the flammability behavior, closed-cellular morphology and mechanical properties of PURCFs by using characterization techniques such as SEM, limiting oxygen index (LOI) tests, compression and torsion testing methods. Based on these various types of characterization methods, authors found out that using the combination of HGM and GF particles in PURCFs could elevate the mechanical properties. On the other hand, it was also reported that incorporation of EG into PURFs improved flame retardant properties of PURCFs [32].

The SEM images for the fractured sections of PURCFs are given in Fig. 2.6. It is clearly seen that most of the cells in pure PURF which is given in Fig. 2.6a are spherical and they have a closed-closed cellular morphology. In addition, the cells in Fig. 2.6a are dispersed homogeneously within the PURF matrix and the cell diameters are close to each other. However, authors reported that in compressed PURCF filled with 16 wt% EG (Fig. 2.6b), the spherical shape of cells became distorted due to the unique morphology of EG and its weaker compatibility with PU chains in the PURCF system [32]. Alternatively, the poor distribution of EG particles within PURF matrix can be eliminated by performing surface modification of EG particles. Thus, the favorable interactions at the interface between EG particles and PU chains should be provided by performing suitable surface modification of EG particles. By this way, poor interfacial interactions between EG particles and PU chains might be eliminated and PURFs with improved physical and mechanical properties can be fabricated. In Fig. 2.6d, HGM particles are added into the PURCF system consisting of EG and GF. It is clearly seen that compatibility of HGM particles with the PURF matrix is much more higher compared to that of GF. The better miscibility of HGM was explained by the authors such that empty interfacial regions between GF and matrix are occupied by HGM particles [32]. Thus, the dispersion of HGM within the interfacial regions between GF and matrix clearly shows that HGM could be able to form favorable interactions with both PU chains and GFs. Finally, it is clearly observed from Fig. 2.6g that the closed-cellular morphology of PURFs is almost collapsed with increasing amounts of HGM and GF.

Yuan et al. investigated the positive effects of using both phosphorus based polyol and nitrogen based polyol in improving flame retardant properties of PURMCFs consisting of EG particles [11]. Authors synthesized these two types of polyols by performing dehydrochlorination and Mannich reactions. The effects of using different amounts of these polyols on flame retardancy and thermal decomposition characteristics of PURMCFs were studied by TGA and LOI tests [11]. Based on these results, authors proved that incorporation of EG particles into PURMCFs consisting of these two polyols improved flame retardant properties [11].

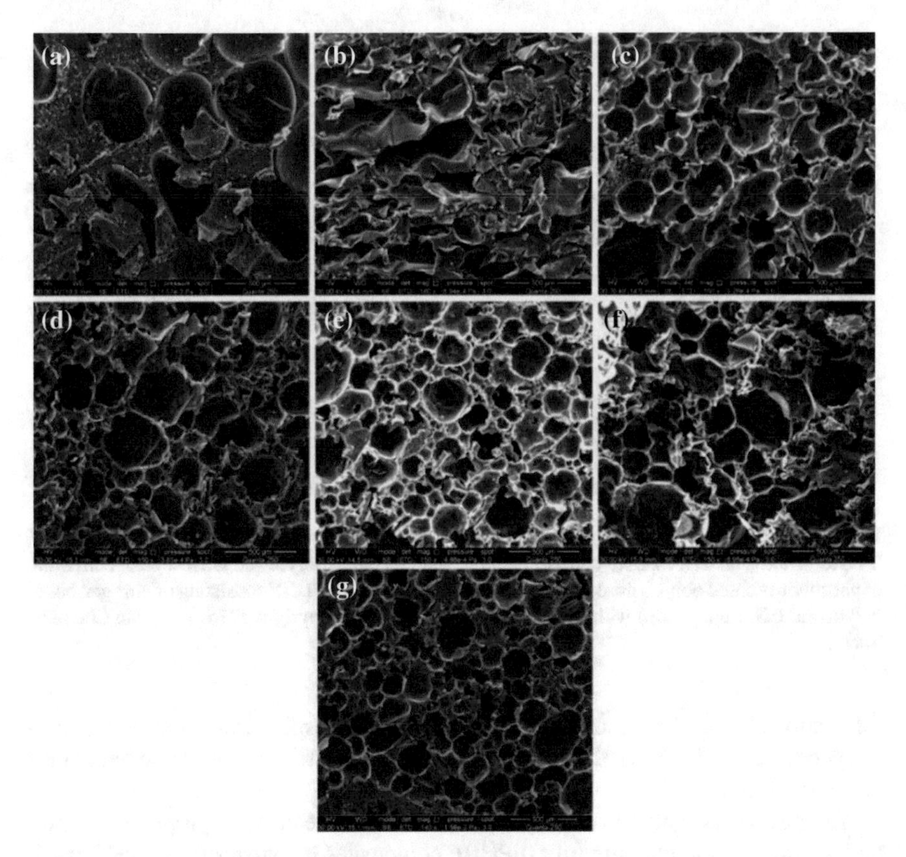

Fig. 2.6 SEM images of pure PURF and compressed PURCFs consisting of diverse contents of EG, GF and HGM. Reproduced with permission from Ref. [32]. Copyright 2013 Wiley Periodicals, Inc.

SEM images of char residues after combustion of pure PURF and PURMCFs consisting of EG and phosphorus and nitrogen based polyols are presented in Fig. 2.7. It was pointed out that large numbers of cracks are visible in pure PURF and PURF/EG which was considered to be a clear proof that EG did not behave as an effective char layer [11]. In agreement with other works [32], it can be concluded that EG particles did not form favorable interactions with PU chains and their distribution within PURF matrix is not very homogenous. Thus, this outcome leads to the conclusion that the addition of EG particles to PURFs does not improve the properties such as combustion and flame retardancy property. On the other hand, it was also pointed out that PURF/phosphorus based polyol/EG sample exhibited a slightly more effective char layer compared to pure PURF due to the extraction of phosphorus based acids from phosphorus based polyol during burning of PURMCFs. In addition, authors also revealed that PURMCFs consisting of phosphorus based polyol, nitrogen based polyol and EG had the most dense and solid char residue and the highest char yield among all the PURMCF samples [11]. Thus, these results clearly showed that

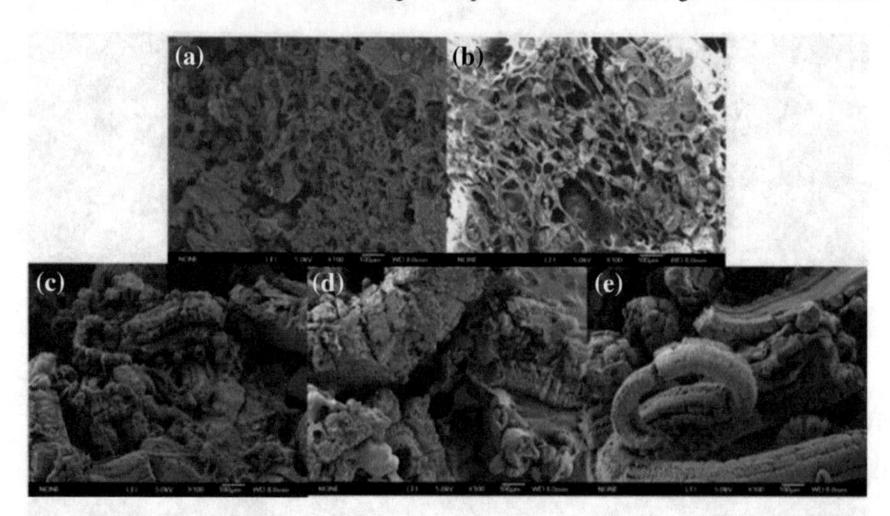

Fig. 2.7 SEM images of char morphology obtained after burning of samples such as **a** pure PURF, **b** PURF filled with EG, **c** PURF consisting of phosphorus based polyol and EG, **d** PURF consisting of phosphorus based polyol, nitrogen based polyol and EG, and **e** PURF consisting of nitrogen based polyol and EG. Reproduced with permission from Ref. [11]. Copyright 2016 American Chemical Society

fabrication of PURFs by using polyols consisting of phosphorus and nitrogen groups is a good strategy in terms of achieving improved properties such as flame retardancy and combustion.

Ye et al. investigated fire delaying behavior and mechanical properties of pulverized expandable graphite (pEG)/PURF composites by encapsulating pEG particles with poly(methyl methacrylate) (PMMA) [9]. Authors confirmed that the pEG-PMMA particles consisting of 22.09 wt% PMMA were successfully synthesized via emulsion polymerization by obtaining experimental data in terms of surface morphology, thermal degradation behaviour and analysis of surface functional groups [9]. Authors emphasized that incorporation of 10 wt% of pEG-PMMA particles into PURF enhanced the values of compression properties and flame retardancy behavior in comparison with the addition of pure pEG to PURFs. The improvement of mechanical properties in PURCFs consisting of pEG-PMMA was explained such that the homogenous distribution of pEG-PMMA within the PURF matrix did not distort closed-cellular morphology due to favorable interfacial interactions between pEG-PMMA and PU chains [9].

Figure 2.8a–c represents the SEM micrographs of PURF consisting of pEG, PURF consisting of pEG-PMMA and PURF consisting of PMMA, respectively. In Fig. 2.8a–c, it is clearly observed that the geometry of cells in the closed-cellular morphology are roughly spherical and isotropic. However, as shown with arrows in Fig. 2.8a, the closed cellular structure is deformed due to unfavorable interactions between pEG particles and PU chains [9]. In previous works [11, 32], the unfavorable interactions between EG particles and PU chains in PURFs were clearly pointed out.

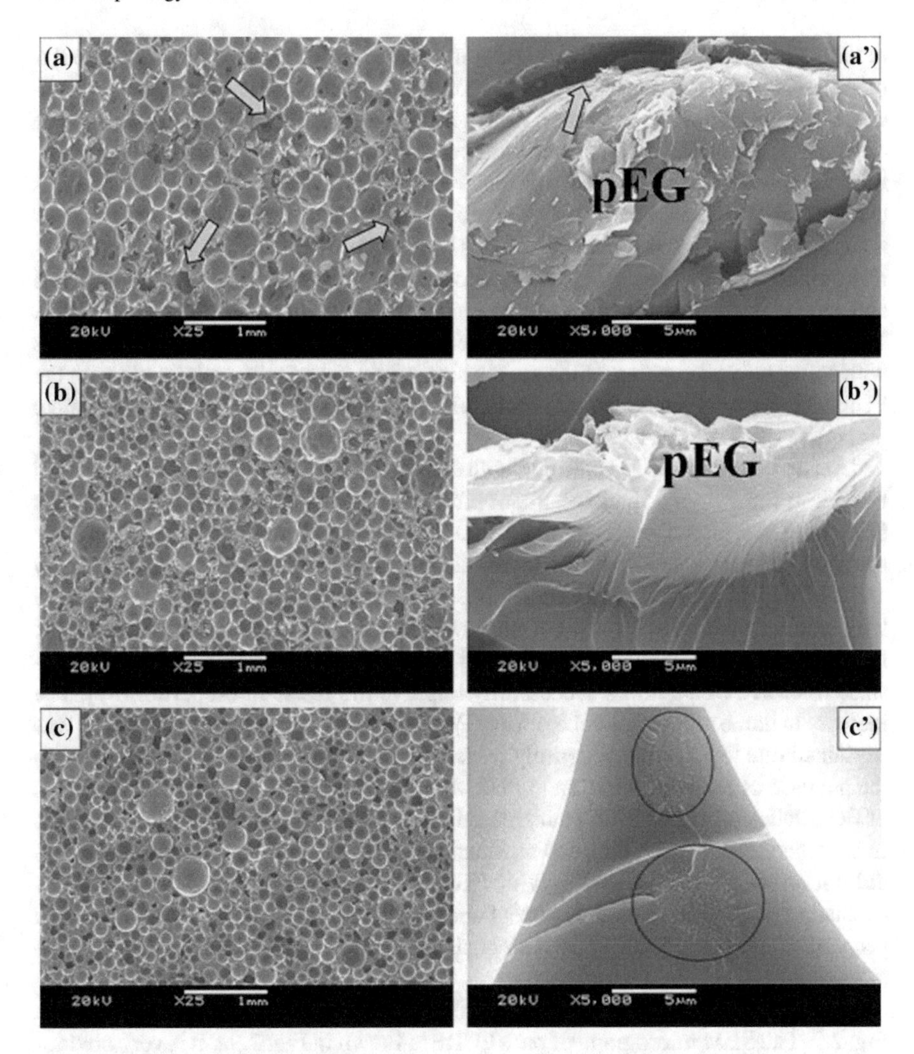

Fig. 2.8 The SEM micrographs consisting of both low and high magnifications: **a** and **a**′ PURF filled with pEG, **b** and **b**′ PURF filled with pEG-PMMA, **c** and **c**′ PURF filled with PMMA. Reproduced with permission from Ref. [9]. Copyright 2009 Elsevier Ltd.

However, the sample pEG-PMMA/PURF in Fig. 2.8b displays cellular morphology in which cells are more uniform and there is less cell deformation. These results clearly showed that surface modification of pEG particles with PMMA leads to the formation of favorable interactions between pEG particles and PU chains [9].

In contrast, based on SEM images in Fig. 2.8, authors also pointed out that highly favorable interactions between pEG-PMMA particles and PU chains is mainly based on positive interfacial adhesion forces between PMMA molecules and PU chains [9]. Thus, in accordance with experimental data, it is precisely noticed that surface modification of rigid particles with softer molecules which are compatible with polymer chains of PURF matrix improves morphology, physical and mechanical properties of PURFs via the formation of beneficial intermolecular interactions. On the other hand, the PURF system filled with only pEG particles does not form the undeformed closed-cellular morphology since the favorable intermolecular interactions between pEG particles and PU chains do not exist. Thus, pEG particles which do not contain any surface functional groups or molecules with suitable physicochemical properties could not lead to the improvement of morphology and thus physical and mechanical properties when they are used as micron-sized fillers in PURFs. Thus, it is clearly understood that it is a requirement to surface modify EG particles with softer polymer chains which are physically and chemically compatible with the PURF matrix to enhance the closed-cellular morphology, physical and mechanical properties of PURMCFs [9].

Li et al. incorporated EG particles with different sizes into water-blown semi-rigid PURFs to improve the flame retardant properties [36]. During sample preparation, EG particles with a size range of 70–960 μm were systematically used to see the beneficial effects of using different particle sizes on thermal stability, foam density and mechanical properties of PURFs. Based on the results, it was revealed that EG with smaller particle sizes did not improve the fire retardant properties positively. In addition, it was shown that using EG with a larger particle size as a micron-sized filler in PURFs might successfully improve fire resistant properties [36]. The increase in flame retardancy behaviour of PURFs with increasing EG particle size was explained due to presence of densely packed segregation layer that was formed with the increase of EG concentration in PURMCFs. It was also reported that LOI values of EG filled semi-rigid PURCFs increased linearly with larger sizes of EG particles. In accordance with TGA data, it was found out that EG particles with different sizes did not affect the thermal stability of PURCFs. Furthermore, semi-rigid PURCFs consisting of EG with a particle size of about 400 μm displayed lower compression properties compared with PURCFs consisting of other EGs with various particle sizes [36].

SEM micrographs of PURCFs consisting of EGs with different sizes are shown in Fig. 2.9. The SEM micrograph of pure PURF is shown in Fig. 2.9a. It is very obvious that the cell geometry is roughly spherical, and there are some open cell structures in the pure PURF matrix. In addition, it was observed that there is no collapse of the cellular morphology which is a clear sign for the uniform and complete cell structure of pure PURF morphology [36]. Also, in Fig. 2.9b–e, SEM micrographs of PURCFs that are filled with EGs with different sizes are depicted. From morphological features in Fig. 2.9b, c, it is clearly noticed that PURF matrix consists of small graphite flakes which are covering the PURF matrix. Furthermore, it was shown that EGs with larger sizes were not found mainly in struts but they are located in regions between cell walls since the average size of EGs is about equal to the size of cells in PURCFs [36]. Other works also revealed the same results in which the particle size of EGs has a direct effect on their locations within the PURF matrix [35, 39, 43]. According to experimental data that was obtained from PURCFs consisting of EGs with different sizes [36], it is clearly noticed that particle sizes of EGs directly influence the

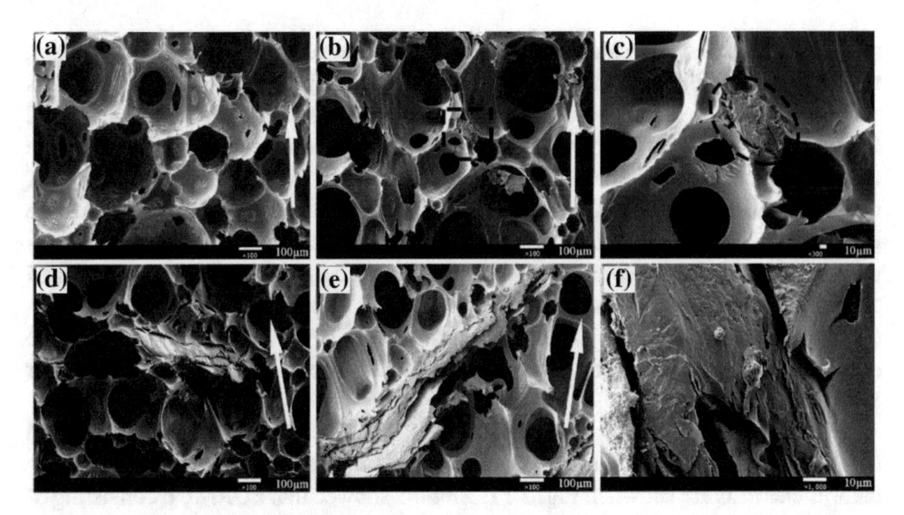

Fig. 2.9 SEM images of pure PURF and PURMCFs consisting of EG particles. Reproduced with permission from Ref. [36]. Copyright 2013 Wiley Periodicals, Inc.

closed-cellular morphology of PURFs. Thus, changes in the morphological features (cell size and shape, particle locations and distribution of particles and cells within PURF matrix) could also lead to changes in PURF properties such as physical, flammability and mechanical properties and thermal stability.

From magnified images of burned layers in Fig. 2.10, it was found out that after burning PURMCF samples, a much larger volume expansion was observed in the foam sample consisting of EG with a larger particle size. This result was reported to be very beneficial in terms of enhancing flame retardant properties in PURMCFs since isolation layer was formed with the inclusion of EG consisting of a larger particle size in PURMCFs [36].

Qian et al. produced flame-retardant PURFs consisting of hexa-phenoxy-cyclotriphosphazene (HPCP) and expandable graphite by using the reactive box-foaming process [37]. Authors used LOI and cone calorimeter tests to systematically analyze the flame retardant properties of PURMCFs. Based on these results, it was shown that the insertion of HPCP into PURFs consisting of EGs enhanced flame

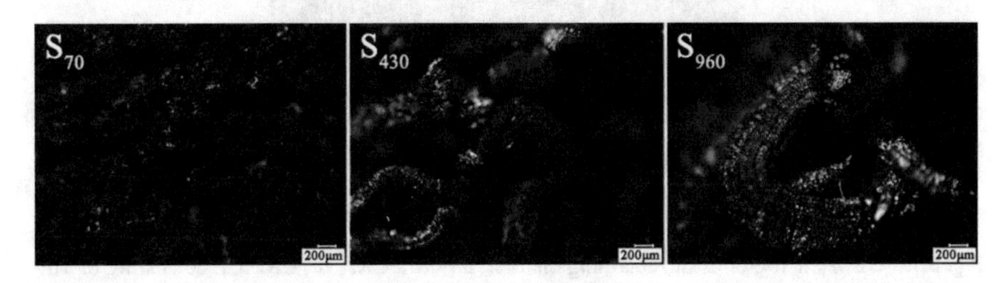

Fig. 2.10 Magnified images of the burned layers (expanded graphite). Reproduced with permission from Ref. [36]. Copyright 2013 Wiley Periodicals, Inc.

retardancy behavior. Besides flame retardant properties, the thermal degradation behavior of HPCP in PURCFs was explored by pyrolysis gas chromatography/mass spectroscopy [37]. Based on these test results, it was shown that HPCP produced PO_2 and phenoxyl free radicals during combustion reaction, and these radicals were used for the quenching of flammable free radicals in the PURF matrix. Furthermore, this observation was shown to be the cause for the flame retardant effect in these PURCFs [37]. After performing the cone calorimeter tests of PURMCFs, morphology and elemental composition of PURMCFs were also characterized by using SEM and energy dispersive X-ray spectroscopy (EDXS). In accordance with experimental data, it was shown that some of phosphorus from HPCP was accumulated in the residual char, and HPCP substantially improved the strength and affinity of the char layer [37].

SEM micrographs of (a) pure PURF, (b) PURMCF consisting of 10% EG, and (c) PURMCF consisting of 10% EG and 10% HPCP which were taken after the cone calorimeter tests are shown in Fig. 2.11. Authors showed that PURMCF consisting of 10% EG and 10% HPCP exhibited a stronger carbon containing layer than PURMCF

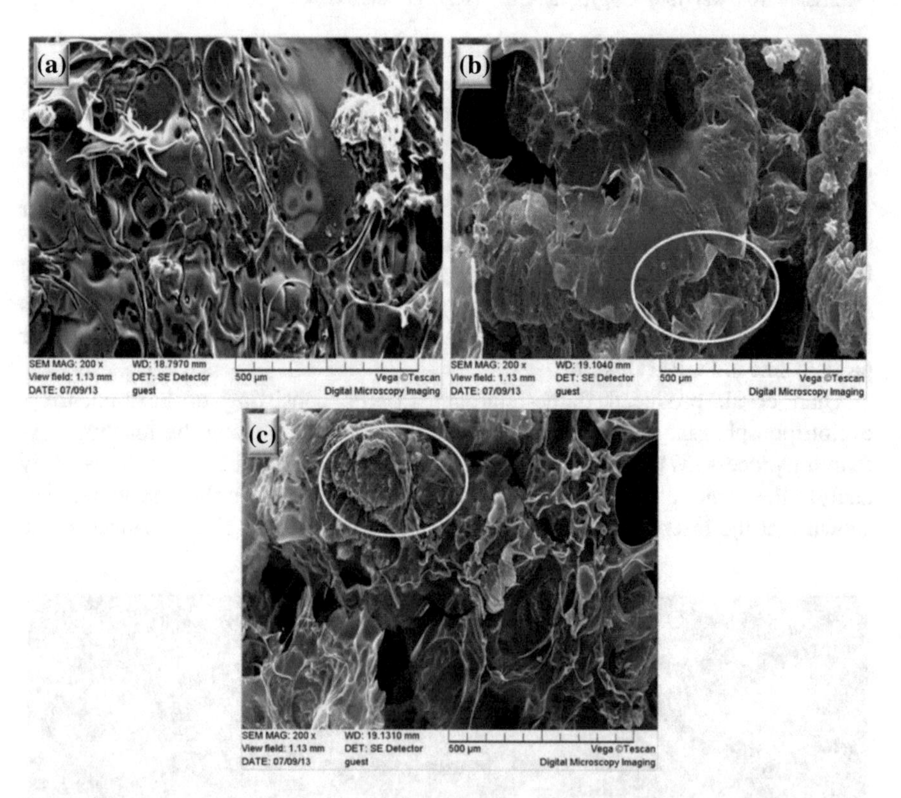

Fig. 2.11 SEM micrographs of remaining chars in **a** pure PURF, **b** PURMCF consisting of 10% EG, and **c** PURMCF consisting of 10% EG and 10% HPCP after cone calorimeter tests. Reproduced with permission from Ref. [37]. Copyright 2013 Elsevier Ltd.

consisting of 10% EG due to the toughness of its residual char. In addition, it was also revealed that PURMCF consisting of 10% EG and 10% HPCP also displayed higher compatibility compared to pure PURF and PURMCF consisting of 10% EG [37]. These experimental findings cleary show that using HPCP in addition EG particles improves the flame retardant behavior of PURFs compared to pure PURF and PURFs consisting of only EG particles. The enhancement of compatibility in PURFs consisting of both EG and HPCP compared to pure PURF and PURFs consisting of only EG particles is a clear sign for the formation of favorable interactions among HPCP, EG and PU chains.

In Fig. 2.11, it is clearly seen that there are some important differences between PURMCF consisting of 10% EG, and PURMCF consisting of 10% EG and 10% HPCP. The morphology of carbon containing layers that surround EG particles had different properties with or without the addition of HPCP into PURFs. It was revealed that without the presence of HPCP in Fig. 2.11b, the EG particles within the residual char had a considerably thin carbon containing layer on their surfaces in PURMCF sample consisting of 10% EG. However, on the other hand, it was shown that with the addition of HPCP as shown in Fig. 2.11c, most of EG particles in the residual char were covered by a thicker carbon containing layer in PURMCF sample consisting of 10% EG and 10% HPCP [37].

Clearly, the main differences in morphological features between PURFs consisting of both HPCP and EG and PURFs consisting of only EG particles in Fig. 2.11 points out and proves the difference of performance of these two foams with regard to fire resistant behaviour. Hence, generation of a thicker carbon containing layer in PURMCF sample consisting of 10% EG and 10% HPCP definitely improves the flame retardancy behavior by acting as a thermal barrier against flammability compared to the much thinner carbonaceous layer that was formed in PURMCF consisting of 10% EG. In addition, the formation of thicker carbonaceous layer in PU/EG foams with the addition of HPCP clearly points out the fact that favorable interactions between EG and HPCP are responsible for the protection of residual char with this thicker carbonaceous layer in PURFs consisting of both EG and HPCP [37]. Thus, these results revealed that using HPCP in PURFs filled with EG particles induced synergistic effects in connection with improving fire resistant behaviour in these PURFs.

Li et al. studied the effects of using different EG particle sizes on favorable fire resistant characteristics between EG and ammonium polyphosphate (APP) in semi-rigid PU foams [39]. In their sample preparation, PURCFs consisting of EGs consisting of three different particle sizes were prepared by adding different mass ratio combinations of EG and APP. The authors used SEM, LOI, TGA and burning tests, to understand synergistic results of using both EG and APP on fire resistant characteristics of PURMCFs. According to flammability data, favorable interactions between EG and APP were observed to be much higher if the EG that was used had a larger particle size. Authors also reported that positive results of using both EG and APP were greatly affected by the tightly-packed structure of protective layer in PURMCF matrix. In addition, TGA and SEM results of the PURMCF system consisting of EG and APP verified the correlation between thermal degradation process and the compactness of protective layer in PURMCFs [39].

Figure 2.12 shows the SEM micrographs of burned PURMCFs consisting of fillers with flame retardant properties. Based on the results of combustion tests, authors showed that the sample series EG3 consisting of EG with the biggest particle exhibited a higher synergistic effect. In Fig. 2.12a, it is clearly seen that many small

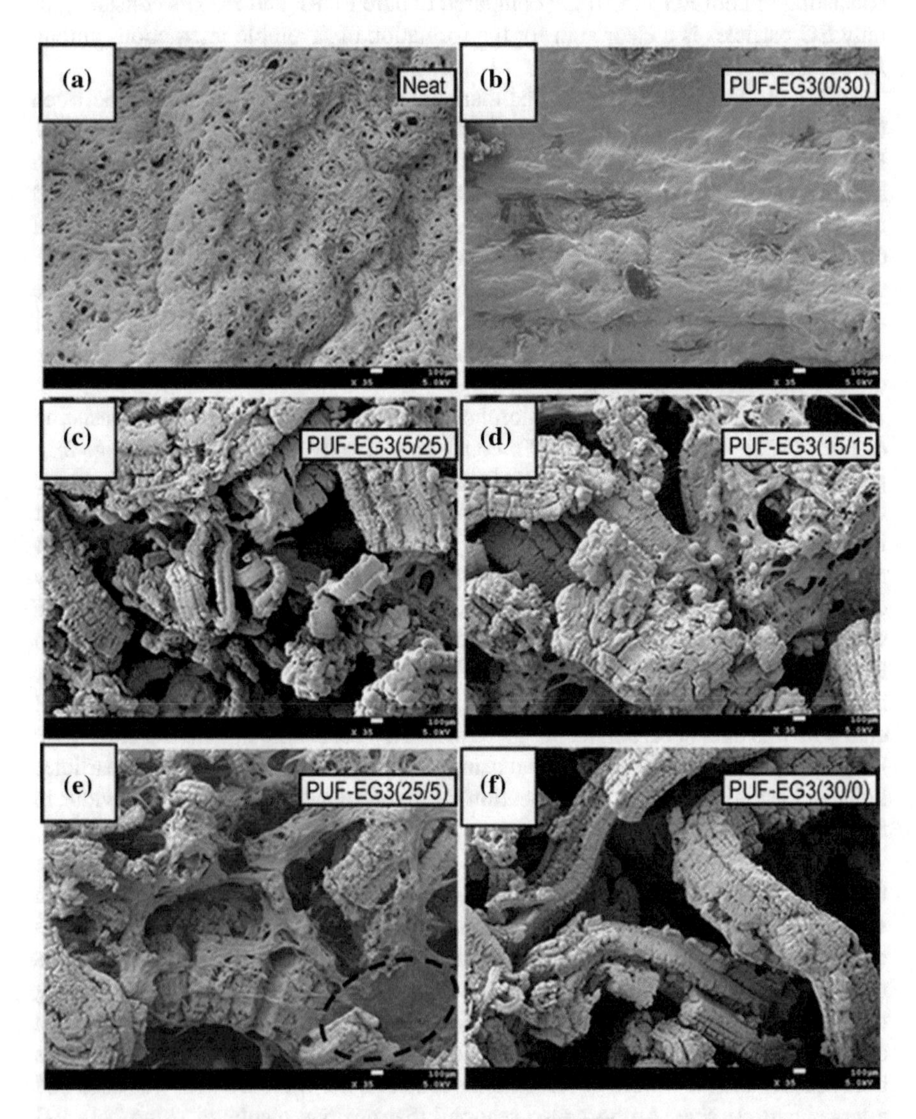

Fig. 2.12 SEM micrographs of burned semi PURFs filled with flame retardants such as EG and ammonium polyphosphate. Reproduced with permission from Ref. [39]. Copyright 2018 Springer-Verlag GmbH Germany

pores exist on the surface of char layer in pure PURF which is not beneficial for the prevention of heat and mass transfer. It was mentioned by the authors that these small holes were formed due to reduced capacity of pure PURF in terms of carbonization and its flame retardant behavior [39]. Authors also reported that PURMCF consisting of 30 pphp APP had a more tightly-packed char layer and a surface with no holes in comparison with the case of pure PURF. This result was explained such that the presence of APP in samples contributed positively in terms of both morphology and flammability properties [39].

In agreement with other works [37], using a second type of flame retardant material such as APP, HPCP, etc. besides EG improves both morphology and flame retardant behavior of PURFs [39]. Thus, the formation of compact char layer and a smooth surface morphology without holes in PURFs containing 30 pphp APP definitely improves the flame retardancy behavior by acting as a barrier against flammability compared to pure PURF. Furthermore, the formation of compact char layer in PU/EG foams with the addition of APP clearly points out the fact that favorable interactions between EG and APP are responsible for the formation of compact char layer in PURFs consisting of both EG and APP. Thus, these results revealed that using a second type of flame retardant material in PURFs filled with EG particles induced synergistic effects in terms of improving the flame retardant behavior in these PURFs.

Liu et al. produced various water-blown rigid polyurethane foams containing EG and diethylene N,N′,N″-tri(diethoxy)phosphoramide (DTP) as flame retardant materials [40]. Based on TGA data, it was shown that using both DTP and EG modified thermal degradation behaviour of PURFs and intensified amount of char residues on PURFs. Based on LOI and combustion calorimeter test results, authors reported that PURMCFs consisting of both EG and DTP exhibited higher LOI and lower heat release rate peak values in comparison with pure PURF case. Based on cone calorimetric test results, it was pointed out that EG/DTP filled PURCF system improved the inhibition of fire amount and decreased the dosage of smoke due to good char forming mechanism in PURCFs. Furthermore, authors performed SEM and EDS experiments to comprehend in detail about the char morphology of foams after combustion. In conclusion, authors proposed that EG and DTP filled PURCFs did form a tightly-packed, steady and partially burned coating full of phosphorus element which acted as an effective thermal barrier against fire during combustion [40].

SEM micrographs of pure PURF and filled PURCFs are shown in Fig. 2.13. It was pointed out that there is discontinuous and small amount of char left in pure PURF after the combustion test due to its higher degree of burning [40]. In addition, it was also reported that there are many chars with holes both on the outer and inner surfaces of burned pure PURF sample in Fig. 2.13a2, a3. On the other hand, in Fig. 2.13b2, b3, PURCF sample consisting of EG/DTP combination exhibited complete and continuous morphology with dense and completely expanded EG char layers due to charring mechanisms of DTP and EG [40].

This study confirmed that using a second type of flame retardant material such as DTP besides EG improves both morphology and flame retardant behavior of PURFs [37, 39, 40]. Thus, the formation of complete and continuous morphology

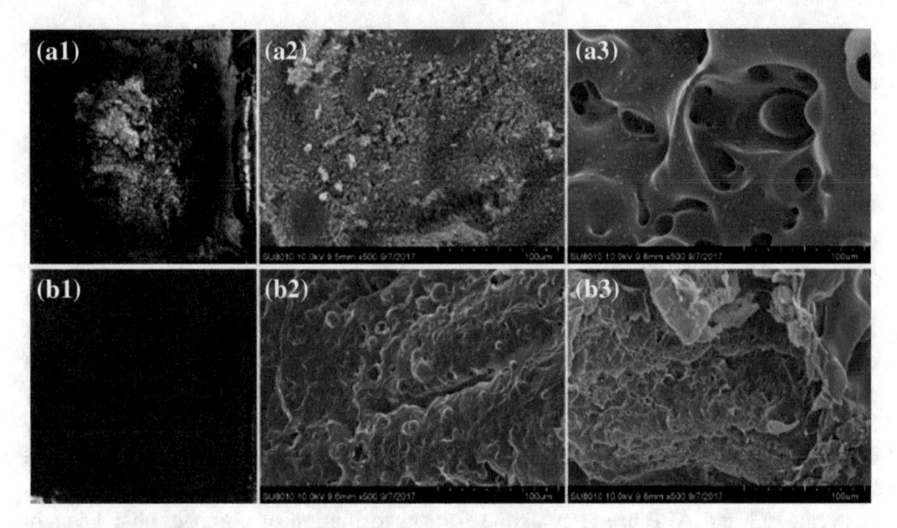

Fig. 2.13 Real space images of char residues that were obtained after cone calorimetry tests: (a1, a2, a3) pure PURF and (b1, b2, b3) PURF consisting of 8% EG and 16% DTP. Reproduced with permission from Ref. [40]. Copyright 2018 Wiley Periodicals, Inc.

with dense and complete expanded EG char layers in PURFs containing both EG and DTP definitely improves the flame retardancy behavior by acting as a barrier against flammability compared to pure PURF. Furthermore, the formation of dense and complete expanded EG char layers in PU/EG foams with the addition of DTP clearly points out the fact that favorable interactions between EG and DTP are responsible for the formation of compact char layer in PURFs consisting of both EG and DTP. Thus, these results revealed that using a second type of flame retardant material in PURFs filled with EG particles induced positive effects in terms of improving the flame retardant behavior in these PURFs.

Shi et al. fabricated PURCFs consisting of EG particles and 10-(2,5-dihydroxy phenyl)-10-hydro-9-oxa-10-phosphorylphenanthrene-10-oxide (HQ) as a second flame retardant material with improved flame retardant properties [41]. Authors used TGA, LOI, burning and cone calorimeter tests to systematically explore fire resistant properties and thermal stability of PURCFs. According to results, it was pointed out that PURF/EG/HQ composites showed higher LOI values, reduced maximum peak heat release rates and increased char yields [41]. Based on SEM results, it was shown that worm-shaped char layer and viscous liquid film in PURCFs were found to be important components in strengthening of fire resistant properties in PURCFs. Based on TG-IR results, it was shown that the extraction of poisonous and easily burnable gaseous compounds from PURMCFs consisting of EG and HQ was considerably reduced in comparison with pure PURF. Authors concluded that inclusion of two diverse flame retardant materials in PURMCFs shows a useful method for fabrication of PURFs with improved fire resistant properties [41].

SEM images of pure PURF and PURCFs consisting of different amounts of EG and HQ are shown in Fig. 2.14. It was pointed out that the cells of pure PURF are

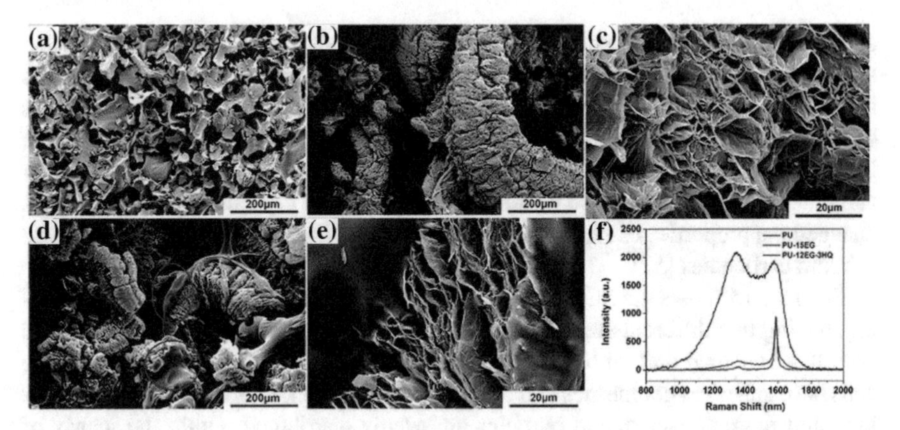

Fig. 2.14 SEM images of pure PURF and PURCFs consisting of different amounts of EG and HQ. Reproduced with permission from Ref. [41]. Copyright 2018 The Royal Society of Chemistry

largely broken with lots of discontinuities on the cell surface [41]. In addition, authors showed that with the addition of 15% EG to PURF, a protective char layer that can be swelled during the heating process was observed in Fig. 2.14c. Furthermore, it was also reported that a concentrated and sticky liquid coating covering the protected and swelled char surface was seen in Fig. 2.14d, e due to the addition of HQ into PURCFs filled with EG [41]. The reason behind the formation of this viscous liquid film was explained due to the fact that the decomposition of HQ into organic acids at high temperatures led to the coverage of the substrate surface. Authors also concluded that the formed viscous liquid film could reinforce the worm-shaped, protected and swelled char layer and enhance the gluing of PURCF with its substrate [41].

According to results of this study [41], it was verified that using two different types of flame retardant materials such as HQ and EG improves both morphology and flame retardant behavior of PURFs [37, 39, 40]. Thus, the formation of a concentrated and sticky liquid coating covering the protected and swelled char layer surface in PURFs containing both EG and HQ definitely improves the flame retardancy behavior by acting as an extra reinforcement compared to pure PURF. Furthermore, the formation of reinforced worm-shaped intumescent char layer in PU/EG foams with the addition of HQ clearly points out the fact that favorable interactions between EG and HQ are responsible for generation of reinforced char residue in PURFs consisting of both EG and HQ. Thus, these results revealed that using two different types of flame retardant materials in PURFs induced positive effects in terms of improving morphology and flame retardant behavior in these PURFs.

Saint-Michel et al. fabricated PURCFs consisting of calcium carbonate fillers with two different sizes and crystallized silica particles with an intermediate size [57]. Basically, authors investigated the effect of using different particle sizes on closed-cellular morphology and mechanical properties of PURCFs. According to SEM data, it was pointed out that samples have a closed-cellular morphology with spherical cell geometries, and the cell size decreased with incorporation of fillers. Based on comparison of mechanical property results with modelling data, it was reported that an accurate explanation of mechanical and viscoelastic properties should be taken into account when the size of fillers is compared to cell wall sizes [57]. Two types of models were established by authors in order to discuss the physical behaviors of PURCFs filled with larger carbonate fillers, and also PURCFs filled with the smaller carbonate fillers. Authors also concluded that the use of crystallized silica particles changed the properties of PURCFs in between that of the PURCFs consisting of two calcium carbonates [57].

Figure 2.15 shows the SEM micrographs of PURCFs filled with calcium carbonates having two different sizes and crystallized silica particles with an intermediate size. It was shown that within the density range of foam samples in Fig. 2.15, the cells have spherical geometries and closed-cellular morphology irrespect of the filler kind and foam densities, and particles are evenly distributed within the matrix of PURCFs [57].

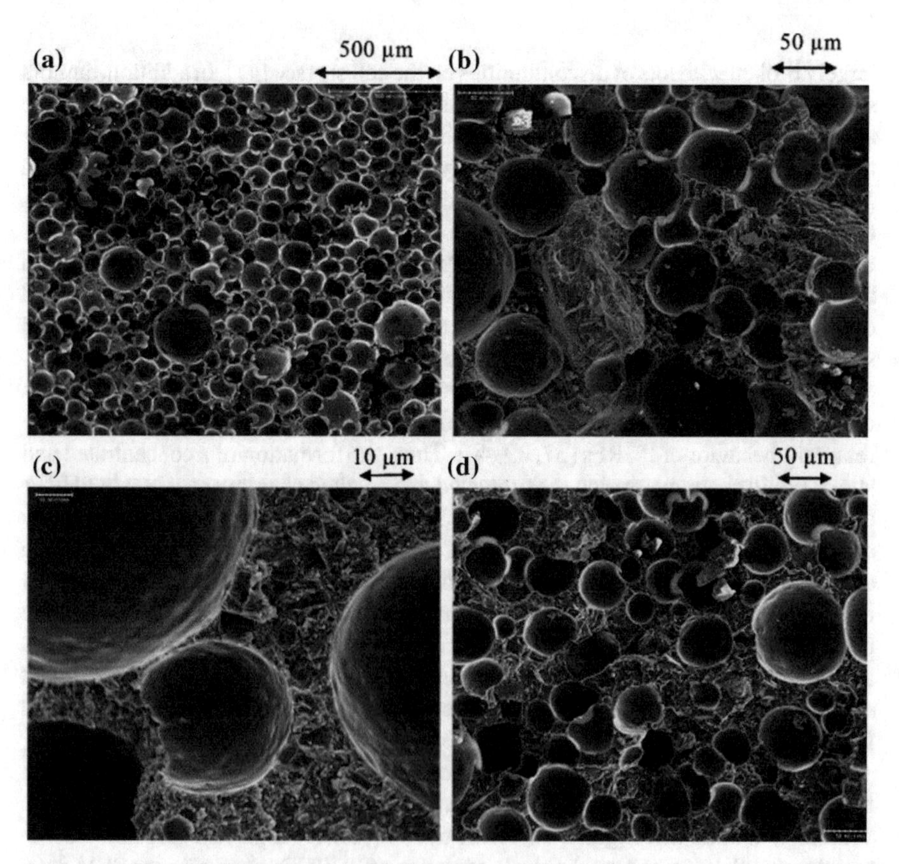

Fig. 2.15 SEM images of PURCFs filled with calcium carbonates having two different sizes and crystallized silica particles with an intermediate size. Reproduced with permission from Ref. [57]. Copyright 2006 Elsevier Ltd.

Bian et al. fabricated PURCFs consisting of EG and whisker silicon (WSi) with different contents which have the same density [53]. Authors showed that PURCFs with the best mechanical properties were obtained with the incorporation of 10 wt% WSi in PURCFs. Moreover, it was also reported that incorporation of 10 wt% WSi into PURFs enhanced the values of compressive strength and modulus. In addition, authors also showed that the inclusion of 10 wt% WSi to PURCFs filled with EG improved the flame retardant and mechanical properties compared to PURCFs filled with only EG [53]. In accordance with TGA data, inclusion of both WSi and EG elevated thermal decomposition behavior in PURCFs. Based on dynamic mechanical (DMA) analysis, authors reported that maximum storage modulus value was obtained in PURCF sample containing 10 wt% EG and 10 wt% WSi because of combined positive effects of using both EG and WSi in PURCFs [53].

Figure 2.16 shows SEM images of burned sample morphology in PURCF consisting of 10 wt% WSi and PURCF consisting of 10 wt% WSi and 10 wt% EG with various magnifications.

Based on SEM results of PURF filled with 10 wt% EG in Fig. 2.16a, authors pointed out the fact that WSi particles and small amounts of burned and blackened

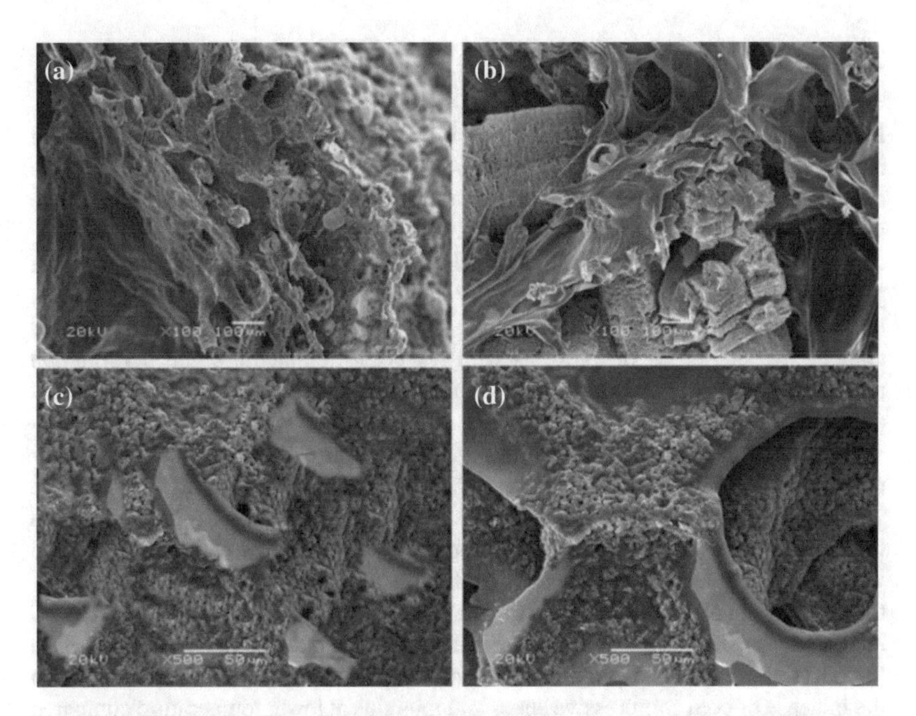

Fig. 2.16 SEM images of burned sample morphology: **a** low magnification of PURF filled with 10 wt% WSi; **b** low magnification of PURF filled with 10 wt% WSi and 10 wt% EG; **c** high magnification of PURF filled with 10 wt% WSi; and **d** high magnification of PURF filled with 10 wt% WSi and 10 wt% EG. Reproduced with permission from Ref. [53]. Copyright 2008 Wiley Periodicals, Inc.

layers can be visualized on micrograph since whisker silicon could not burn and did not possess fire resistant properties [53]. Furthermore, authors also reported that in a higher magnified SEM micrograph, more burnt and blackened layers and less amounts of WSi particles can be found in PURF filled with 10 wt% WSi and 10 wt% EG compared to that of PURF filled with 10 wt% EG. In conclusion, authors proposed that EG particles are assisted by WSi particles for the coverage of the burned surface in PURCF matrix [53].

Based on these results of this study [53], it was shown that using WSi in addition to EG improves both morphology and flame retardant behavior of PURFs [37, 39–41]. Thus, the formation of more charred layers which are covered by EG and WSi particles in PURFs containing both EG and WSi definitely improves the flame retardancy behavior by acting as an extra barrier compared to pure PURF. Furthermore, the formation of more char layers in PU/EG foams with the addition of WSi and EG clearly points out the fact that favorable interactions between EG and WSi are responsible for generation of protective burned sheets in PURFs consisting of both EG and WSi.

2.3 Mechanical Properties

PURCFs consisting of glass fibers were fabricated via reactive foaming by using blowing agents such as CO_2^- and HFC-245fa. Authors studied mechanical properties of PURCFs with respect to changing temperature and diverse deformation rate values [5]. In accordance with data obtained from mechanical properties and closed-cellular morphology, it was revealed that closed cellular morphology and mechanical properties of PURCFs were substantially influenced by the type of blowing agent that was used during the reactive foaming process of PURCFs. Authors obtained compressive properties of PURCFs with respect to temperature and deformation rate to systematically understand the influences of these parameters on mechanical properties [5].

According to experimental data in Fig. 2.17, it was shown that decreasing temperature values in PURCFs containing either one of the blowing agents enhanced the values of both compressive stress and elastic modulus. For instance; it was reported that the compressive stress of PURF sample which was blown with CO_2 increased with decreasing deformation speeds at −163 °C compared to that at room temperature. In addition, it was also added that the elastic modulus of PURCFs at −163 °C increased with decreasing deformation speeds compared to that at room temperature [5]. Thus, these results clearly showed that deformation speed is a critical factor in the increase of both compressive stress and modulus at lower temperature compared to room temperature. Also, it is very critical that choosing the most suitable blowing agent with proper physical and chemical properties is very important in the design of advanced PURFs with improved mechanical properties.

In accordance with experimental data, at temperatures above the freezing point of blowing agent, mechanical testing data of PURCFs elevated with increasing

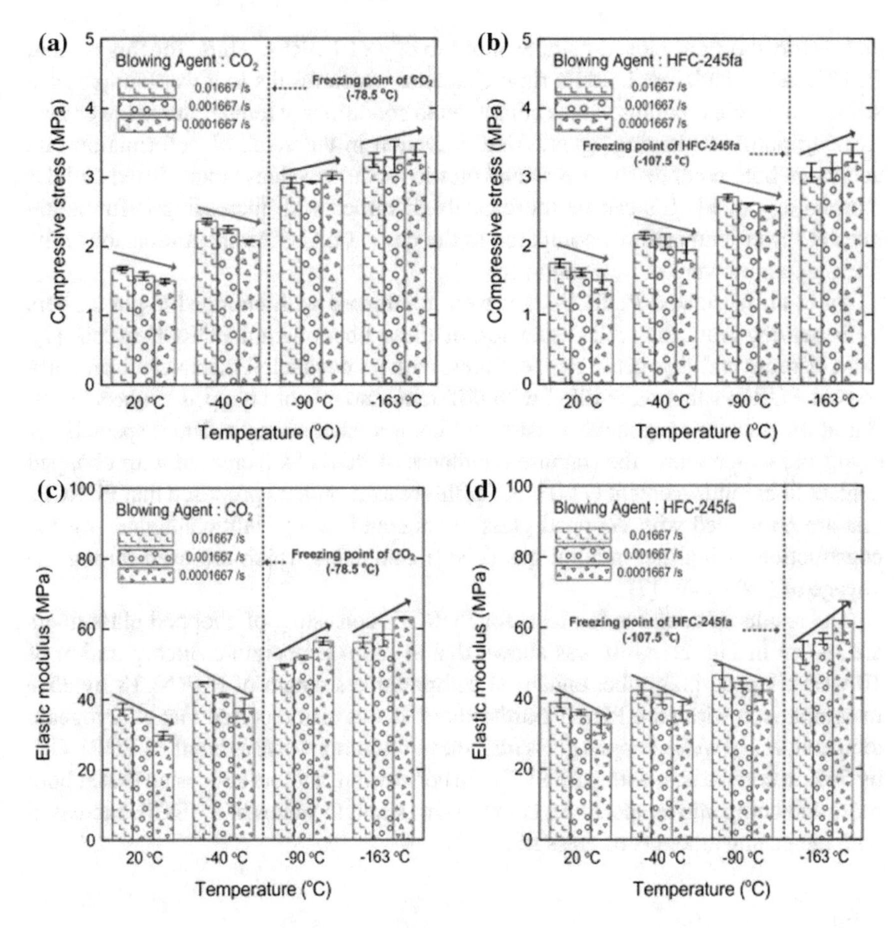

Fig. 2.17 Compression test results of PURFs as a function of temperature: compressive stress values and elastic modulus data of PURCFs that are blown with blowing agents. Reproduced with permission from Ref. [5]. Copyright 2016 Elsevier Ltd.

deformation rate if the data of compressive strength and elastic modulus are compared with respect to deformation rate value. On the other hand, it was also shown that at temperatures below the freezing point of blowing agent, the values of compressive stress and elastic modulus diminished with increasing deformation rate [5]. These results clearly show that the freezing temperature of blowing agents acts as a clear transition temperature between opposite observations of elastic modulus and compressive stress values as a function of deformation rate value. Basically, at temperatures lower than freezing temperature, the blowing agent remains inside cells without disturbing the closed-cellular morphology of PURFs. Thus, for this reason, PURF matrix maintains its rigid character which leads to higher compressive stress and elastic modulus values. On the other hand, at temperatures higher than freezing point, blowing agents much more likely escape from the inner regions of cells and distort the rigid closed-cellular morphology of PURFs. Thus, for this reason, PURF matrix could not hold its rigid character which results in lower compressive stress and elastic modulus values compared to conditions at temperatures lower than freezing point of blowing agent. Also, elevation in the value of deformation rate decreases both compressive stress and elastic modulus values since closed cellular morphology of PURFs can be more easily disturbed with increasing deformation rates at lower temperatures compared to the case at higher temperatures above the freezing temperature of blowing agent [5].

Yu et al. fabricated PURCFs that were reinforced with chopped E-glass fibers by benefiting from the well dispersion of these fibers within PURCF matrix [7]. Both at room and cryogenic temperatures, authors conducted mechanical property tests of PURCFs that were filled with different amounts of chopped E-glass fibers. Based on fracture toughness, tensile and compressive strength data, especially at cryogenic temperature, the fracture toughness of PURCFs increased with chopped E-glass fiber reinforcement [7]. Thus, for this reason, authors proposed that PURCFs that are reinforced with chopped glass fibers could be the right candidates for the construction of liquified natural gas (LNG) containers which are used during the voyage of LNG ships [7].

The results of compression tests for PURCFs consisting of chopped glass fibers are shown in Fig. 2.18a. It was shown that at room temperature, incorporation of 10 wt% chopped glass fiber enhanced compressive strength of PURNCFs by 35% in comparison with pure PURF. Furthermore, it was also reported that at cryogenic temperature, 10 wt% chopped glass fiber elevated compressive strength of PURNCFs by 54% in comparison with pure PURF. In conclusion, authors suggested that at both room and cryogenic temperature, compressive strength values of PURCFs increased with increasing amounts of glass fibers [7].

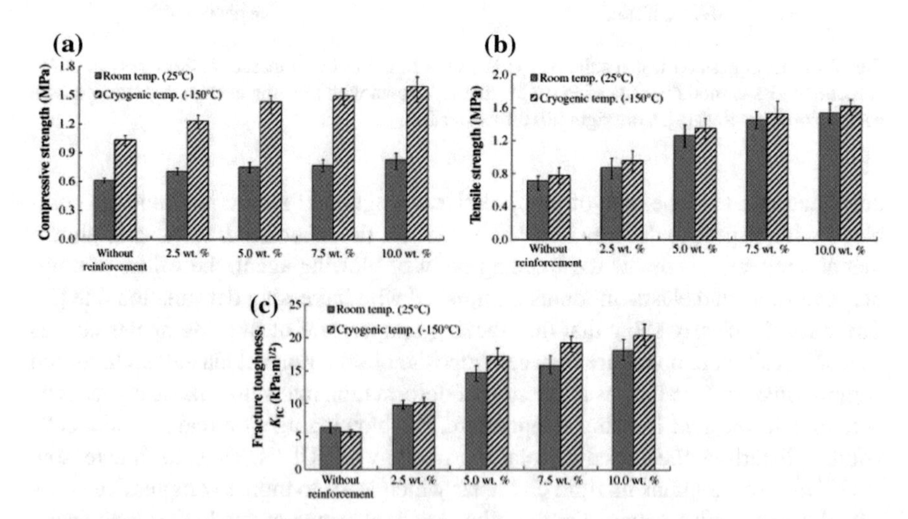

Fig. 2.18 Mechanical properties of PURCFs filled with chopped glass fibers. Reproduced with permission from Ref. [7]. Copyright 2014 Elsevier Ltd.

Tension test results for PURCFs consisting of different amounts of chopped glass fibers are shown in Fig. 2.18b. It was shown that at room temperature, the value of tensile strength increased about 117% in maximum when the amount of chopped glass fibers was adjusted as 10 wt% compared to that of pure PURF. In addition, it was also reported that at cryogenic temperature, tensile strength increased by 108% in maximum compared to that of pure PURF when the amount of chopped glass fibers was 10 wt%. In conclusion, authors proposed that at both room and cryogenic temperature, tensile strength values of PURCFs increased with increasing amounts of glass fibers [7].

Fracture toughness test results for PURCFs consisting of chopped glass fibers are shown in Fig. 2.18c. It was shown that at room temperature, incorporation of 10 wt% chopped glass fiber enhanced fracture toughness of PURNCFs by 290% in comparison with pure PURF. In addition, it was also reported that at cryogenic temperature, incorporation of 10 wt% chopped glass fiber enhanced fracture toughness of PURNCFs by 360% in comparison with pure PURF. In conclusion, authors suggested that at both room and cryogenic temperature, fracture toughness values of PURCFs increased with increasing amounts of glass fibers [7].

Previously, rigid polyurethane foams consisting of lignin in which some of the synthetic polyol was exchanged with different amounts of lignin were systematically fabricated and investigated [6]. In that study, PURCFs consisting of 37.19% lignin was also mixed with 5 wt% pulp fiber in the maximum amount. According to experimental data obtained from lignin containing PURCFs, authors systematically investigated closed cellular morphology, chemical structure, mechanical and thermal properties to perceive influences of partial replacement of polyol with lignin on performance of PURCFs [6].

The values of density and compressive strength of PURCFs as a function of lignin content are given in Fig. 2.19. Based on these results, authors showed that the density values of samples consisting of 1 and 2% lignin decreased drastically in comparison with that of PURF sample containing no lignin. However, on the other hand, it was also noted that with more additions of lignin content in PURCFs, foam densities did not decline substantially [6]. In addition, as shown in Fig. 2.19, it was expressed that in PURCFs consisting of 5% lignin content but different amounts of pulp fiber, densities of PURCFs decreased with increasing pulp fiber content. Based on these results, authors proposed that pulp fiber addition to PURFs might affect the foam expansion and density and chemical reaction of polyol and isocyanate in the PURCF system [6].

In accordance with experimental data, it was revealed that compressive strength values of PURCFs filled with lignin was significantly reduced with increasing amounts of lignin compared to pure PURF foam. The decrease of compressive strength in PURCFs filled as a function of increasing amount of lignin was explained such that relatively low amounts of hydroxyl groups in lignin diminish the values of cross-linking density and compressive strength in PURCFs in comparison with the pure polyol system [6]. Moreover, authors also showed that further increasing the amount of lignin content did not decrease the compressive strength of PURCFs. Based on mechanical testing results, on the overall, authors suggested that very high

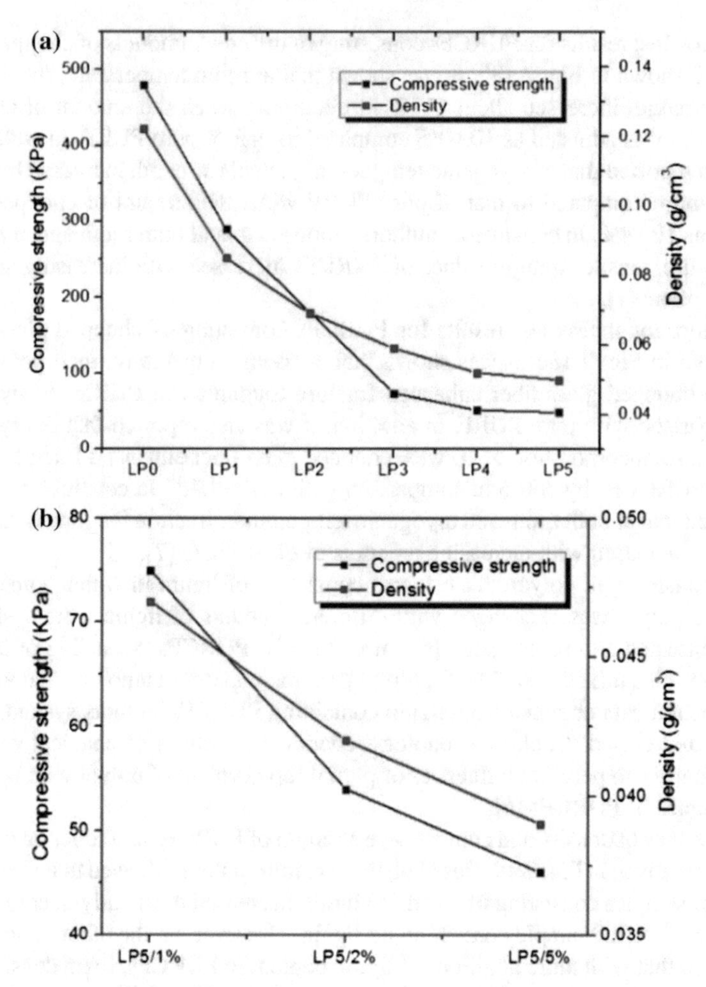

Fig. 2.19 Foam density and compressive strength data of PURCFs with respect to lignin and pulp fiber content. Reproduced with permission from Ref. [6]. Copyright 2014 American Chemical Society

amounts of lignin in PURCFs could form aggregations among themselves due to their unfavorable intermolecular interactions with PU chains [6].

Aranguren et al. fabricated and characterized PURCFs containing pine wood-fibers or hemp for using these materials in applications such as acoustic panels and automotive industry [4]. PURFs with different foaming levels were adjusted, and they were fabricated from crosslinked PUs consisting of a castor oil type of polyol. In addition, PURCFs with anisotropic alignment properties were prepared with the inclusion of long hemp fibers. On the contrary, PURCF samples with uneven distribution of short hemp and wood fibers were also fabricated. Authors used SEM to analyze the closed-cellular morphology, and they also used three point bending and dynamic mechanical tests to understand mechanical and thermomechanical properties of PURCFs [4]. The flexural properties of PURCFs with different aging times are shown in Fig. 2.20.

According to experimental data in Fig. 2.20, it was shown that the flexural properties such as maximum stress, deformation up to the rupturing point and modulus of PURCFs are affected by the aging time of PURCFs. In addition, authors showed that the values of maximum stress and bending modulus of PURCFs elevated in regard to time. In contrast, it was reported that the maximum deformation of PURCFs was not reduced at the breaking point. Moreover, authors showed that the maximum deformation at the breaking point increased in most of PURCFs [4].

Yu et al. investigated the values of impact energy of PURCFs consisting of chopped E-glass fibers as a function of increasing concentration of fibers [3]. Authors performed the impact tests at −196 °C by conditioning the foam samples in liquid nitrogen. Based on drop test results, authors measured parameters such as permanent strain, damage factor, maximum impact pressure and consequently the critical impact energy values of PURCFs. Based on results from measurements, authors suggested that the impact resistance values of PURCFs at very low temperatures substantially were enhanced with the addition of chopped fibers [3].

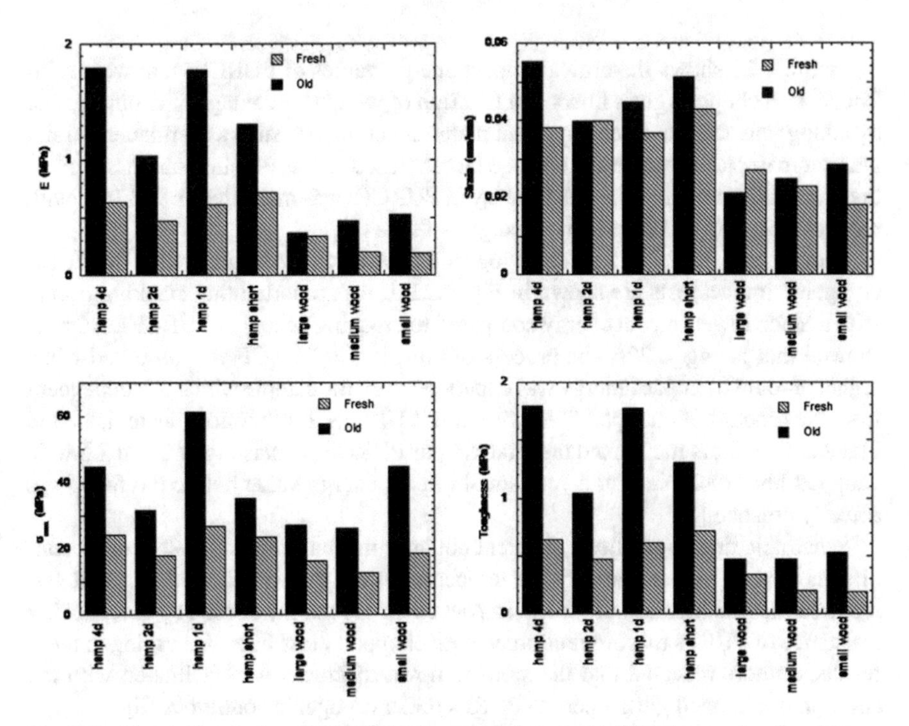

Fig. 2.20 Flexural properties of PURCFs that were reinforced with fibers in regard to time. Reproduced with permission from Ref. [4]. Copyright 2007 Wiley Periodicals, Inc.

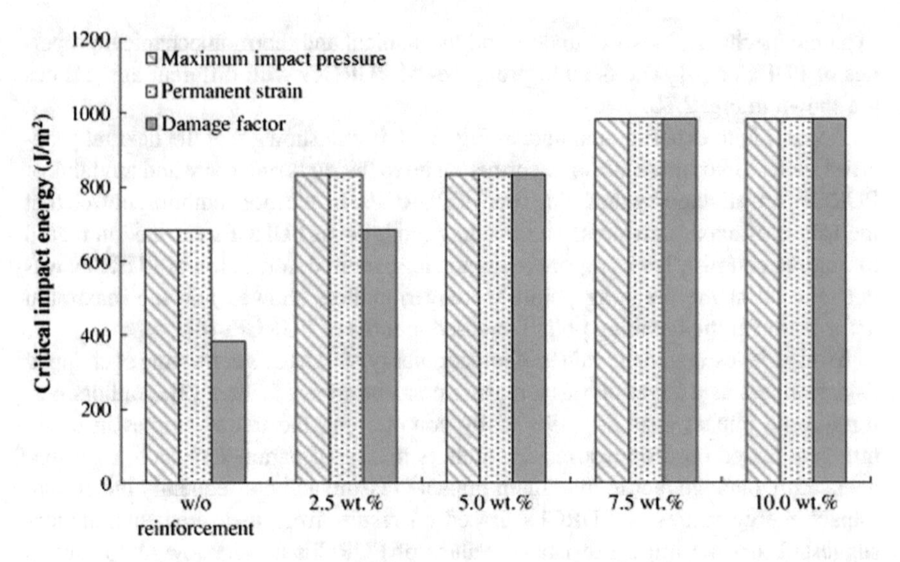

Fig. 2.21 Critical impact energies of PURCFs that were reinforced with chopped fibers. Reproduced with permission from Ref. [3]. Copyright 2015 Elsevier Ltd.

Figure 2.21 shows the critical impact energy values of PURCFs that were reinforced with chopped glass fibers as a function of weight percentage of chopped fiber by taking into account three important material parameters such as permanent strain, maximum impact pressure and damage factor. Based on these results, authors showed that the maximum critical impact energy in PURCFs was measured as 983 J/m^2 with the incorporation of 10 wt% chopped glass fibers [3].

Photographs of PURCFs consisting of 2.5 wt% chopped glass fibers after the cryogenic impact tests are shown in Fig. 2.22. It was reported that cracking occurs with a much lower impact energy compared to crushing occurs in PURCFs. Authors showed that in Fig. 2.22b, the process of crushing in PURCF was observed when higher amount of impact energy was experienced by the sample which is independent from the amount of chopped fiber present in PURCFs. Furthermore, as indicated in Fig. 2.22a, authors mentioned that cracking of PURCFs consisting of 0 and 2.5 wt% chopped fiber took place at a very small impact energy value before the foam was actually crushed [3].

Schematic drawings of two different collapse mechanisms in PURFs under continuous impact pressures at very low temperatures are shown in Figs. 2.23a, b. It was reported that the increase of critical impact energy is mainly caused by the cracking resistance of PURFs that are reinforced with chopped glass fiber. According to these results, authors revealed that the slowing down of cracks was facilitated with the addition of chopped glass fibers to PURFs under cryogenic conditions [3].

Polyurethane rigid composite foams (PURCFs) consisting of industrial residue of blanched cellulose fibers as micron-sized cylindrical fillers were fabricated recently [27]. Authors investigated the effects of using different amounts of cellulose fibers on closed-cellular morphology, mechanical and thermal properties of rigid foams.

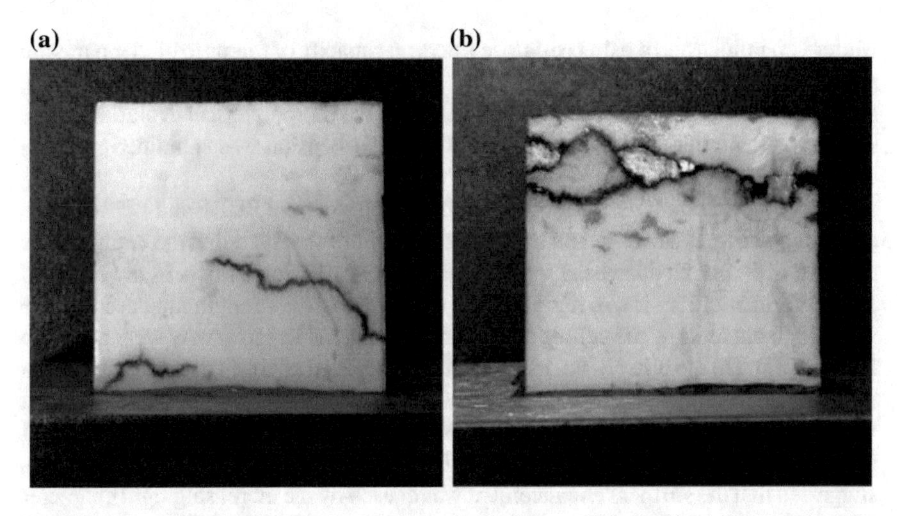

Fig. 2.22 Photographs of **a** cracked and **b** crushed PURCFs containing 2.5 wt% chopped fiber after the cryogenic impact tests. Reproduced with permission from Ref. [3]. Copyright 2015 Elsevier Ltd.

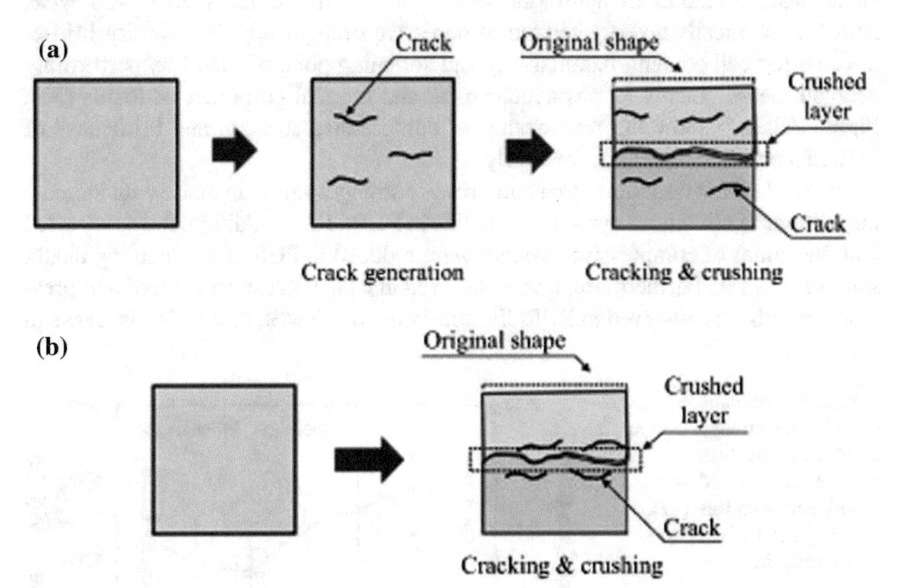

Fig. 2.23 Schematic drawings of collapse mechanisms of PURCFs under repeated impact pressures at the cryogenic temperature: **a** no addition of chopped fiber and **b** the addition of chopped fiber. Reproduced with permission from Ref. [3]. Copyright 2015 Elsevier Ltd.

Moreover, influences of using cellulose fibers on capacity of foams to hinder parasitic assault of organisms were systematically examined [27]. For this specific purpose, authors devised a particular experiment in which the morphological evolution of the surface of foam that was in contact with fungus suspension was examined within a certain period of time [27].

The values of compressive modulus and strength for pure PURF and PURCFs are presented in Fig. 2.24. Based on these results, authors observed that the compressive strength of PURCFs decreased with the addition of cellulosic fibers in the range of 1–8 wt%. In addition, it was reported that the compressive strength increased a little when high contents of the cellulose fiber in the range of 12–16% w/w were added to PURFs [27]. On the other hand, in the overall, a detailed analysis of the compression modulus curve with respect to fiber amount was performed by the authors, and it was shown that the compression modulus of PURCFs increased slightly when it is compared with the compression modulus of pure PURF. In conclusion, authors suggested that the stiffness enhancement was caused by the increasing rigid character of foam because of the inclusion of cellulose fibers in PURCFs [27].

Czuprynski et al. produced PURCFs consisting of various fillers such as talc, chalk, starch, aluminum hydroxide and borax in the amounts of 2.5–20 wt%. Authors specifically investigated the compressive strength, apparent density, brittleness, closed cell content, flammability and softening point of foams by performing detailed measurements on physicochemical and thermal properties of foams [30]. Figure 2.25a, b show the relationship of compressive strength and brittleness of PURCFs on filler content, respectively.

In Fig. 2.25a, it was shown that compressive strength values increase with increasing amount of aluminum hydroxide and talc in PURCFs. In addition, it was reported that the values of compressive strength were reduced in PURCFs containing chalk, starch and borax. Furthermore, it was observed that the highest decrease of compressive strength was observed in PURCFs consisting of 20 wt% borax. The increase in

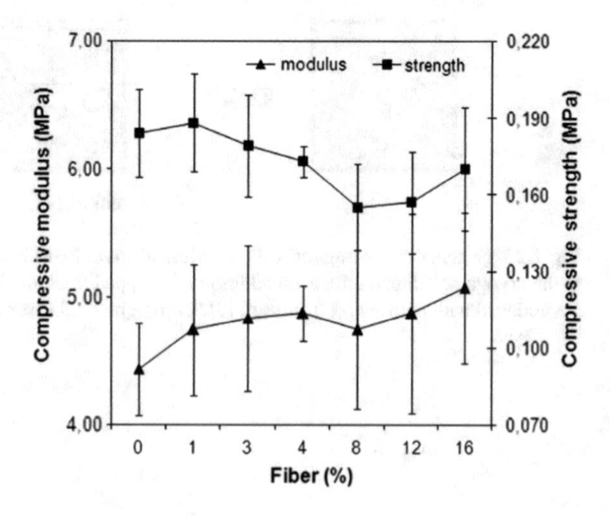

Fig. 2.24 The compressive modulus and strength data of PURCFs versus fiber amount. Reproduced with permission from Ref. [27]. Copyright 2010 Wiley Periodicals, Inc.

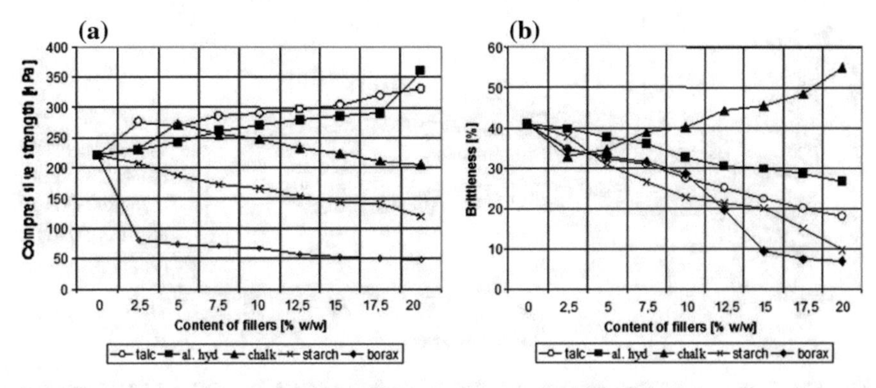

Fig. 2.25 Dependence of **a** compressive strength and **b** brittleness on content of fillers in foams. Reproduced with permission from Ref. [30]. Copyright 2009 Wiley Periodicals, Inc.

compressive strength was attributed to the increase in apparent density of PURCFs consisting of talc and aluminum hydroxide. In conclusion, authors proposed that increases in foam density and compressive strength was caused by toughening of the cell structure by addition of fillers such as talc and aluminum hydroxide [30]. The increase of compressive stress values with incorporation of talc and aluminum hydroxide to PURCFs proves the fact that favorable intermolecular interactions are formed between these particles and PU chains. In addition, these favorable interactions could lead to the formation of a much more compact and finer closed cellular morphology in PURCFs which ultimately enhances the values of compressive stress values in these PURCFs [30].

In Fig. 2.25b, it was shown that the decrease in brittleness of PURCFs in comparison with pure PURF occurred in PURCFs containing talc, aluminum hydroxide starch and borax. Moreover, it was shown that brittleness increased to about 55% in PURCFs consisting of 20 wt% chalk [30]. In conclusion, authors proposed that PURCFs consisting of talc and aluminum hydroxide could be used in applications such as building and heat engineering [30]. These results clearly show that chalk is a much better reinforcing agent in the enhancement of brittleness of foams compared to fillers such as talc, aluminum hydroxide starch and borax.

PURCFs containing expandable graphite (EG), HGMs and GFs were prepared, recently [32]. Authors basically investigated the flammability behavior, closed-cellular morphology and mechanical properties of PURCFs by using characterization techniques such as SEM, limiting oxygen index (LOI) tests, compression and torsion testing methods. Based on these various types of characterization methods, authors found out that using the combination of HGM and GF particles in PURCFs could elevate the mechanical properties [32].

Figure 2.26 shows the compressive strength values of pure PURF and PURCFs filled with EG and HGM. According to compression test results, it was shown that compressive strength values of PURCFs increased slowly when small amount of inorganic fillers were added to the system. It was also reported that among the

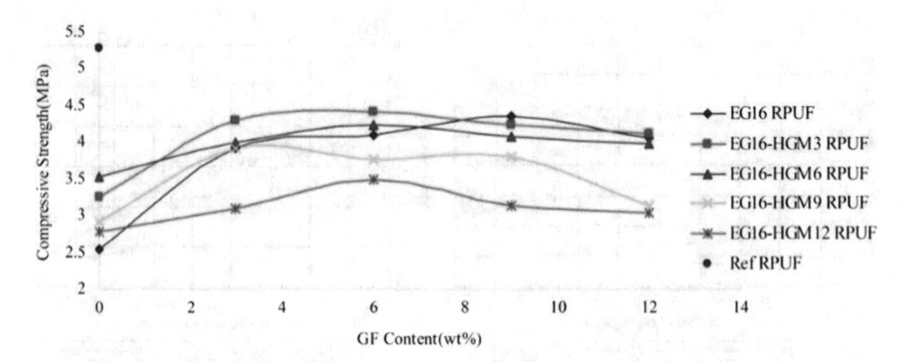

Fig. 2.26 Compressive strength data of PURCFs consisting of different GF and HGM contents. Reproduced with permission from Ref. [32]. Copyright 2013 Wiley Periodicals, Inc.

additives, HGM which has a spherical structure and high resistance to stress is the best micron-sized filler in terms of enhancing the mechanical properties [32]. However, it was shown that compressive strength values decrease slowly with increasing amount of inorganic fillers due to the aggregation behavior and collection of fillers in the PURF matrix [32]. In addition, it was further noted that too many additives weaken the intermolecular interactions between fillers and polymer chains, also cause thinning of cell walls in PURCF matrix. Thus, based on this analysis, it was shown that consequently the closed-cellular structure of PURCFs is destroyed by collapsing of the actual closed-cellular structure, and the PURCFs cannot hold the applied stress any more due to the weakening of mechanical properties [32]. If the optimum content of fillers is exceeded in PURCFs, unfavorable filler-filler interactions basically dominate over favorable polymer-filler interactions. Consequently, the mechanical properties are negatively affected by the increase of unfavorable filler-filler interactions at high filler contents. Thus, it is very important to find the optimum amount of fillers that should be used in the fabrication of PURCFs with improved morphology and mechanical properties.

The mechanical properties and flame retardant behavior of PURCFs consisting of pulverized expandable graphite (pEG) and pulverized EG-poly(methyl methacrylate) (PMMA) were systematically analyzed [9]. Authors confirmed that pulverized EG-PMMA particles consisting of 22.09 wt% PMMA were successfully synthesized via emulsion polymerization by obtaining experimental data in terms of surface morphology, thermal degradation behaviour and analysis of surface functional groups [9]. Authors emphasized that incorporation of 10 wt% of pulverized EG–PMMA particles into PURFs enhanced the values of compression properties and flame retardancy behavior in comparison with the addition of pure pulverized EG to PURFs. The improvement of mechanical properties in PURCFs consisting of pulverized EG–PMMA was explained such that the homogenous distribution of pulverized EG-PMMA within the PURF matrix did not distort closed-cellular morphology due to favorable interfacial interactions between pulverized EG-PMMA and PU chains [9].

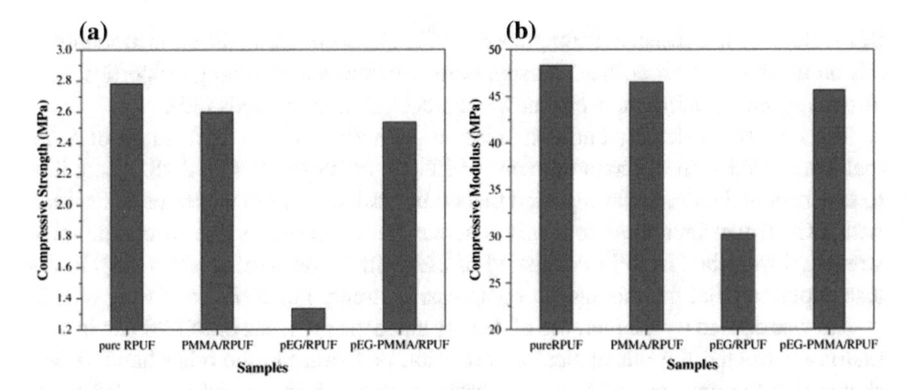

Fig. 2.27 Compressive **a** strength and **b** modulus of filled and unfilled PURCFs. Reproduced with permission from Ref. [9]. Copyright 2009 Elsevier Ltd.

Compression testing results of filled and unfilled PURFs are displayed in Fig. 2.27. Based on the compression test results, authors showed that in PURF sample consisting of pulverized EG, the compressive strength and modulus substantially decrease due to unfavorable interactions between pulverized EG particles and PURF matrix [9]. However, on the other hand, it was unveiled that compressive strength and modulus values of PURCFs increased in great extent when the pulverized EG particles were encapsulated by PMMA. The considerable improvement of compressive strength and modulus values in PURCFs consisting of pulverized EG–PMMA particles was explained such that carboxylic acid functional groups of pulverized EG–PMMA particles could possibly react with urethane functional groups of PURFs, thus increasing the favorable intermolecular interactions between fillers and PU chains. On the other hand, these type favorable interactions are not present in pulverized EG/PURF system since pulverized EG particles are not surface modified with PMMA molecules. In addition, in pulverized EG/PURF system, filler-filler interactions might dominate over polymer-filler type of interactions. Consequently, for this reason, the mechanical properties of pulverized EG–PMMA/PURF were enhanced in a great extent in comparison with those of pEG/PURF [9].

Luo et al. investigated the effects of using EG with different particle sizes and concentrations on the characteristics of semi-rigid polyurethane foams (SPFs) which were blown with water during reactive foaming [35]. In accordance with experimental data, it was shown that increasing EG amount in SPFs leads to generation of a much effective blockade layer and exhibited improved flame retardant properties compared to lower EG contents. Based on horizontal burning test results, it was shown that using EGs with particles sizes of 430 and 960 μm with different amounts extensively improved the flame retardancy behavior of SPFs. However, on the other hand, it was shown that using EG particles with a size of 70 μm did not elevate the fire resistance of SPFs due to the generation of a less efficient burnt and blackened layer on the surface of SPFs. According to TGA data, it was revealed that using EG as a flame retardant positively enhanced the thermal decomposition behaviour of

SPFs. Besides fire resistant properties of SPFs, effects of using different amounts of EG on mechanical properties, density, pore structure and damping properties were also analyzed by using the well-known characterization methods [35].

The results for density and compression strength values at 25% strain of SPFs that were filled with different amounts of EG are displayed in Fig. 2.28. According to experimental data, it was reported that compressive strength values of SPFs filled with EG-70 μm increased suddenly whereas the numerical value of compressive strength diminished in SPFs consisting of EGs with larger particle sizes [35]. It was also explained that the increase in compressive strength of SPFs consisting of EG-70 μm was caused by the increase of density since the presence of EG-70 μm in SPF matrix obstructed the mechanical deformation of foam. On the other hand, it was shown that the presence of EGs with larger particle sizes formed many defects in SPFs due to unfavorable interactions between these EGs and SPF matrix [35]. Thus, based on these results, it is concluded that EGs with smaller particle sizes are much more effective plate-like micron-sized particles in terms of enhancing the density and compressive strength values of PURCFs.

Thirumal et al. fabricated water-blown polyurethane rigid composite foams (PURCFs) consisting of two different particle sizes and concentrations of EG [8]. Authors investigated in detail the effects of EG particle size on morphological, mechanical, thermal, water absorption, and flame-retardant properties of PURCFs. In accordance with experimental data, it was pointed out that inclusion of higher amounts of EG deteriorated mechanical properties of PURCFs for all of EG particles. Based on the water absorption and SEM results, it was reported that water absorption increased with increasing addition of EG due to the disturbance of closed-cellular morphology of PURCFs. The thermal conductivity results showed that thermal insulation values of PURCFs consisting of EGs diminished with growing EG loadings. It was reported that increasing EG loadings enhanced the flame retardant behavior of PURCFs. In conclusion, enhanced fire retardant and mechanical properties were observed in

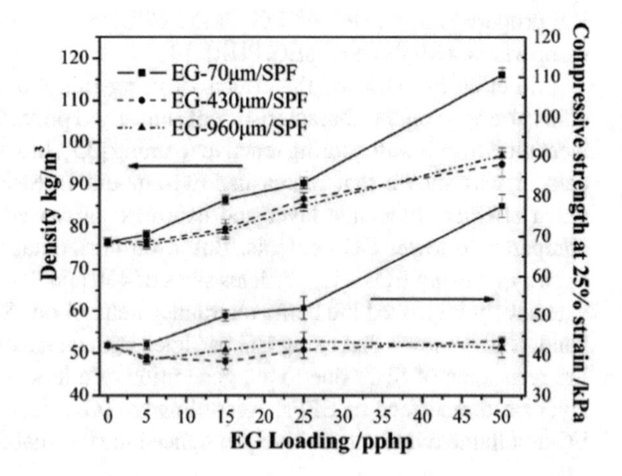

Fig. 2.28 Foam density and compressive strength data of SPFs containing diverse amounts of expandable graphite. Reproduced with permission from Ref. [35]. Copyright 2014 The Royal Society of Chemistry

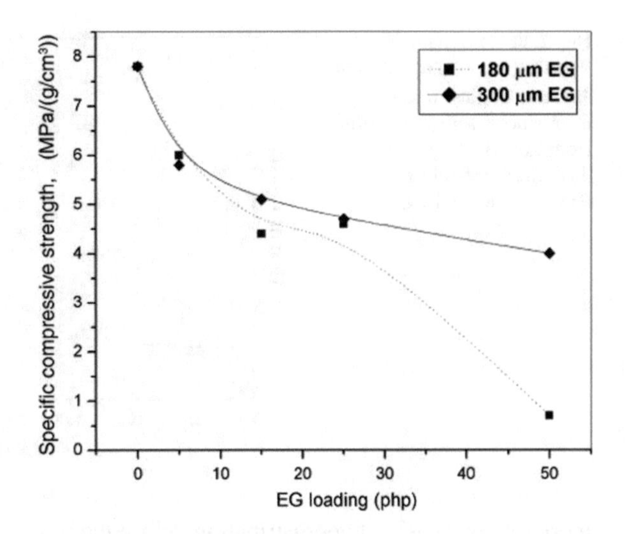

Fig. 2.29 Specific compressive strength data of PURCFs consisting of different amounts of EG. Reproduced with permission from Ref. [8]. Copyright 2008 Wiley Periodicals, Inc.

PURCFs consisting of EGs with larger particle sizes compared to PURCFs consisting of EGs with smaller particle sizes [8].

Figure 2.29 shows the effects of using EG particles with two different sizes on compression property results of PURCFs. Based on these findings, it was shown that the specific compressive values at 10% strain decayed with increasing amounts of EG [8]. This experimental finding could be explained such that at higher EG loadings, the aggregation behavior of EG particles and domination of unfavorable filler-filler interactions are generally the main reasons for the decrease of specific compressive strength values. However, in Fig. 2.29, the compressive strength values of PURCFs consisting of EG with larger particle size are much higher than those of PURCFs consisting of EGs with smaller particle sizes. This finding was explained such that EG particles with smaller sizes were unevenly distributed and this property led to the formation of aggregates and non-uniform closed cellular morphology in PURCFs [8].

2.4 Thermal and Thermomechanical Properties

Kim et al. fabricated polyurethane rigid foam (PURF)/glass fiber composites from PMDI, polypropylene oxide and glass fibers by using HFC 365mfc as the blowing agent [1]. Based on SEM, DSC (differential scanning calorimetry), compression and tensile test, density, thermal conductivity and TGA results, it was shown that incorporation of glass fibers into PURFs enhanced thermal conductivity, mechanical properties, glass transition and decomposition temperatures. Furthermore, authors compared thermal decomposition temperatures, glass transition temperatures and thermal conductivity data of PURCFs with the results obtained from closed-cellular

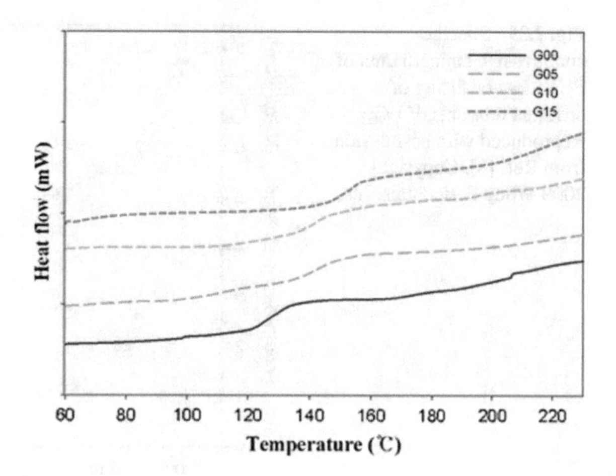

Fig. 2.30 DSC data of PURCFs containing different amounts of glass fibers. Reproduced with permission from Ref. [1]. Copyright 2010 Springer Science + Business Media, LLC

morphology, density, fiber distribution within the foam matrix and a simple model that explains the heat transfer mechanism of PURCFs [1].

Figure 2.30 shows the DSC thermograms of PURCFs consisting of 0, 5, 10 and 15 wt% glass fiber. Based on DSC results, it was well noted that T_g of PURCFs intensified with higher contents of glass fiber from 0 to 20 wt%. Moreover, it was pointed out that inclusion of 15 wt% glass fiber enhanced glass transition temperature by 20 °C in PURCFs in comparison with pure PURF. The basis behind increase of T_g with the incorporation of glass fiber was explained such that the motion of polymer chains are slowed down because of the existence of glass fibers in PURCFs [1]. Thus, the mobility slow-down of polymer chains in the vicinity of glass fibers is the primary mechanism for the increase of T_g in PURCFs. This result is very common in polymer composites filled with rigid particles in which glass transition temperature generally increases due to favorable interactions at the interface between polymer chains and rigid particles. Thus, these favorable interactions decrease movement of polymer molecules which are close to rigid particles, and eventually this mechanism elevates T_g of PURCF system.

Figure 2.31 shows the thermal conductivity behavior of PURCFs containing different amounts of glass fiber. Based on thermal conductivity results, it was shown that thermal conductivity values of PURCFs consisting of glass fibers elevate with growing amount of glass fibers because of higher value of coefficient of thermal conductivity associated with glass fibers [1]. It is very well-known that lower thermal conductivity values in PURFs are achieved by producing foam structures with smaller cell sizes [61, 62]. However, in Fig. 2.31, the numerical value of thermal conductivity increases not that much with increasing amount of glass fiber up to 15 wt%. Moreover, PURCF doped with 15 wt% glass fiber mat exhibited an elevation of not more than 8% regarding thermal conductivity compared to pure PURF. Thus, since PURFs are typically used for thermal insulation applications, the PURCF sample filled with glass fibers should be selected such that its mechanical properties should be superior but its thermal conductivity should not be that high at the same time in order to meet the criteria of thermal insulation applications [1].

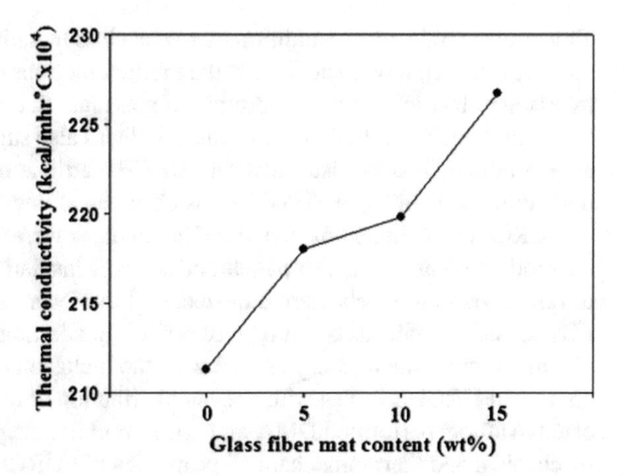

Fig. 2.31 Thermal conductivity data of PURCFs consisting of different amounts of glass fiber mat. Reproduced with permission from Ref. [1]. Copyright 2010 Springer Science + Business Media, LLC

Previously, glass fiber and silica reinforced PURCFs and also only glass fiber filled PURCFs were prepared from polymeric diphenylmethane diisocyanate (PMDI) and polyol by using HFC 365mfc as the blowing agent [17]. Authors systematically investigated thermal properties such as glass transition and decomposition temperatures and thermal conductivity, closed-cellular morphology, mechanical properties and foam reaction kinetics of PURCFs. Based on the results, it was found out that using both glass fibers and silica in PURCFs induced synergistic improvements in terms of thermal conductivity, thermomechanical properties [17].

Figure 2.32 shows the DSC thermograms of PURCFs consisting of glass fibers with respect to glass mat fiber content. According to experimental data, it was disclosed that glass transition temperatures of PURCFs got bigger with increasing amounts of glass fiber mat [17]. In addition, in Fig. 2.32, the maximum value of

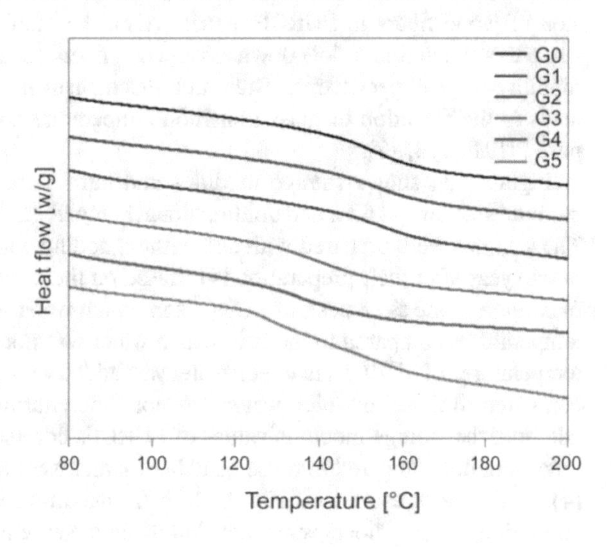

Fig. 2.32 DSC data of PURCFs containing different amounts of glass fiber mat. Reproduced with permission from Ref. [17]. Copyright 2017 BME-PT

T_g was observed with the addition of 2 wt% glass mat fibers. In agreement with previous results [1], it was shown that the thermal motions of PU chains in the PURCF matrix are obstructed due the addition of glass mat fibers. Thus, this behavior led to the increase of T_g in PURCFs. In addition, it was also suggested that glass mat fibers are not homogenously distributed in PURCF matrix at high fiber contents since the maximum value of T_g in PURCFs was observed at low glass mat contents [17].

PURCFs containing pine wood and hemp fibers were fabricated and characterized the production of car interior panels and acoustic insulation panels [4]. PURCFs with different foaming levels were adjusted, and they were fabricated from crosslinked PUs consisting of a castor oil type of polyol. In addition, PURCFs with anisotropic alignment properties were prepared with the inclusion of long hemp fibers. On the contrary, PURCF samples with uneven distribution of natural fibers were also fabricated. Authors performed DMA and three point bending experiments to understand mechanical and thermomechanical properties of PURCFs [4].

Figure 2.33 shows DMA results such as storage modulus and tan δ data for pure PURF and PURCFs consisting of pine wood fibers with different sizes. In accordance with DMA data, glass transition temperature values were found to be in the range of 98–108 °C for PURCFs reinforced with fibers, while it was found to be around 70–73 °C for the pure PURF sample [4]. In addition, it was also unveiled that reduction of storage modulus with respect to growing temperature in pure PURF was much more pronounced than PURCFs consisting of fibers since the reinforcement effect of fibers puts an obstacle against the decrease of storage modulus. The large differences between DMA results of reinforced and unreinforced PURFs were explained such that in reinforced PURCFs, hydroxyl functionalities on cellulosic fibers could be chemically bonded with isocyanate functionalities in PUs. Thus, this behavior could increase the modulus because of chemical bonding interactions between PU chains and wood fibers [4]. Thus, in agreement with other works [1, 17], the favorable interactions between wood fibers and PU chains contribute to homogenous dispersion of wood fibers in PURCF matrix. Also, the mobility of PU chains near the surface of wood fibers slow down since wood fibers act as rigid surfaces against the mobility of polymer chains. Thus, this mechanism in wood fiber filled PURCFs leads to the elevation of glass transition temperature compared to the situation in pure PURF [1, 4, 17].

Figure 2.34 shows storage modulus and tan δ data of PURCFs consisting of medium sized wood fiber and unidirectional hemp fibers with respect to temperature. The samples were prepared with and without adding water and they were analyzed in one year after their preparation [4]. Based on these results, it was shown that the maximum in the tan δ peak of PURCFs in which water was added appears at lower temperatures compared to those in which water was not added. So, glass transition temperatures of PURCFs in which water was added were observed to be much lower compared to those in which water was not added during the foam preparation. In addition, the storage modulus values of PURCFs consisting of water appears to be much smaller compared to those of PURFCs that were prepared without any water [4]. Furthermore, as shown in Fig. 2.34a, b, the maximum in the tan δ peak of PURCFs consisting of hemp fibers was located at much lower temperatures compared to that

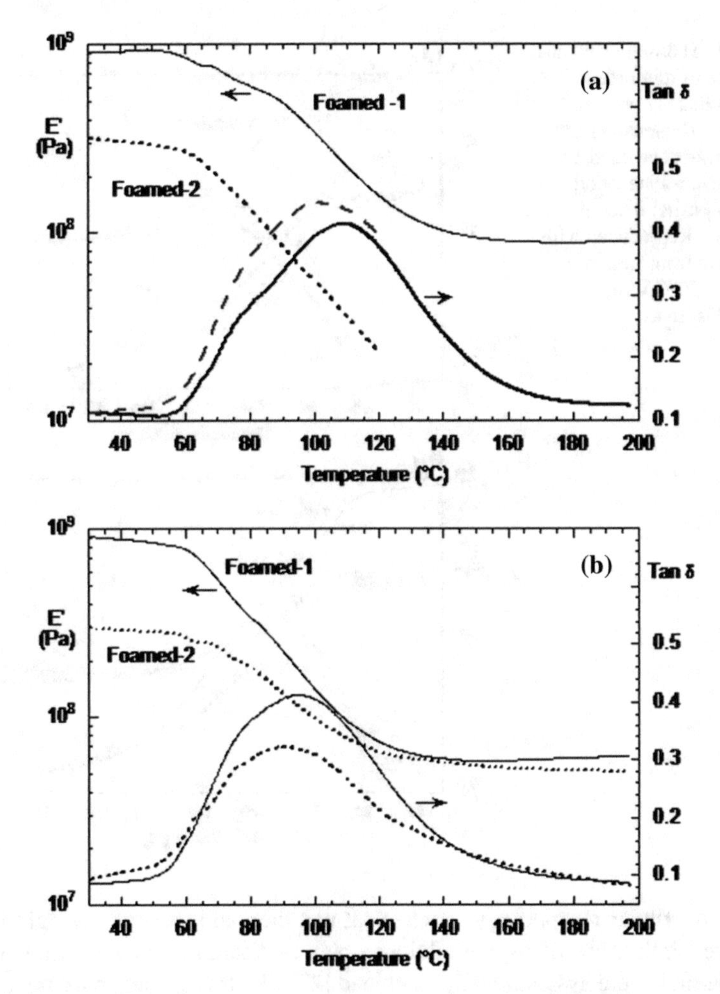

Fig. 2.33 Storage modulus and tan δ data of pure PURF and PURCFs consisting of pine wood fibers with different sizes. Reproduced with permission from Ref. [4]. Copyright 2007 Wiley Periodicals, Inc.

of PURCFs consisting of pine wood fibers [4]. Thus, in hemp fiber filled PURCFs, the mobility of polymer chains is not drastically reduced compared to PURCFs reinforced with pine wood fibers. The reason for this observation can be explained such that hemp fibers in PURCF matrix might be less dispersed due to their longer continuous lengths compared to pine wood fibers. Thus, this property of hemp fibers might lead to the decrease of favorable fiber-polymer interactions and also glass transition temperature.

Recently, polyurethane rigid composite foams (PURCFs) consisting of industrial residue of blanched cellulose fibers as micron-sized cylindrical fillers were fabricated [27]. Authors investigated the effects of using different amounts of cellulose fibers

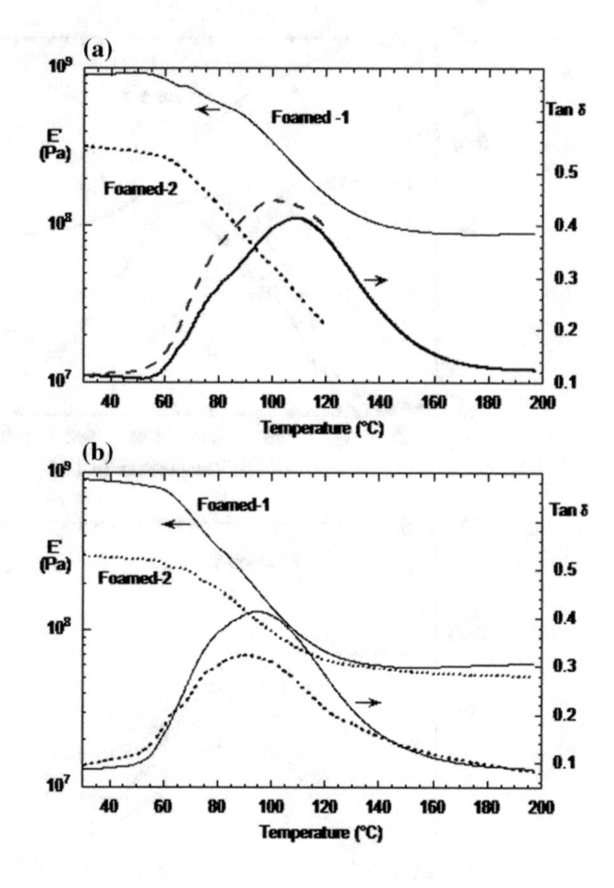

Fig. 2.34 Thermomechanical properties of reinforced PURCFs that are prepared with water (Foamed-2) and without water (Foamed-1) after 1 year. **a** Pine wood fiber composite; **b** hemp composite. Reproduced with permission from Ref. [4]. Copyright 2007 Wiley Periodicals, Inc.

on closed-cellular morphology, mechanical and thermal properties of rigid foams. Moreover, influences of using cellulose fibers on foam capacity to counteract to fungal assault were systematically examined [27]. For this specific purpose, authors devised a particular experiment in which the morphological evolution of the surface of foam that was in contact with fungus suspension was examined within a certain period of time [27].

Figure 2.35 shows the thermal conductivity and average cell diameter results of pure PURF and PURCFs consisting of industrial residue of blanched cellulose fibers with respect to fiber amount. According to experimental data, it was disclosed that thermal conductivity reduced about 32% with the addition of 16 wt% cellulose fiber to pure PURF [27]. In addition, as shown in Fig. 2.35, it was also reported that thermal conductivity and average cell diameter curves followed the same trends with increasing fiber content. Thus, it was suggested that this result is a clear proof for the strong influence of the average cell size on the thermal conductivity behavior of PURFs [27]. Based on the results from other works [35, 36, 39, 43, 61, 62], it is very-well known that there is a directly proportional relationship between the average cell size and thermal conductivity behavior in PURFs.

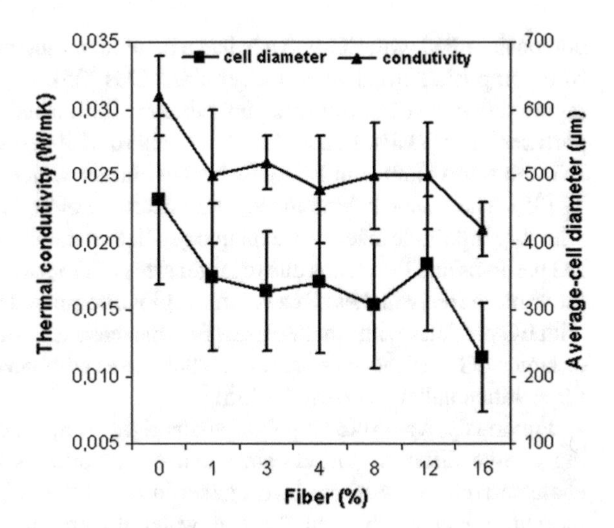

Fig. 2.35 Thermal conductivity and average cell sizes in the vertical direction of pure PURF and PURCFs consisting of different amounts of fiber. Reproduced with permission from Ref. [27]. Copyright 2010 Wiley Periodicals, Inc.

Influences of using EG with different particle sizes and concentrations on the characteristics of semi-rigid polyurethane foams (SPFs) which were blown with water during reactive foaming were investigated in detail [35]. In accordance with experimental data, it was shown that increasing EG amount in SPFs leads to generation of a much effective blockade layer and exhibited enhanced fire resistant characteristics compared to lower EG contents. Besides fire resistant properties of SPFs, effects of using different amounts of EG on mechanical properties, density, pore structure and damping properties were also analyzed by using the well-known characterization methods [35].

Figure 2.36 shows the dynamic mechanical properties (tan δ vs. temperature) of pure SPF and SPF composites filled with EG particles consisting of different sizes. As specified by DMA data, EG incorporation led to small amount of increase on the damping properties [35]. In addition, it was reported that with increasing EG content, EG particles obstructed the deformation of SPFs which led to the decay of tan δ peak area or damping behavior. Moreover, authors also showed that the SPF

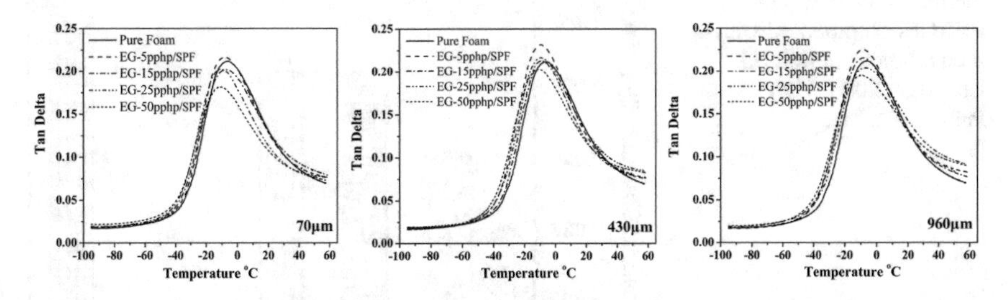

Fig. 2.36 Dynamic mechanical properties of SPF composites filled with different amount of EG particles. Reproduced with permission from Ref. [35]. Copyright 2014 The Royal Society of Chemistry

composite filled with EG-960 μm had a better damping property in comparison with SPF composites filled with smaller sized EGs [35]. As shown in Fig. 2.36, it was reported that the glass transition temperature diminished with higher amounts of EG particles in PURCFs. In addition, it was also added that with increasing amounts of EG, the free volumes in PURCFs became larger which led to further reduction of T_g [35]. The reason behind much larger decrease of T_g in PURCFs consisting of EG with larger particle sizes was explained such that there is more free volume between EG particles and PU chains due to larger sizes of EG particles. Thus, PU chains could move more freely in these free volume regions compared to the PURCF system filled with EG particles with smaller sizes. For this reason, T_g of PURCF system consisting of larger EG particles was observed much lower compared to that of PURCF system filled with smaller EG particles [35].

Previously, water-blown polyurethane rigid composite foams (PURCFs) consisting of two different particle sizes and concentrations of EG were prepared and characterized [8]. Authors investigated in detail the effects of EG particle size on morphological, mechanical, thermal, water absorption, and flame-retardant properties of PURCFs. In accordance with experimental data, it was shown that increasing amounts of EG deteriorated mechanical properties of PURCFs for all types of EG particles. The thermal conductivity results showed that thermal insulation values of PURCFs consisting of EGs diminished with growing EG loadings [8].

Figure 2.37 shows the effects of using different EG loadings on the thermal conductivity behavior of PURCFs. Based on thermal conductivity results, it was pointed out that thermal conductivity values elevated with higher amounts of EG loading in PURCFs because of the higher thermal conductivity coefficient of EG [8]. In addition, authors explained the reason behind higher thermal conductivity in PURCFs in comparison with pure PURF such that increased thermal conductivity is caused by the increase of average cell size and damaged cell morphology in EG-filled PURCFs [8]. Based on the results from other works [27, 35, 36, 39, 43, 61, 62], it is very well known that the disturbance and transform of closed-cellular morphology into open

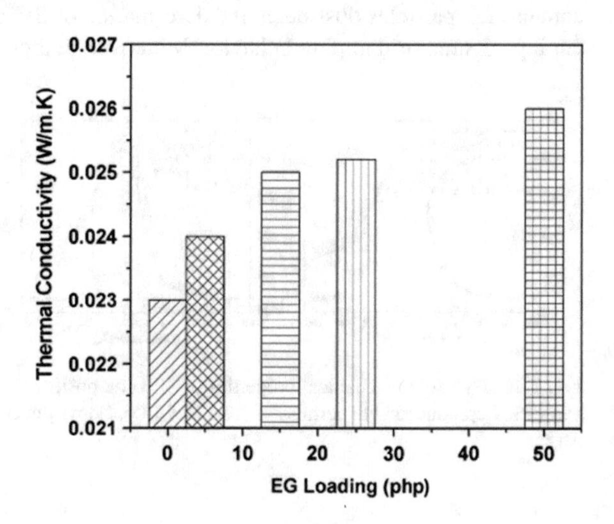

Fig. 2.37 Effect of EG with a size of 300 μm on the thermal conductivity of PURCFs. Reproduced with permission from Ref. [8]. Copyright 2008 Wiley Periodicals, Inc.

cell structure, broken cell walls and non-uniform distribution of closed cells and also the increase of cell size generally leads to the increase of thermal conductivity behavior in PURFs.

Recently, PURCFs consisting of calcium carbonate fillers with two different sizes and crystallized silica particles with an intermediate size were studied [57]. Basically, authors investigated the effect of using different particle sizes on closed-cellular morphology and mechanical properties of PURCFs. As claimed by comparison of mechanical property results with modelling data, it was reported that an accurate explanation of mechanical and viscoelastic properties should be taken into account when the size of fillers is compared to cell wall sizes [57]. Two types of models were established by authors in order to discuss the physical behaviors of PURCFs filled with larger carbonate fillers, and also PURCFs filled with the smaller carbonate fillers. Authors also concluded that the use of crystallized silica particles changed the properties of PURCFs in between that of the PURCFs consisting of two calcium carbonates [57].

Figure 2.38 shows the change of storage (G′) and loss (G″) moduli of PURCFs consisting of calcium carbonate with two different sizes with respect to temperature. In compliance with DMA data, it was unveiled that storage and loss moduli curves

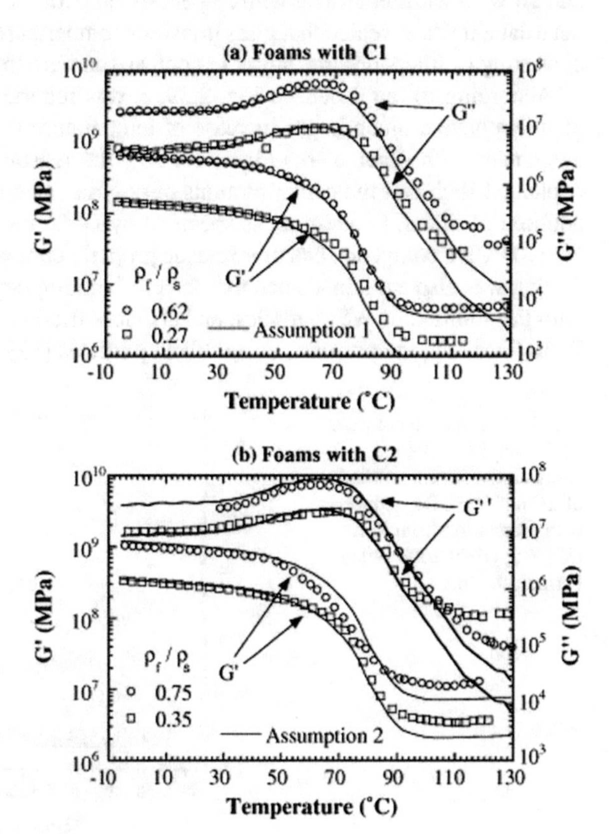

Fig. 2.38 Elastic (G′) and loss (G″) modulus data with respect to temperature for PURCFs reinforced with **a** calcium carbonate C1 with a larger size and **b** calcium carbonate C2 with a smaller size. Reproduced with permission from Ref. [57]. Copyright 2006 Elsevier Ltd.

associated with PURCFs containing different densities were parallel to each other and only shifted along the y-axis of graphs [57]. In addition, it was also disclosed that reduction of foam density diminished storage modulus values in PURCFs independent of filler type. Moreover, it was also noted that the elevation of storage modulus with incorporation of fillers was observed for higher density PURCFs. Based on the experimental results from DMA experiments, the effects of using different sizes of fillers on the elastic properties of PURCFs was clearly revealed in such a way that the storage modulus of PURCFs increases with decreasing average size of fillers [57].

Previously, PURCFs consisting of EG and whisker silicon with different contents which have the same density were systematically investigated [53]. Authors showed that PURCFs with the best mechanical properties were obtained with the incorporation of 10 wt% WSi in PURCFs. Moreover, it was also reported that incorporation of 10 wt% WSi into PURFs enhanced the values of compressive strength and modulus. According to DMA data, highest storage modulus value was obtained in PURCF sample containing 10 wt% EG and 10 wt% WSi because of combined positive effects of using both EG and WSi in PURCFs [53].

Figure 2.39 shows tan δ curves of (a) pure PURF, PURCF consisting of (b) 10 wt% EG, (c) PURCF consisting of 10 wt% WSi and (d) PURCF consisting of 10 wt% EG and 10 wt% whisker silicon with respect to temperature. In accordance with DMA tan δ data, it was revealed that glass transition temperatures associated with PURCFs consisting of fillers became larger in comparison with that of pure PURF [53].

According to tan δ data in Fig. 2.39, it was reported that PURCFs consisting of fillers have a much larger increase of temperature range for damping effect. In agreement with other works [1, 4, 17, 35], the reason behind this behavior was explained such that molecular motions of polymer chains were obstructed with the addition of fillers. In addition, as specified by DMA data, it was also unveiled that WSi10 PURF composite had more elastic property compared to other PURCFs. This finding was also explained such that the crosslinking degree of PURCFs decreased with the addition of WSi particles, and also less friction or loss modulus occured in PURCFs due to the presence of rigid WSi particles [53].

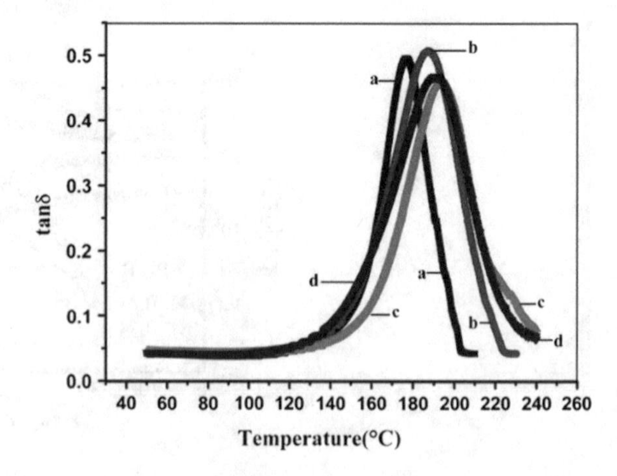

Fig. 2.39 Tan δ data of pure PURF and PURCFs containing different amounts of EG and WSi. Reproduced with permission from Ref. [53]. Copyright 2008 Wiley Periodicals, Inc.

2.5 Thermal Degradation and Flammability

Polyurethane rigid foam (PURF)/glass fiber composites from glass fiber, polymeric diphenylmethane diisocyanate (PMDI) and polypropylene glycols (PPGs) by were fabricated by using HFC 365mfc as the blowing agent [1]. Based on SEM, DSC (differential scanning calorimetry), compression and tensile test, density, thermal conductivity and TGA results, it was shown that incorporation of glass fibers into PURFs enhanced thermal conductivity, mechanical properties, glass transition and decomposition temperatures. Furthermore, authors compared thermal decomposition and glass transition temperatures, thermal decomposition behavior and thermal conductivity data of PURCFs with the results obtained from closed-cellular morphology, density, fiber distribution within the foam matrix and a simple model that explained the heat transfer mechanism of PURCFs [1].

Figure 2.40 shows TGA thermograms of glass fiber reinforced PURCFs versus different amounts of glass fiber. In compliance with TGA data, it was disclosed that decomposition temperatures associated with PURCFs elevated with incorporation of glass fiber to pure PURF [1]. In addition, it was also reported that ultimate weight loss corresponding to glass fiber reinforced PURCFs diminished by about 20% compared to pure PURF. The thermal stability and thermal decomposition behavior enhancement in PURCFs was explained such that glass fibers in PURCFs acted as an obstacle against the diffusion of oxygen and other volatile compounds [1]. In agreement with other works [6, 11, 17, 34, 35], the presence of rigid particles such as glass fibers, pulp fibers and expandable graphite prevented movements of polymer chains near their surfaces which led to elevation of glass transition temperatures as well as decomposition temperatures of PURCFs. In addition, these rigid particles also slow down the mobility of gaseous compounds that emerge during the decomposition process of PURCFs [1].

Previously, glass fiber and silica reinforced PURCFs and also only glass fiber filled PURCFs were prepared from polymeric diphenylmethane diisocyanate (PMDI) and polyol by using HFC 365mfc as the blowing agent [17]. Authors systematically

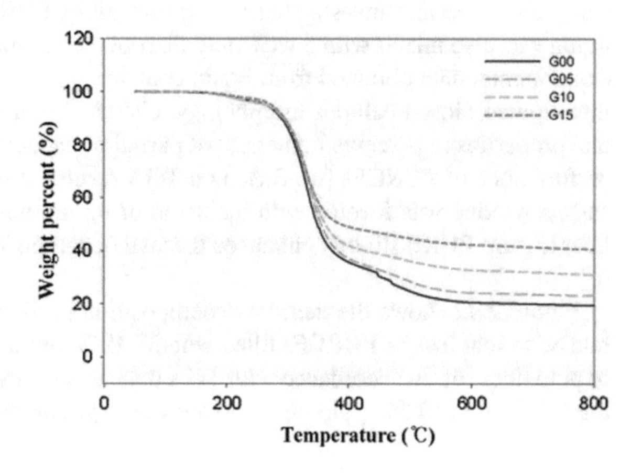

Fig. 2.40 TGA data of PURCFs consisting of different amounts of glass fiber. Reproduced with permission from Ref. [1]. Copyright 2010 Springer Science + Business Media, LLC

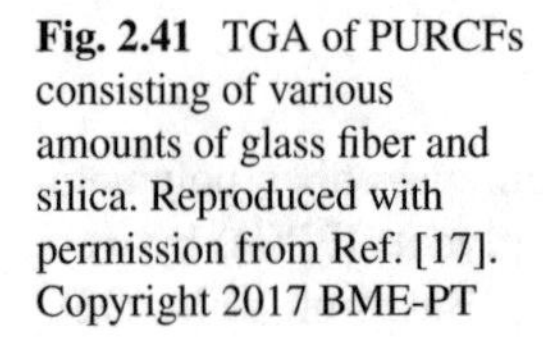

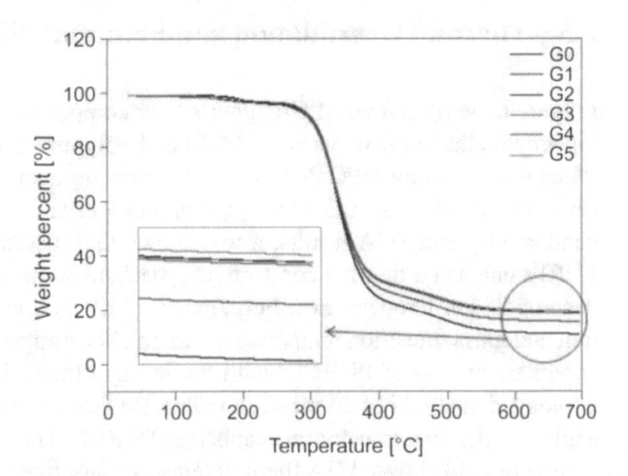

Fig. 2.41 TGA of PURCFs consisting of various amounts of glass fiber and silica. Reproduced with permission from Ref. [17]. Copyright 2017 BME-PT

investigated the closed-cellular morphology, mechanical properties, foam reaction kinetics, glass transition temperature, thermal conductivity and thermal degradation temperatures of PURCFs. Based on the results, it was found out that using both glass fibers and silica in PURCFs induced synergistic improvements in terms of mechanical properties, thermal conductivity, thermomechanical and thermal decomposition properties [17].

Figure 2.41 shows the thermal decomposition and thermal stability behaviours of PURF/glass fiber/silica and PURF/glass fiber composites. According to TGA data, highest thermal degradation temperatures associated with 10 and 50% weight loss were obtained in PURCFs consisting of 2% glass mat composition. In addition, in agreement with previous fiber reinforced PURCFs works [1, 6], the enhancement of thermal stability and thermal decomposition of PURCFs with the addition of glass fibers was explained such that glass fiber mats acted as barriers and reduced the flow of oxygen and evaporative products throughout the PURCF matrix [1, 6, 17].

Previously, rigid polyurethane foams consisting of lignin in which some of the synthetic polyol was exchanged with different amounts of lignin were systematically fabricated and investigated [6]. In that study, PURCFs consisting of 37.19% lignin was also mixed with 5 wt% pulp fiber in the maximum amount. According to experimental data obtained from lignin containing PURCFs, authors systematically investigated closed cellular morphology, chemical structure, mechanical and thermal properties to perceive influences of partial replacement of polyol with lignin on performance of PURCFs [6]. Based on TGA results, it was demonstrated that high carbon residue was formed with inclusion of lignin, however introduction of pulp fiber to pure PURF slightly enhanced thermal decomposition behaviour of PURCFs [6].

Figure 2.42 shows the thermal decomposition profiles, thermal stability and the rate of weight loss of PURCFs filled with 37.19% lignin and various concentrations of pulp fiber [6]. In accordance with TGA data, it was unveiled that PURCF containing 5% lignin and 2% pulp fiber had a lower degradation pace compared to sample

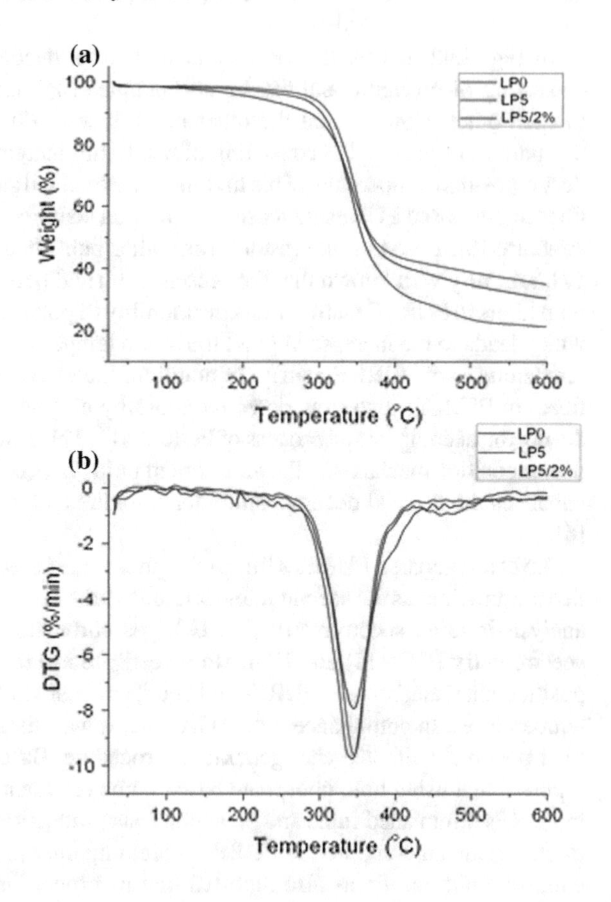

Fig. 2.42 TGA results of PURCFs consisting of different amounts of pulp fiber and lignin. **a** The weight loss curves and **b** the rate of weight loss curves. Reproduced with permission from Ref. [6]. Copyright 2014 American Chemical Society

filled with only 5% lignin. Authors explained motivation behind this observation such that the pulp fiber has less number of interactions with the polymer chains of PURF during the reactive polymer foaming process [6]. In agreement with the results that were published in other works [1, 11, 17, 34, 35], the decrease in the number of favorable interactions between particles and PU chains generally decreases thermal stability, mechanical and thermomechanical properties of PURCFs. The reason behind this decrease in properties can be explained as in the following sentences. The particles which are not surface-modified with compatible and specific surface functional groups more likely form aggregated structures within the PURCF matrix. Also, the homogenous distribution of these particles becomes less likely due to unfavorable particle-particle interactions. Thus, during the reactive foaming process, these aggregated particles cannot form favorable interactions with PU chains in terms of reducing the cell size and producing a uniform distribution of closed-cellular morphology in PURFs. Thus, the fabrication of PURFs consisting of larger cell sizes and non-uniform distribution of closed-cellular morphology generally exhibit poor performance in terms of thermal stability, mechanical and thermomechanical properties [1, 6, 11, 17, 34, 35].

In Fig. 2.42, it was also noted that maximum decomposition rates of PURCF consisting of no lignin and 5% lignin (samples LP0 and LP5) were almost equal to each other. However, on the other hand, it was concluded that the addition of 2% pulp fiber into PURF consisting of 5% lignin increases the value of maximum decomposition temperature. This finding was explained such that the addition of pulp fiber to bio-based PURFs increases the thermal stability of the closed-cellular foam structure [6]. Based on the results from other published works in the literature [1, 17], it is very well known that the presence of rigid particles such as glass fibers or pulp fibers in PURCF matrix limits the mobility of polymer chains near fiber surfaces which leads to the increase of glass transition temperature as well as decomposition temperatures of PURCF matrix. In addition, the existence of pulp fibers and glass fibers in PURCFs also slow down the mobility of gaseous compounds that emerge during the decomposition process of PURCFs [1, 17]. Thus, as a result of this thermal decomposition mechanism, the inclusion of pulp fibers in the amount of 2% basically enhances the thermal decomposition temperature and thermal stability of PURCFs [6].

Xu et al. prepared PURCFs filled with phosphate based flame retardant and modified aramid fiber as the second flame retardant [26]. Authors used thermogravimetric analysis/infrared spectrometry (TG-IR), gas chromatography combine with mass spectrometry (GC-MS) and FT-IR to investigate fire resistant and thermal decomposition characteristics of PURFs and details of char residues in PURCFs at various temperatures. In compliance with TGA data, it was disclosed that modified aramid fiber positively affected char generation procedure. Based on TG-IR results, it was reported that using both phosphate based flame retardant and modified aramid fiber in PURFs diminished fume and poisonous gas production and increased the amount of char remnant compared to PURFs containing only phosphate based flame retardant. In addition, it was also reported that incorporation of both phosphate based flame retardant and modified aramid fiber in PURFs reduced poisonous gases such as hydrogen cyanide and increased the amounts of CO_2 and water. Based on GC-MS results, it was also found out that incorporation of both phosphate based flame retardant and modified aramid fiber in PURFs induced barrier and quenching effects in the PURF system. Based on these findings from experimental tests, authors proposed a potential thermal degradation mechanism in order to explain the experimental observations in a more systematic way [26].

Figure 2.43 shows the TGA and DTG data of PURF and PURFs consisting of phosphate based flame retardant and modified aramid fiber as the flame retardants. In accordance with this experimental data, it was shown that thermal decomposition temperature at 5% weight loss for PURF-2 was diminished by about 60 °C in comparison with that of PURF-1 [26]. This result was explained such that the presence of phosphate based flame retardant in sample PURF-2 led to low decomposition and volatilization temperature in this foam composite. In addition, it was also reported that the temperature corresponding to maximum rate of weight loss in sample PURF-4 was increased compared to that of PURF-1 due to the improved thermal stability by the positive effect of modified aramid fiber presence in the system [26].

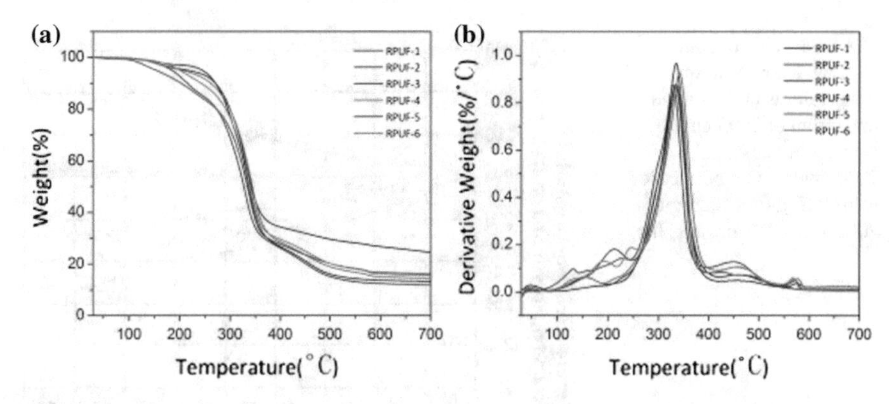

Fig. 2.43 Thermal degradation data of pure PURF and PURFs consisting of various amounts of flame retardants. Reproduced with permission from Ref. [26]. Copyright 2018 Elsevier Ltd.

As shown in Fig. 2.43, the degradation temperature associated with highest rate of weight loss in PURF-5 was observed to be much higher compared to those in PURF-2 and PURF-3. Authors explained this result in such a way that the dense char residue was formed by the synergistic effects of degraded phosphate group and aramid fiber which acted as the main components to prevent PURF from further degradation. Thus, it was suggested that using both phosphate based flame retardant and modified aramid fiber in PURFs enhanced the thermal stability, and effectively acted as the major mechanism against the reduction of thermal stability of flame retardant PURFs [26].

Based on these results of this study [26], it was verified that using two types of flame retardant materials such as phosphate based flame retardant and modified aramid fiber improves both morphology and thermal stability behavior of PURFs [37, 39, 40]. Thus, the formation of a dense char residue in PURFs containing both phosphate based flame retardant and modified aramid fiber definitely improves the thermal stability behavior by acting as an extra reinforcement compared to pure PURF. Furthermore, the formation of a dense char layer in PU/modified aramid fiber foams with the addition of phosphate based flame retardant clearly points out the fact that favorable interactions exist between phosphate based flame retardant and modified aramid fiber in PURFs consisting of both phosphate based flame retardant and modified aramid fiber. Thus, these results revealed that using two different types of flame retardant materials in PURFs induced positive effects in terms of improving morphology, thermal stability and flame retardancy behavior in these PURCFs [26, 37, 39, 40].

Previously, PURMCFs consisting of various fillers such as talc, chalk, starch, aluminum hydroxide and borax in the amounts of 2.5–20 wt% were prepared. Authors specifically investigated the compressive strength, apparent density, brittleness, closed cell content, flammability and softening point of foams by performing detailed measurements on physicochemical and thermal properties of foams [30].

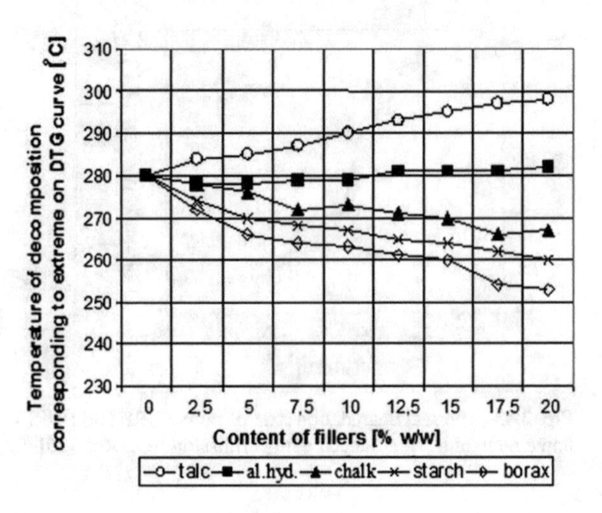

Fig. 2.44 The maximum thermal decomposition temperatures based on the maximum of DTG curves versus filler amount. Reproduced with permission from Ref. [30]. Copyright 2009 Wiley Periodicals, Inc.

Figure 2.44 shows the dependence between the values of maximum thermal decomposition temperature and amounts of fillers in PURMCFs. According to experimental data, it was shown that maximum thermal decomposition temperature on DTG curve for pure PURF was observed to be around 280 °C. In addition, it was also reported that the maximum thermal decomposition temperature on DTG curve for PURCF consisting of 20 wt% talc was observed to be around 298 °C. However, it was also shown that the maximum thermal decomposition temperatures on DTG curve for PURCFs consisting of 20 wt% chalk, starch and borax decreased to 270, 260 and 253 °C, respectively [30]. In conclusion, it was found out that the incorporation of fillers such as chalk, starch, talc, aluminum hydroxide and borax into the PURFs affected both physical and mechanical properties of foams in addition to their heat and thermal resistance. It was reported that inclusion of talc and aluminum hydroxide into PURMCFs from 2.5 to 20 wt% improved their functional properties compared to pure PURF consisting of no filler. Authors suggested that PUR-PIR foams consisting of talc and aluminum hydroxide could possibly be used industrial applications such as building and heat engineering [30].

Recently, beneficial effects of using both phosphorus based polyol and nitrogen based polyol in improving flame retardant properties of PURMCFs consisting of EG particles were investigated [11]. The effects of using different amounts of these polyols on flame retardancy and thermal decomposition characteristics of PURMCFs were studied by TGA and LOI tests [11]. Based on these results, authors proved that incorporation of EG particles into PURMCFs consisting of phosphorus and nitrogen based polyols improved fire resistant properties [11]. In addition, authors also reported that the incorporation of EG particles into PURF consisting of both phosphorus based polyol and nitrogen based polyol improved flame retardant properties of PURCFs. Moreover, it was pointed out that the LOI value of PURCFs approached to ~34% when the concentration of EG was 15 wt% [11].

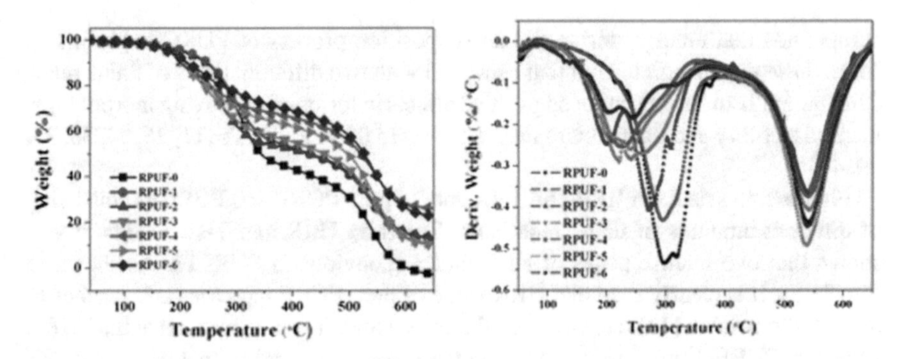

Fig. 2.45 TG and DTG curves of pure PURF and PURCFs consisting of different amounts of flame retardants in air environment. Reproduced with permission from Ref. [11]. Copyright 2016 American Chemical Society

Figure 2.45 shows thermal degradation data of pure PURF and PURMCFs consisting of flame retardant materials in air environment. Based on the data in Fig. 2.45a, it was reported that the sample PURF-4 containing 1:1 ratio of phosphorus based polyol and nitrogen based polyol had the best synergetic effect in terms of improving the thermal stability of PURCFs filled with EG [11]. Furthermore, the reason behind enhanced thermal stability in PURCFs was explained such that using both phosphorus based polyol and nitrogen based polyol in PURCFs loaded with EG improved the thermal stability of flame retardant PURFs via char formation by suppressing the mass loss rates that were observed from DTG curves [11]. These results clearly showed that using both EG, phosphorus based polyol and nitrogen based polyol in PURCFs induced synergistic effects in terms of improving thermal stability and flame retardancy behavior in these PURCFs.

Based on the results from other similar works [26, 34, 35, 37, 39, 40], it was confirmed that using two types of flame retardant materials such as phosphorus based polyol and nitrogen based polyol in addition to EG improves morphology, thermal stability and flame retardancy behavior of PURCFs. Thus, the char formation in PURCFs containing both phosphorus based polyol and nitrogen based polyol besides EG definitely improves the thermal decomposition and flame retardant properties by acting as an extra reinforcement compared to pure PURF [11]. Furthermore, the formation of a char layer in phosphorus based polyol and nitrogen based polyol based PURMCFs with the addition of EG clearly points out the fact that favorable interactions between EG and flame retardants are responsible for generation of reinforced and burned coating in PURFs consisting of phosphorus based polyol and nitrogen based polyol and EG. Based on the results from other published works in the literature [1, 6, 17], it is very well known that the presence of particles in PURCF matrix limits the mobility of polymer chains near particle surfaces which leads to the increase of glass transition temperature as well as decomposition temperatures of PURCF matrix. In addition, the existence of EG with phosphorus based polyol and nitrogen based polyol in PURCFs could also decrease the mobility of gaseous compounds that emerge during the decomposition process of PURCFs [1, 6, 17]. Thus, these results pointed out that using EG with two different types of flame retardant materials in PURFs induced positive effects in terms of improving morphology, thermal stability and flame retardancy behavior in these PURCFs [11, 26, 34, 35, 37, 39, 40].

Figure 2.46 sets forth HRR and THR data of pure PURF and PURCFs consisting of different amounts of flame retardants. Based on HRR and THR results, it was shown that two intense peaks of HRR data are obvious in PURCFs. As shown in Fig. 2.46a, it is clearly seen that HRR value of pure PURF increases and reaches to its maximum value [11]. However, on the other hand, it was reported that the PHRR values of PURF-1 and PURF-2 were much lower compared to that of pure PURF due to the partial burning effects of phosphorus based polyol during the burning process. Moreover, it was also shown that the inclusion of nitrogen based polyol led to substantial amounts of decrease in PHRR values of PURCFs. In conclusion, it was demonstrated that using both phosphorus based polyol and nitrogen based polyol as flame retardant materials increases the LOI values and enhances the char layer generation substantially in PURCFs [11].

Gao et al. fabricated PURMCFs from sustainable rosin based polyester polyol, EG and melamine polyphosphate (MPP) as swollen flame retardants and layered double hydroxide (LDH) as micron-sized filler [34]. Authors investigated closed-cellular morphology, mechanical properties, thermal conductivity, thermal decomposition behaviour, fire and flame retardant properties of pure PURF and PURCFs consisting of different amounts of fillers. In addition, authors explored the effect of synergy between LDH and flame retardant materials on the fire properties of PURCFs. According to XRD data, it was disclosed that LDH had the exfoliated structure and well distributed in rosin-based PURFs. In compliance with TGA data, it was noted that incorporation of EG, MPP or more addition of LDH into PURFs reduced the initial and second maximum-rate decomposition temperatures while increasing

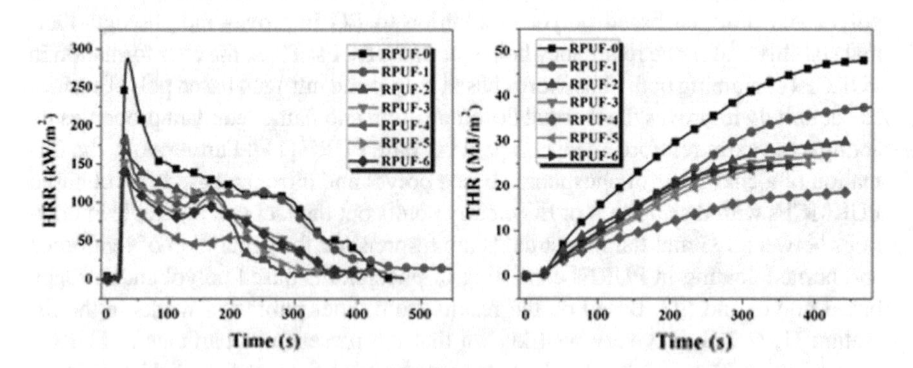

Fig. 2.46 HRR and THR data of pure PURF and PURCFs consisting of different amounts of flame retardants. Reproduced with permission from Ref. [11]. Copyright 2016 American Chemical Society

the amounts of char residue in PURFs at higher temperatures. Furthermore, it was concluded that adding EG, MPP and LDH into PURFs could extensively enhance mechanical, flame retardant, and fire resistance properties of rosin-based PURFs [34].

Figure 2.47 shows (a) TGA and (b) differential TGA of pure PURF, PURMCF consisting of 10% EG and 10% MPP and PURMCF consisting of 10% EG and 10% MPP and 3% LDH in air environment. As shown in Fig. 2.47, it was reported that in both PURMCF consisting of 10% EG and 10% MPP and PURMCF consisting of 10% EG and 10% MPP and 3% LDH, the first, second and third step thermal decomposition temperatures diminished compared to that of pure PURF which indicated that both of these PURCFs had lower thermal stability compared to pure PURF [34]. Thus, considerable decrease in first step and second step maximum degradation temperatures of PURFs occurred due to incorporation of EG and MPP or inclusion

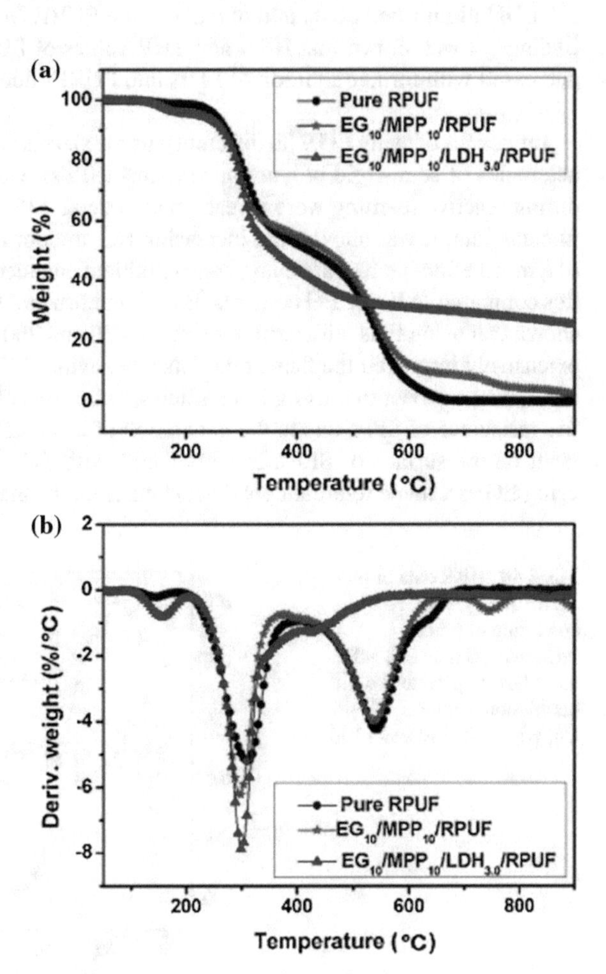

Fig. 2.47 Thermal degradation data of pure PURF, PURMCFs containing different amounts of EG, MPP and LDH in air environment. Reproduced with permission from Ref. [34]. Copyright 2013 Elsevier Ltd.

of higher LDH amounts into PURFs. However, in Fig. 2.47a, the char residues of PURCFs consisting of MPP, EG and LDH are much higher compared to pure PURF revealing the fact that flame retardant properties of PURFs are more superior than that of pure PURF [34].

HRR data of pure PURF, PURMCF consisting of 10% EG and 10% MPP and PURMCF consisting of 10% EG and 10% MPP and 3% LDH is depicted in Fig. 2.48. HRR curves of PURMCF consisting of 10% EG and 10% MPP and PURMCF consisting of 10% EG and 10% MPP and 3% LDH display two characteristic peaks instead of only one peak [34]. The reason behind this change was reported such that oxygen amount surrounding PURFs diminished due to presence of ammonia and other nonflammable gases that were degraded from EG, MPP and LDH. In addition, it was also reported that the peak and average heat release rates of PURMCF consisting of 10% EG and 10% MPP and PURMCF consisting of 10% EG and 10% MPP and 3% LDH diminished compared to that of pure PURF. In conclusion, based on these findings, it was shown that HRR and THR values of PURCFs could be effectively decreased with further addition of LDH into PURFs due to the thermal blockade of LDH [34].

Influences of using EG with different particle sizes and concentrations on the characteristics of semi-rigid polyurethane foams (SPFs) which were blown with water during reactive foaming were investigated in detail [35]. In accordance with experimental data, it was shown that increasing EG amount in SPFs leads to generation of a much effective blockade layer and exhibited enhanced fire resistant characteristics compared to lower EG contents. Based on horizontal burning test results, it was shown that using EGs with particles sizes of 430 and 960 μm with different amounts extensively improved the flame retardancy behavior of SPFs. However, on the other hand, it was shown that using EG particles with a size of 70 μm did not elevate the fire resistance of SPFs due to the generation of a less efficient burnt and blackened layer on the surface of SPFs. In compliance with TGA data, it was disclosed that using EG as a flame retardant positively enhanced the thermal stability of SPFs [35].

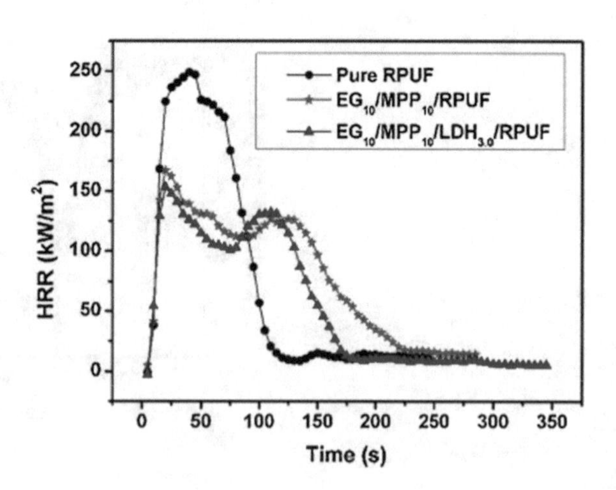

Fig. 2.48 HRR data of pure PURF and PURMCFs consisting of flame retardants such as EG, MPP and LDH. Reproduced with permission from Ref. [34]. Copyright 2013 Elsevier Ltd.

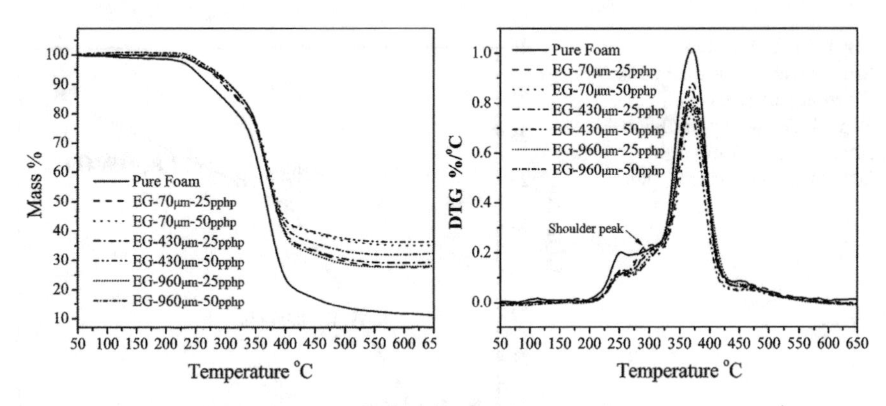

Fig. 2.49 TG and DTG curves of pure SPF and SPFs consisting of EGs with different sizes. Reproduced with permission from Ref. [35]. Copyright 2014 The Royal Society of Chemistry

Figure 2.49 shows TGA and DTG data of pure semi rigid polyurethane foam (SPF) and SPFs consisting of EGs having three different particle sizes in nitrogen environment. In accordance with experimental data, it was revealed that using higher amount of EG with larger particle sizes extensively increased $T_{5wt\%}$ of SPF systems. This finding was explained such that using EG with higher amount and a larger particle size in SPFs affected positively the morphology of isolation layer, and led to higher thermal decomposition behaviour. Based on DTG thermogram results, it was shown that the values of T_{max} slightly changed with the addition of EGs to SPFs. In addition, it was concluded that SPFs consisting of EGs and also higher amount of EGs exhibited higher amounts of char residues compared to pure foam at high temperatures [35]. In agreement with other works [34], the presence of higher amounts of char residues in EG filled PURCF system clearly shows that these materials have improved flame retardant properties compared to pure PURF.

Various EG particles containing different sizes were included into water-blown semi-rigid PURFs to improve the flame retardant properties [36]. According to experimental data, it was unveiled that using EG with smaller particle sizes did not improve fire retardant properties of PURMCFs. In addition, it was shown that using EG with a larger particle size as a micron-sized filler in PURMCFs might efficiently intensify fire retardant characteristics [36]. Elevation of flame retardancy in PURMCFs with increasing EG particle size was explained due to presence of densely packed segregation layer that was formed with the increase of EG concentration in PURMCFs. It was also reported that LOI values of EG filled semi-rigid PURMCFs increased linearly with the increase in sizes of EGs. In accordance with TGA data, it was emphasized that EG particles with different sizes did not affect the thermal stability of PURMCFs [36].

Figure 2.50 shows the LOI results of SPFs consisting of EGs (20 pphp) with different particle sizes. Based on results, it was verified that EGs with larger particle sizes are more favourable than smaller ones in terms of enhancing the flame retardant properties of SPFs [36]. As shown in Fig. 2.50, using EGs with particle sizes less

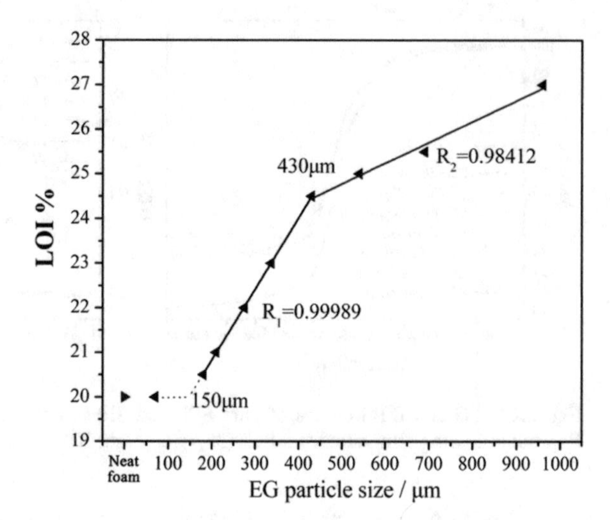

Fig. 2.50 LOI results of SPFs consisting of EGs with different particle sizes. Reproduced with permission from Ref. [36]. Copyright 2013 Wiley Periodicals, Inc.

than 150 μm in SPFs did not increase the LOI values. However, the addition of EGs with particle sizes greater than 150 μm to foams dramatically increases the LOI values of SPFs. Thus, this finding was explained such that EGs with smaller sizes might not create an efficient blockade against heat penetration whereas EGs with particle sizes bigger than 150 μm increased the flame retardancy due to their larger expanded volumes [36].

Previously, flame-retardant PURFs consisting of hexa-phenoxy-cyclotriphosphazene (HPCP) and EG were prepared by using the reactive box-foaming process [37]. In accordance with LOI and cone calorimetry test results, HPCP incorporation into PURFs consisting of EGs enhanced flame retardant properties. The thermal degradation behavior of HPCP in PURCFs was explored by pyrolysis gas chromatography/mass spectroscopy [37]. Based on these test results, it was shown that HPCP produced PO_2 and phenoxyl free radicals during combustion reaction, and these radicals were used for the quenching of flammable free radicals in the PURF matrix. Furthermore, this observation was shown to be the cause for the flame retardant effect in these PURCFs [37]. In compliance with EDS and SEM data, it was disclosed that some of phosphorus from HPCP was accumulated in the residual char, and HPCP substantially improved the strength and affinity of the char layer [37].

Figure 2.51 shows HRR curves of pure PURF and PURCFs consisting of various proportions of HPCP and EG. Based on these results, it was verified that EG acts as an efficient barrier against heat and oxygen, and it also reduces the combustion degree of PURCF matrix [37]. In addition, it was also reported that when the amount of HPCP in PURCFs approached to 15 wt%, the peak heat release rate decreased much more than those of pure PURF and PURCF consisting of only 10 wt% EG. In accordance with experimental data, combustion intensity and spreading speed of flame in PURCFs were extensively inhibited by the incorporation of HPCP to PURFs [37]. In agreement with previously published works [34], the substantial decrease of

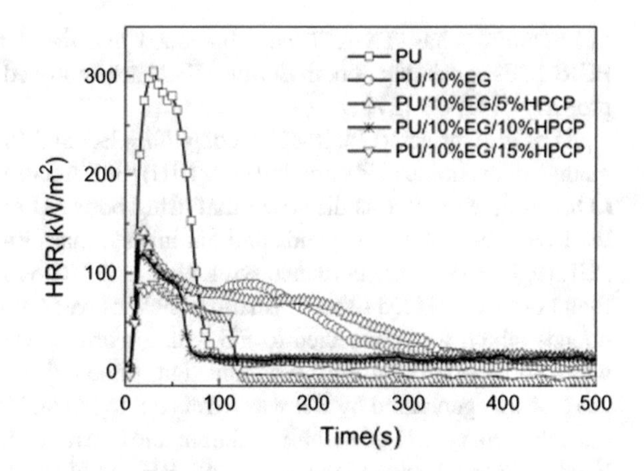

Fig. 2.51 HRR data of pure PURF and PURMCFs consisting of different amounts of EG and HPCP. Reproduced with permission from Ref. [37]. Copyright 2013 Elsevier Ltd.

peak heat release rate as a function of increasing EG content implies that EG particles are very effective flame retardant materials in terms of improving the flammability properties of PURCFs.

The flame retardant mechanism of using both HPCP and EG in PURMCFs is shown in Fig. 2.52. Basically, substantial fire delaying characteristics of PURMCFs consisting of EG and HPCP could be related to flame retardant effect of HPCP which have two distinct phases [37]. As shown in Fig. 2.52, it was reported that during the combustion of PURCFs, the flammable free radicals coming from substrate were quenched by phenoxyl and PO_2 free radicals that were produced by HPCP, and thus this process inhibited the multiple stages of combustion reaction and decreased the burning rate of PURMCFs. On the other hand, authors also stated that the decomposed products of EG filled PURMCF system reacted with phosphorus coming from HPCP, and they formed a strong carbonaceous layer which enhanced the blockade impact

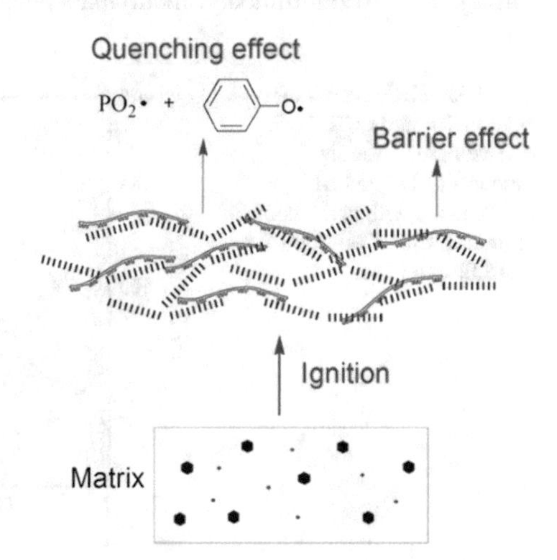

Fig. 2.52 The mechanism of fire delaying process of using both HPCP and EG in PURMCFs. Reproduced with permission from Ref. [37]. Copyright 2013 Elsevier Ltd.

of remaining char. Thus, it was concluded that the flame retardant properties of PURMCFs consisting of both EG and HPCP are improved by bi-phase flameretardant property of HPCP [37].

Xi et al. fabricated PURMCFs containing EG and [bis(2-hydroxyethyl)amino]-methyl-phosphonic acid dimethyl ester (BH) [38]. According to cone calorimetry and LOI testing data, it was disclosed that using both BH and EG in PURFs increased the LOI values and char yields and diminished mass loss and heat release rates of PURMCFs [38]. In accordance with SEM and GC-MS experimental data, it was found out that BH led to the formation of char layer consisting of phosphorus compounds which was connected to EG with a worm-shaped geometry. In addition, it was also reported that during combustion, dimethyl methylphosphonate (DMMP) gas that was generated by BH was pyrolyzed to PO and PO_2 free radicals, and these radicals were used in quenching of flammable free radicals within the PURCF matrix. Finally, it was concluded that using both BH and EG as flame retardants in PURFs is more efficient in providing better flame retardancy than the case in which only one of them is used as a flame retardant [38].

Figure 2.53 shows HRR curves of pure PURF and PURMCFs consisting of different proportions of EG and BH. Based on the results, it was shown that the peak heat release rate of PURMCF consisting of 8% EG is much lower compared to that of pure PURF [38]. Based on this finding, it was confirmed that during combustion, the burning intensity of PURMCF matrix was effectively reduced by the expanded graphite of EG. Furthermore, it was reported that the peak heat release rate of PURMCF consisting of 18% BH and 8% EG was found to be 32.9% lower than that of PURMCF consisting of 8% EG. Based on these results, it was clearly shown that using the combination of EG and BH greatly inhibits the degree of combustion by exerting their substantial flame retardant properties in PURMCFs [38].

The flame-retardant mechanism that was generated by incorporating BH and EG into PURFs is shown in Fig. 2.54. Based on this mechanism, it was reported that during combustion, dimethyl methylphosphonate (DMMP) gas that was generated

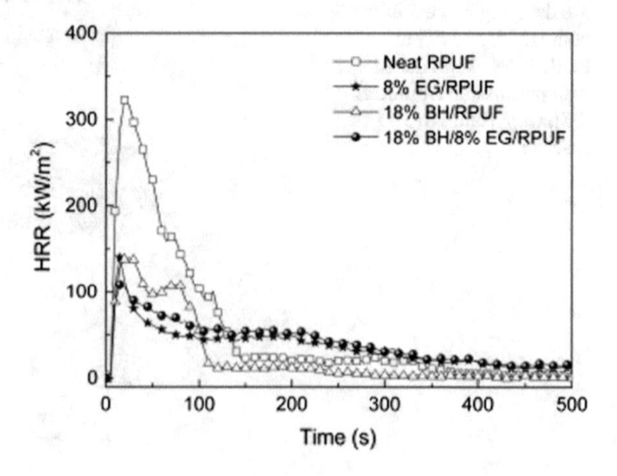

Fig. 2.53 HRR curves of pure PURF and PURMCFs consisting of different amounts of EG and BH. Reproduced with permission from Ref. [38]. Copyright 2015 Elsevier Ltd.

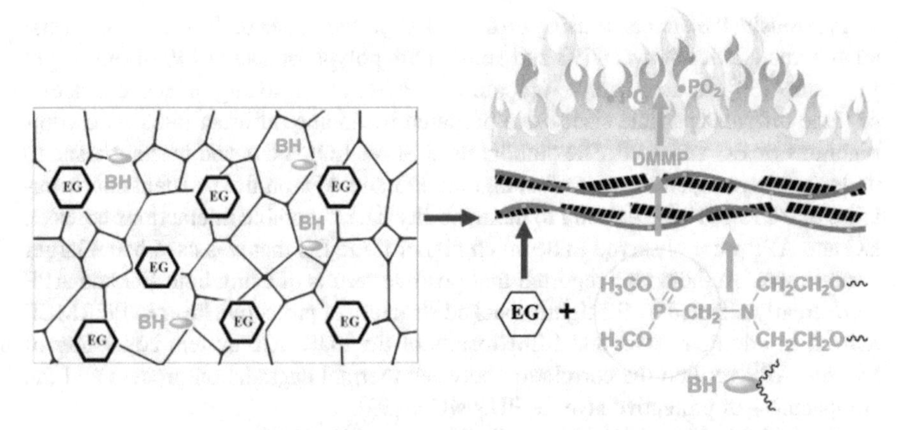

Fig. 2.54 The fire delaying process of using both BH and EG in PURMCFs. Reproduced with permission from Ref. [38]. Copyright 2015 Elsevier Ltd.

by BH was pyrolyzed to PO and PO_2 free radicals, and these radicals were used in quenching the flammable free radicals within the PURCF matrix [38]. In addition, it was also reported that BH provided the formation of char layer containing phosphorus compounds which was connected to the worm-shaped EG with substantial hindrance against heat and fire. Thus, it was concluded that using EG/BH flame retardant system in PURFs gave rise to beneficial flame retardant effects which improved the flame retardant properties of PURMCFs [38].

In accordance with results from other similar works [11, 26, 34, 35, 37, 39, 40], it was confirmed that using two types of flame retardant materials such as BH and EG in improves morphology, thermal stability and flame retardancy behavior of PURCFs. Thus, the formation of phosphorus-containing char layer in PURCFs containing both BH and EG definitely improves flame retardant properties and thermal stability behavior by acting as an extra reinforcement compared to pure PURF [11, 35, 37, 39, 40]. Furthermore, the char layer consisting of phosphorous compounds which was connected to the worm-shaped EG in PURF/BH/EG foams clearly points out the fact that favorable interactions between EG and BH are responsible for the formation of reinforced char layer in PURFs consisting of BH and EG. Based on the results from other published works in the literature [1, 6, 17], it is very well known that the presence of particles in PURCF matrix limits the mobility of polymer chains near particle surfaces which leads to the increase of glass transition temperature as well as decomposition temperatures of PURCF matrix. In addition, the existence of EG with BH could also decrease the mobility of gaseous compounds that emerge during the decomposition process of PURCFs [1, 6, 17]. Thus, these results clearly show that using EG with two different types of flame retardant materials in PURFs induced positive effects in terms of improving morphological and flame retardant properties and thermal stability behavior in these PURMCFs [11, 26, 34, 35, 37–40].

Previously, the effects of using different EG particle sizes on favorable fire resistant characteristics between EG and ammonium polyphosphate (APP) in semi-rigid PU foams [39]. In their sample preparation, PURCFs consisting of EGs consisting of three different particle sizes were prepared by adding different mass ratio combinations of EG and APP. The authors used SEM, LOI, TGA and burning tests, to understand synergistic results of using both EG and APP on fire resistant characteristics of PURMCFs. According to flammability data, favorable interactions between EG and APP were observed to be much higher if the EG that was used had a larger particle size. Authors also reported that positive results of using both EG and APP were greatly affected by the tightly-packed structure of protective layer in PURMCF matrix. In addition, TGA and SEM results of the PURMCF system consisting of EG and APP verified the correlation between thermal degradation process and the compactness of protective layer in PURMCFs [39].

Figure 2.55 shows LOI test results for three different flame retardant systems consisting of EGs with three different particle sizes (EG1 = 70 μm; EG2 = 430 μm; EG3 = 960 μm). Based on these results, it was verified that using both EG3 and APP provides the best synergistic effect in PURFs [39]. It was also noted that EG1 particle showed small degree of positive flame retardant effect with APP due to smaller particle size of EG1. In addition, it was also reported that EG2 and EG3 reached to their maximum LOI values when EG/APP was adjusted to 25/5. Consequently, LOI results established that the favourable effect of flame retardant system was found to be much larger if the particle size of EG that was used in PURFs was much larger [39].

Based on the results from other similar works [35, 36], it was confirmed that using larger sized EGs in comparison with smaller ones considerably improves morphology, thermal stability and flame retardancy behavior of PURCFs. Thus, using EG with higher amounts and a larger particle size in PURCFs influences positively the favorable interactions between PU chains and EG particles, and leads to higher thermal stability and flame retardant properties [35, 36, 39]. In addition, using smaller

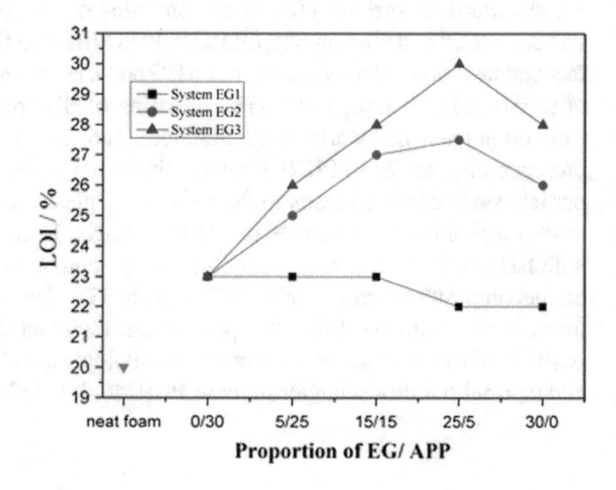

Fig. 2.55 LOI test results of PURCFs consisting of three different flame retardant systems. Reproduced with permission from Ref. [39]. Copyright 2018 Springer-Verlag GmbH Germany

EG particles did not increase the LOI values of PURCFs since EGs with smaller sizes could not generate an adequate protective blockade layer against flame penetration. Furthermore, it was also reported that EGs with larger particle sizes substantially improved the flame retardancy of PURCFs due to their larger expanded volumes compared to smaller EG particles [35, 36, 39].

Figure 2.56 provides a simplified flame retardant representation for combustion properties of PURCFs consisting of APP and EG with different particle sizes [39]. Based on the proposed flame retardant model, it was shown that a more tightly-packed insulation char sheet was formed since the three-dimensional network morphology occupied the regions between the expanded graphite layers of EG. In addition, it was also reported that changing the numerical ratio of EG/APP could modify the volume of three-dimensional network morphology. It was concluded that a very efficient protective layer could be formed when the particle size of EG is suitable with the amount of APP in PURCFs [39].

In agreement with the results from other similar works [26, 34, 35, 37, 40, 41], using two types of flame retardant materials such as EG and APP improves morphology, flame retardant properties and thermal stability of PURCFs. Thus, a very effective protective char layer formation in PURCFs containing both EG and APP definitely improves the thermal stability and flame retardancy behavior by acting like an extra reinforcement barrier compared to pure PURF [11, 37, 38, 41]. Furthermore, the formation of a very effective protective char layer in PURF/EG/APP foams clearly points out the fact that favorable interactions between EG and APP are responsible for generation of reinforced and partially burned shielding in PURFs

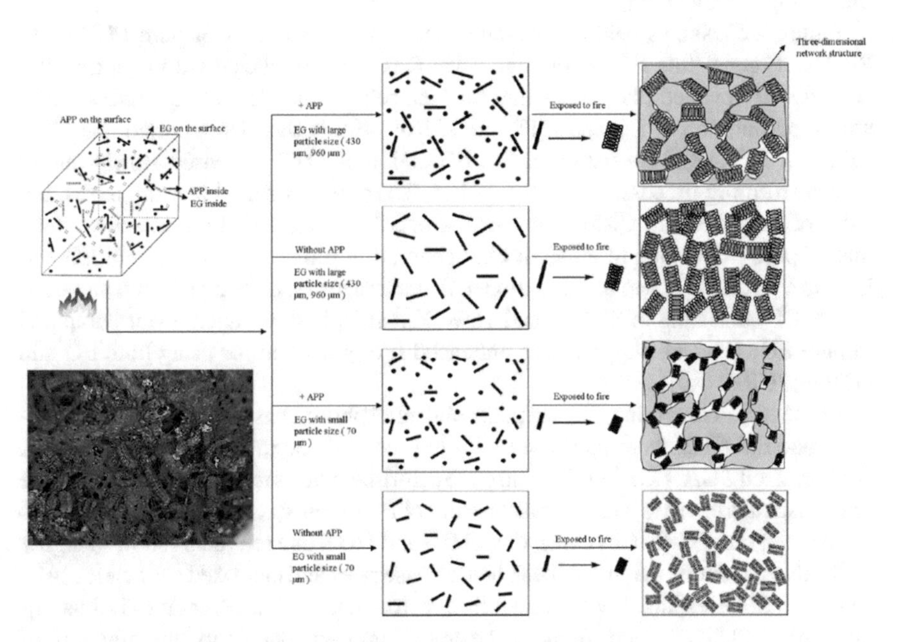

Fig. 2.56 A simplified flame retardant representation for PURCFs consisting of APP and EG. Reproduced with permission from Ref. [39]. Copyright 2018 Springer-Verlag GmbH Germany

containing APP and EG. Based on the results from other published works in the literature [1, 6, 17], it is very well known that the presence of particles in PURCF matrix limits the mobility of polymer chains near particle surfaces which leads to the increase of glass transition temperature as well as decomposition temperatures of PURCF matrix. In addition, the existence of EG with APP in PURCFs could also decrease the mobility of gaseous compounds that emerge during the decomposition process of PURCFs [1, 6, 17]. Thus, these results show that using EG with different types of flame retardant materials in PURFs leads to synergistic effects in terms of improving morphology, flame retardant properties and thermal stability in these PURCFs [11, 26, 34, 35, 37–41].

Previously, various water-blown rigid polyurethane foams containing EG and DTP as flame retardant materials were fabricated [40]. Based on TGA data, it was shown that using both DTP and EG modified thermal degradation behaviour of PURFs and intensified amount of char residues on PURFs. Based on LOI and combustion calorimeter test results, authors reported that PURMCFs consisting of both EG and DTP exhibited higher LOI and lower heat release rate peak values in comparison with pure PURF case. Based on cone calorimetry testing results, it was pointed out that EG/DTP filled PURCF system improved the inhibition of fire amount and decreased the dosage of smoke due to good char forming mechanism in PURCFs. Furthermore, authors performed SEM and EDS experiments to comprehend in detail about the char morphology of foams after combustion. In conclusion, authors proposed that EG and DTP filled PURCFs did form a tightly-packed, steady and insulation shielding sheet filled with phosphorus element which acted as an effective thermal barrier against fire during combustion [40].

Figure 2.57 shows heat release rates and mass loss curves of pure PURF and PURMCF consisting of 8% EG and 16% DTP. In accordance with experimental data, it was unveiled that peak value for heat release rate decreased by about 48% with incorporation of EG and DTP into PURF [40]. It was also reported that HRR values of PURMCF consisting of 8% EG and 16% DTP decreased slowly due to its char forming process. As shown in Fig. 2.57b, it was stated that the maximum value of THR data in PURMCF consisting of 8% EG and 16% DTP was lower than that of pure PURF due to efficient char generation property of using both EG and DTP in PURMCF. It was also mentioned that average rate of heat emission values of PURMCF consisting of 8% EG and 16% DTP sample were much lower compared to pure PURF which suggested an enhanced fire protection for using both EG and DTP in PURFs [40].

PURMCFs containing EG particles and phosphorus based second flame retardant material (HQ) were fabricated with improved flame retardant properties [41]. Authors used TGA, LOI, burning and cone calorimeter tests to systematically explore fire resistant properties and thermal stability of PURCFs. According to results, it was pointed out that PURMCFs consisting of EG and HQ flame retardants showed higher LOI values, reduced maximum peak heat release rates and increased char yields [41]. Based on SEM results, it was shown that worm-shaped char layer and viscous liquid film in PURCFs were found to be important components in strengthening of fire resistant properties in PURCFs. Based on TG-IR results, it was shown that the

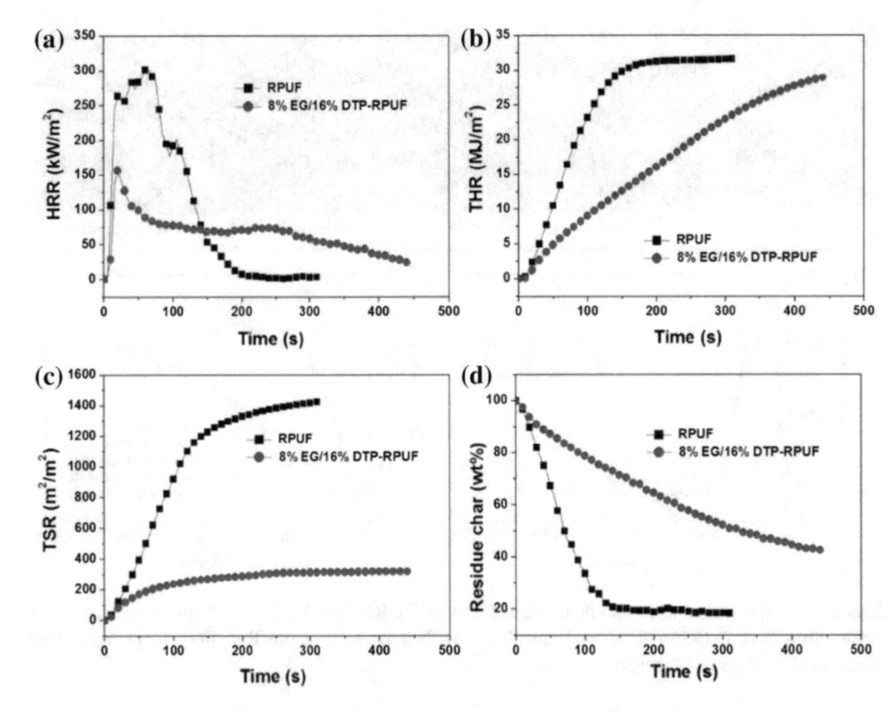

Fig. 2.57 Cone calorimetry data of PURF and PURMCF consisting of 8% EG and 16% DTP. Reproduced with permission from Ref. [40]. Copyright 2018 Wiley Periodicals, Inc.

extraction of poisonous and easily burnable gaseous compounds from PURMCFs consisting of EG and HQ was considerably reduced in comparison with pure PURF. Authors concluded that inclusion of two diverse fire delaying materials in PURM-CFs shows a useful method for fabrication of PURFs with improved fire resistant characteristics [41].

Figure 2.58 shows (a) optical microscopy images of PURMCFs containing EG and HQ after completing cone calorimeter tests; (b) HRR, and (c) THR curves of PURMCFs consisting of flame retardants. Based on optical images, it was shown that very tiny content of flat and partially burned remaining material was left in pure PURF, whereas long, cylindrically shaped, swollen and tightly-packed burned sheet was left after incorporation of EG and HQ to PURF [41]. Based on HRR results, it was unveiled that incorporation of EG and HQ to PURF reduced the time to reach the peak value for heat release rate due to the higher degradation effect of EG and HQ. Moreover, it was reported that the PHRR value of PURMCF containing 10% EG and 5% HQ decreased by 58.5% compared to pure PURF. It was concluded that the formation of worm-shaped intumescent char layer in PURMCFs consisting of EG and HQ leads to decay of the peak value for heat release rate which leads to improved fire-resistant properties and inhibition of heat [41].

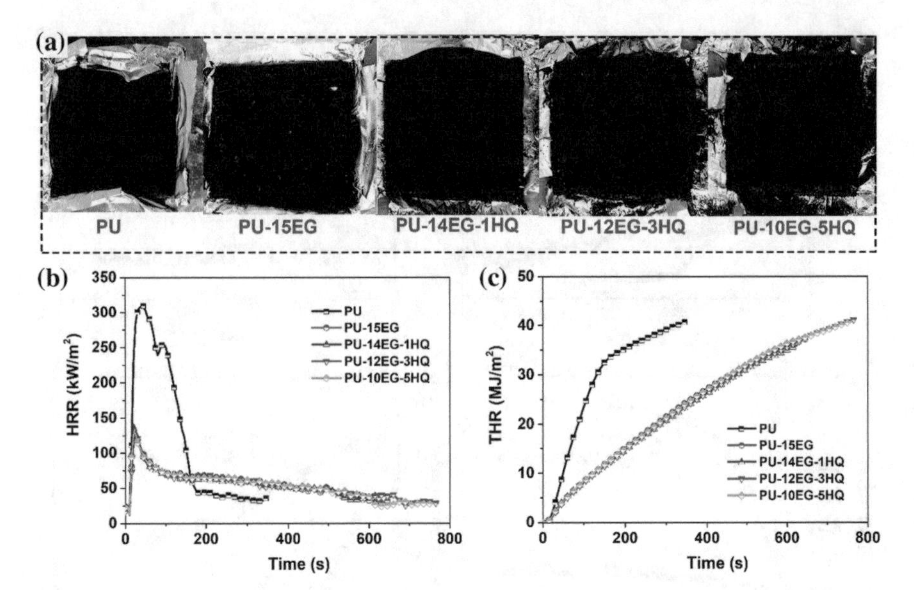

Fig. 2.58 **a** Optical images and heat release data of PURMCFs containing flame retardants after completing cone calorimeter tests. Reproduced with permission from Ref. [41]. Copyright 2018 The Royal Society of Chemistry

Previously, water-blown polyurethane rigid microcomposite foams (PURMCFs) consisting of two different particle sizes and concentrations of EG were fabricated [8]. Authors investigated in detail the effects of EG particle size on morphological, mechanical, thermal, water absorption, and flame-retardant properties of PURMCFs. The thermal conductivity results showed that thermal insulation values of PURCFs consisting of EGs diminished with growing EG loadings. It was reported that increasing EG loadings enhanced the flame retardant behavior of PURCFs. In conclusion, enhanced fire retardant and mechanical properties were observed in PURCFs consisting of EGs with larger particle sizes compared to PURCFs consisting of EGs with smaller particle sizes [8].

Figure 2.59 shows the dependence of LOI values of PURMCFs on the particle size and concentration of EG. In accordance with experimental data, it was emphasized that LOI values elevated directly proportionally in the company of growing amounts of EG in PURMCFs. It was also reported that at high temperatures, graphite which is intercalated between the sheets of EG reacts with sulfuric acid and produces gaseous compounds such as H_2O, CO_2 and SO_2. Thus, it was concluded that LOI values increased in PURMCFs since the oxygen concentration around fire region was reduced by these gases during burning [8].

PURMCFs containing EG and whisker silicon (WSi) with different contents which had the same density were prepared recently [53]. Authors showed that the inclusion of 10 wt% WSi to PURCFs filled with EG improved the flame retardant and

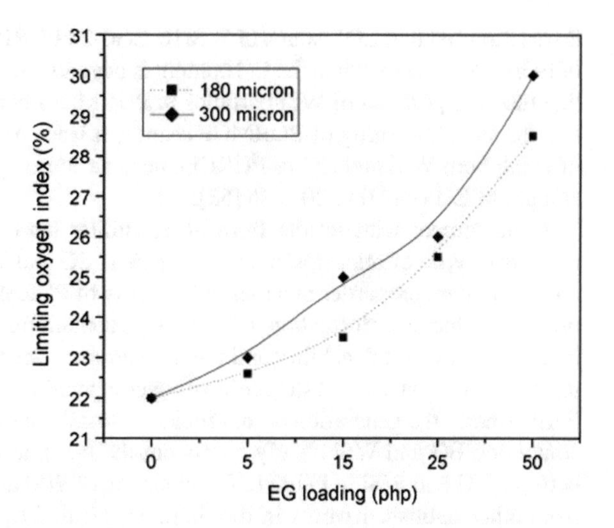

Fig. 2.59 Effect of EG concentration and particle size on LOI of PURMCFs. Reproduced with permission from Ref. [8]. Copyright 2008 Wiley Periodicals, Inc.

mechanical properties compared to PURCFs filled with only EG [53]. According to TGA data, incorporation of these micron-sized particles improved thermal decomposition behavior of PURMCFs [53].

Figure 2.60 shows the LOI values of PURMCFs consisting of only EG in the amount of 0–20 wt% (EG0-20 PURF), PURMCFs consisting of only WSi in the amount of 0–30 wt% (WSi0-30 PURF), PURMCFs consisting of 10 wt% WSi and EGs in the amount of 0–20 wt%, (WSi10-EG0-20 PURF). According to experimental data, LOI values in EG/PURF system intensified linearly with increasing amounts of EG from 0 to 20 wt% [53]. In addition, it was shown that LOI values of PURMCFs consisting of WSi did not change that much with increasing amounts of WSi. However, it was shown that LOI values of WSi10-EG0-20 PURF exhibited much higher values compared to EG0-20 PURF at the same EG content. For instance;

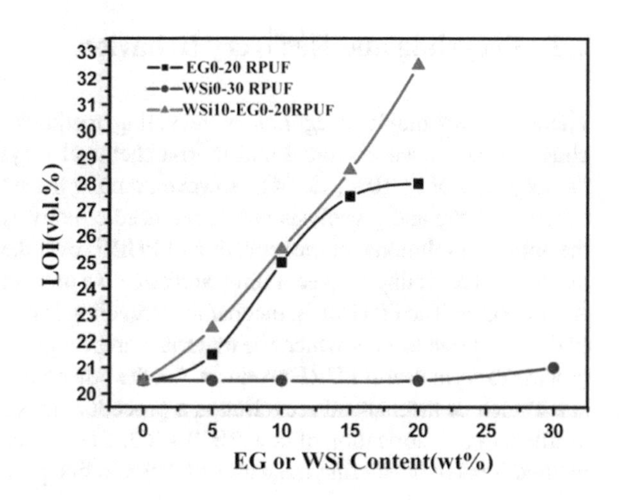

Fig. 2.60 LOI values of PURMCFs containing different amounts of EG or WSi. Reproduced with permission from Ref. [53]. Copyright 2008 Wiley Periodicals, Inc.

it was reported that LOI values of WSi10-EG0-20 PURF enhanced more than those of EG/PURF system when the EG content is beyond 5 wt% [53]. It was also reported that the incorporation of WSi particles in PURMCFs consisting of EG could facilitate the surface-coating of PURMCF matrix. It was concluded that positive effects of using both WSi and EG in PURCFs became more apparent with increasing the amount of EG from 0 to 20 wt% [53].

In agreement with results from other similar works [26, 34, 35, 37, 40, 41], using two types of micron-sized fillers such as EG and WSi improves morphology, flame retardant properties and thermal stability of PURMCFs. Thus, a very effective protective char layer formation in PURMCFs containing both EG and WSi definitely improves the thermal stability and flame retardancy behavior by acting like an extra reinforcement barrier and surface-coating compared to pure PURF [11, 37, 38, 41]. Furthermore, the generation of an efficient partially burned shielding in PURMCFs containing EG and WSi clearly points out the fact that favorable interactions exist between EG and WSi in PURMCFs consisting of WSi and EG. Based on the results from other published works in the literature [1, 6, 17], it is very well known that the presence of particles in PURCF matrix acts as thermal barrier and limits the diffusion of volatile gases during decomposition process which leads to the increase of decomposition temperatures of PURCF matrix. In addition, the existence of EG with WSi in PURCFs could also decrease the mobility of gaseous compounds that emerge during the decomposition process of PURCFs [1, 6, 17]. Thus, these results show that using EG with different types of flame retardant materials in PURFs leads to synergistic effects in terms of improving morphology, flame retardant properties and thermal stability in these PURCFs [11, 26, 34, 35, 37–41, 53].

In conclusion, using both EG and WSi in PURCFs provides better flame retardancy behavior compared to the case of using only one of the fillers. Using the synergetic effect of binary fillers in PURCFs not only improves the flame retardant property and thermal stability but also enhances the mechanical properties of PURCFs [53].

2.6 Recycling and Recovery Behavior

There are two major categories of recycling methods such as mechanical recycling and advanced chemical and thermochemical recycling that can be used for the recycling of PURFs [63, 64]. However, among these two major recycling methods, regrinding and glycolysis have been used effectively and economically to find the optimum solutions for the recycling of PURFs over the years [64]. The method of mechanical recycling is an economic and effective procedure for recycling of PURFs and PURCFs. The first step in mechanical recycling is to decrease the size of original PURFs to a size level at which the material can be reprocessed by using a secondary process to form useful PURF products. On the other hand, regrinding method which is a branch of mechanical recycling is a procedure that uses waste PURF products as fillers in the fabrication of new PURFs [63, 64]. Typically, the steps of regrinding method involve powdering the waste PURF into fine powder with a size of less than

100 μm, and blending this powder with polyol for the production of new PURFs [61, 62]. Typically, for grinding of PURFs, ball mills were used to obtain very small powders with sizes of 85 μm [64].

Glycolysis is a type of chemical processing method in which PURFs are reacted with diols at temperatures above 200 °C [65]. Glycolysis is the most commonly used chemical recycling method mainly for PURFs. The mostly preferred and effective method for recycling and recovery of PURFs are chemical recycling processes due to higher thermal stability of PURFs. Among these chemical recycling methods, the glycolysis has been studied in a great extent both in academia and industry for the recycling of PURFs [65]. The aim of glycolysis is to enable the recycle of polyols for the generation of new PU materials [65, 66].

Previously, the glycolysis method was used to recycle PUs which were fabricated via reaction injection molding manufacturing method, and these PUs were reinforced with glass fibers that were obtained from the automotive industry. It was shown that these PU waste products can be directly recycled by using the glycolysis method for production of new rigid foam insulating materials [67]. Very recently, PURFs consisting of thermoregulating microcapsules were neatly recycled by using a single-phase glycolysis method which included unpurified glycerol as the transesterification agent. At the end of glycolysis, the recovered polyol was obtained with 71% of purity [66]. By using this method, both microcapsules and polyols were recovered, and this way of recycling demonstrated the successful realization of chemically conversion of these types of composites into reusable starting components of PUs for the first time. Lastly, the recovered polyols were successfully used for the replacement of unused polyether polyols for fabrication of fresh PURFs [66].

Besides optimization of experimental conditions for the fabrication of PURFs, the proper methods for recycling and recovery of these materials should be systematically determined in order to recover various types of waste PURFs, reduce fossil fuel consumption, and consequently increase the energy value and sustainability of PURFs. However, the recycling and recovery of PURFs into useful daily life consumer products should be executed without giving any damage to environment in order to provide both energy conservation and sustainability. The first aim of recycling methods is to decrease the size of PURF waste products such that they can be reused for the fabrication of new PURFs. Based on lots of research data both in academia and industry, researchers have found a few number of recycling and recovery processes for PURFs which are beneficial in terms of economically and environmentally [64, 65].

Some of the recycling and recovery methods that were published as patents or articles by researchers both in industry and academia in the past have been used in the establishment of new factories where waste PURF products are efficiently recycled and used as raw materials in the fabrication of new useful PURF products. For example; Aprithane which is located at Abtsgmünd, Germany conducted a glycolysis operation of PURFs and successfully used this final single-phase product in the synthesis of new PURFs [65, 68]. However, this factory does not operate any more since the glycolysis process consisting of expensive glycolysis agents is not profitable [69]. On the other hand, Troy Polymers has developed a single-phase glycolysis operation of PU foams and successfully achieved in commercializing polyols for synthesis of PURFs. In addition, this recovered polyol has been used in the production of PURFs for the automotive industry since 2002 [65, 70].

These examples showed that there are still some economical and environmental feasibility problems associated with glycolysis in order to use the glycolysis method efficiently for the recycling and recovery of PURFs in the industrial scale. Thus, for this reason, researchers both in academia and industry should try to find new alternatives or modified methods of glycolysis to conduct the industrial scale recycling method more efficiently, economically and environmental-friendly for producing high standard recovered PURF components that will be used in the fabrication of new PURF products.

References

1. Kim, S.H., Park, H.C., Jeong, H.M., Kim, B.K.: Glass fiber reinforced rigid polyurethane foams. J. Mater. Sci. **45**(10), 2675–2680 (2010)
2. Burgaz, E.: Thermomechanical analysis of polymer nanocomposites. In: Huang, X., Zhi, C. (eds.) Polymer Nanocomposites. Springer, Cham (2016)
3. Yu, Y.H., Nam, S., Lee, D., Lee, D.G.: Cryogenic impact resistance of chopped fiber reinforced polyurethane foam. Compos. Struct. **132**, 12–19 (2015)
4. Aranguren, M.I., Racz, I., Marcovich, N.E.: Microfoams based on castor oil polyurethanes and vegetable fibers. J. Appl. Polym. Sci. **105**(5), 2791–2800 (2007)
5. Park, S.B., Choi, S.W., Kim, J.H., Bang, C.S., Lee, J.M.: Effect of the blowing agent on the low-temperature mechanical properties of CO_2^- and HFC-245fa-blown glass-fiber-reinforced polyurethane foams. Compos. Part B Eng. **93**, 317–327 (2016)
6. Xue, B.L., Wen, J.L., Sun, R.C.: Lignin-based rigid polyurethane foam reinforced with pulp fiber: synthesis and characterization. ACS Sustain. Chem. Eng. **2**(6), 1474–1480 (2014)
7. Yu, Y.H., Choi, I., Nam, S., Lee, D.G.: Cryogenic characteristics of chopped glass fiber reinforced polyurethane foam. Compos. Struct. **107**, 476–481 (2014)
8. Thirumal, M., Khastgir, D., Singha, N.K., Manjunath, B.S., Naik, Y.P.: Effect of expandable graphite on the properties of intumescent flame-retardant polyurethane foam. J. Appl. Polym. Sci. **110**(5), 2586–2594 (2008)
9. Ye, L., Meng, X.Y., Ji, X., Li, Z.M., Tang, J.H.: Synthesis and characterization of expandable graphite-poly(methyl methacrylate) composite particles and their application to flame retardation of rigid polyurethane foams. Polym. Degrad. Stabil. **94**(6), 971–979 (2009)
10. Cheng, J.J., Shi, B.B., Zhou, F.B., Chen, X.Y.: Effects of inorganic fillers on the flame-retardant and mechanical properties of rigid polyurethane foams. J. Appl. Polym. Sci. **131**(10), 40253 (2014)
11. Yuan, Y., Yang, H.Y., Yu, B., Shi, Y.Q., Wang, W., Song, L., Hu, Y., Zhang, Y.M.: Phosphorus and nitrogen-containing polyols: synergistic effect on the thermal property and flame retardancy of rigid polyurethane foam composites. Ind. Eng. Chem. Res. **55**(41), 10813–10822 (2016)
12. Kargerkocsis, J., Harmia, T., Czigany, T.: Comparison of the fracture and failure behavior of polypropylene composites reinforced by long glass-fibers and by glass mats. Compos. Sci. Technol. **54**(3), 287–298 (1995)
13. Wilberforce, S., Hashemi, S.: Effect of fibre concentration, deformation rate and weldline on mechanical properties of injection-moulded short glass fibre reinforced thermoplastic polyurethane. J. Mater. Sci. **44**(5), 1333–1343 (2009)
14. Cotrgreave, T.C., Shorttall, J.B.: The mechanism of reinforcement of polyurethane foam by high-modulus chopped fibres. J. Mater. Sci. **12**, 708–717 (1997)

15. Yang, Z.G., Zhao, B., Qin, S.L., Hu, Z.F., Jin, Z.K., Wang, J.H.: Study on the mechanical properties of hybrid reinforced rigid polyurethane composite foam. J. Appl. Polym. Sci. **92**(3), 1493–1500 (2004)
16. Kim, S.H., Park, H.C., Jeong, H.M., Kim, B.K.: Glass fiber reinforced rigid polyurethane foams. J. Mater. Sci. **45**, 2675–2680 (2010)
17. Kim, M.W., Kwon, S.H., Park, H., Kim, B.K.: Glass fiber and silica reinforced rigid polyurethane foams. Express Polym. Lett. **11**(5), 374–382 (2017)
18. Taniguchi, T., Okamura, K.: New films produced from microfibrillated natural fibres. Polym. Int. **47**(3), 291–294 (1998)
19. Chang, L.C., Xue, Y., Hsieh, F.H.: Comparative study of physical properties of water-blown rigid polyurethane foams extended with commercial soy flours. J. Appl. Polym. Sci. **80**(1), 10–19 (2001)
20. Teixeira, M.J., Fernandes, A.C., Saramago, B., Rosa, M.E., Bordado, J.C.: Influence of the wetting properties of polymeric adhesives on the mechanical behaviour of cork agglomerates. J. Adhes. Sci. Technol. **10**(11), 1111–1127 (1996)
21. Banik, I., Sain, M.M.: Structure of glycerol and cellulose fiber modified water-blown soy polyol-based polyurethane foams. J. Reinf. Plast. Comp. **27**(16–17), 1745–1758 (2008)
22. Gu, R.J., Sain, M.M., Konar, S.K.: A feasibility study of polyurethane composite foam with added hardwood pulp. Ind. Crop. Prod. **42**, 273–279 (2013)
23. Rials, T.G., Wolcott, M.P., Nassar, J.M.: Interfacial contributions in lignocellulosic fiber-reinforced polyurethane composites. J. Appl. Polym. Sci. **80**(4), 546–555 (2001)
24. Ryabov, S.V., Kercha, Y.Y., Kotel'nikova, N.E., Gaiduk, R.L., Shtompel', V.I., Kosenko, L.A., Yakovenko, A.G., Kobrina, L.V.: Biodegradable polymer composites based on polyurethane and microcrystalline cellulose. Polym. Sci. Ser. A **43**(12), 1256–1260 (2001)
25. Johnson, M., Shivkumar, S.: Filamentous green algae additions to isocyanate based foams. J. Appl. Polym. Sci. **93**(5), 2469–2477 (2004)
26. Xu, D.F., Yu, K.J., Qian, K.: Thermal degradation study of rigid polyurethane foams containing tris(1-chloro-2-propyl)phosphate and modified aramid fiber. Polym. Test. **67**, 159–168 (2018)
27. Silva, M.C., Takahashi, J.A., Chaussy, D., Belgacem, M.N., Silva, G.G.: Composites of rigid polyurethane foam and cellulose fiber residue. J. Appl. Polym. Sci. **117**(6), 3665–3672 (2010)
28. Bledzki, A.K., Zhang, W.Y., Chate, A.: Natural-fibre-reinforced polyurethane microfoams. Compos. Sci. Technol. **61**(16), 2405–2411 (2001)
29. Silva, M.C., Lopes, O.R., Colodette, J.L., Porto, A.O., Rieumont, J., Chaussy, D., Belgacem, M.N., Silva, G.G.: Characterization of three non-product materials from a bleached eucalyptus kraft pulp mill, in view of valorising them as a source of cellulose fibres. Ind. Crop. Prod. **27**(3), 288–295 (2008)
30. Czuprynski, B., Paciorek-Sadowska, J., Liszkowska, J.: Properties of rigid polyurethane-polyisocyanurate foams modified with the selected fillers. J. Appl. Polym. Sci. **115**(4), 2460–2469 (2010)
31. Nikje, M.M.A., Garmarudi, A.B., Haghshenas, M.: Effect of talc filler on physical properties of polyurethane rigid foams. Polym. Plast. Technol. **45**(11), 1213–1217 (2006)
32. Cheng, J.J., Shi, B.B., Zhou, F.B., Chen, X.Y.: Effects of inorganic fillers on the flame-retardant and mechanical properties of rigid polyurethane foams. J. Appl. Polym. Sci. **131**(10), 40253 (2014)
33. Kirpluks, M., Cabulis, U., Zeltins, V., Stiebra, L., Avots, A.: Rigid polyurethane foam thermal insulation protected with mineral intumescent mat. Autex Res. J. **14**(4), 259–269 (2014)
34. Gao, L.P., Zheng, G.Y., Zhou, Y.G., Hu, L.H., Feng, G.D., Xie, Y.L.: Synergistic effect of expandable graphite, melamine polyphosphate and layered double hydroxide on improving the fire behavior of rosin-based rigid polyurethane foam. Ind. Crop. Prod. **50**, 638–647 (2013)
35. Luo, W., Li, Y., Zou, H.W., Liang, M.: Study of different-sized sulfur-free expandable graphite on morphology and properties of water-blown semi-rigid polyurethane foams. RSC Adv. **4**(70), 37302–37310 (2014)
36. Li, Y., Zou, J., Zhou, S.T., Chen, Y., Zou, H.W., Liang, M., Luo, W.Z.: Effect of expandable graphite particle size on the flame retardant, mechanical, and thermal properties of water-blown semi-rigid polyurethane foam. J. Appl. Polym. Sci. **131**(3), 39885 (2014)

37. Qian, L.J., Feng, F.F., Tang, S.: Bi-phase flame-retardant effect of hexa-phenoxy-cyclotriphosphazene on rigid polyurethane foams containing expandable graphite. Polymer **55**(1), 95–101 (2014)
38. Xi, W., Qian, L.J., Chen, Y.J., Wang, J.Y., Liu, X.X.: Addition flame-retardant behaviors of expandable graphite and [bis(2-hydroxyethyl)amino]-methyl-phosphonic acid dimethyl ester in rigid polyurethane foams. Polym. Degrad. Stabil. **122**, 36–43 (2015)
39. Li, J., Mo, X.H., Li, Y., Zou, H.W., Liang, M., Chen, Y.: Influence of expandable graphite particle size on the synergy flame retardant property between expandable graphite and ammonium polyphosphate in semi-rigid polyurethane foam. Polym. Bull. **75**(11), 5287–5304 (2018)
40. Liu, D.Y., Zhao, B., Wang, J.S., Liu, P.W., Liu, Y.Q.: Flame retardation and thermal stability of novel phosphoramide/expandable graphite in rigid polyurethane foam. J. Appl. Polym. Sci. **135**(27), 46434 (2018)
41. Shi, X.X., Jiang, S.H., Zhu, J.Y., Li, G.H., Peng, X.F.: Establishment of a highly efficient flame-retardant system for rigid polyurethane foams based on bi-phase flame-retardant actions. RSC Adv. **8**(18), 9985–9995 (2018)
42. Zhang, L.Q., Zhang, M., Zhou, Y.H., Hu, L.H.: The study of mechanical behavior and flame retardancy of castor oil phosphate-based rigid polyurethane foam composites containing expanded graphite and triethyl phosphate. Polym. Degrad. Stabil. **98**(12), 2784–2794 (2013)
43. Shi, L., Li, Z.M., Xie, B.H., Wang, J.H., Tian, C.R., Yang, M.B.: Flame retardancy of different-sized expandable graphite particles for high-density rigid polyurethane foams. Polym. Int. **55**(8), 862–871 (2006)
44. Duquesne, S., Delobel, R., Le Bras, M., Camino, G.: A comparative study of the mechanism of action of ammonium polyphosphate and expandable graphite in polyurethane. Polym. Degrad. Stabil. **77**(2), 333–344 (2002)
45. Hu, X.M., Wang, D.M.: Enhanced fire behavior of rigid polyurethane foam by intumescent flame retardants. J. Appl. Polym. Sci. **129**(1), 238–246 (2013)
46. Meng, X.Y., Ye, L., Zhang, X.G., Tang, P.M., Tang, J.H., Ji, X., Li, Z.M.: Effects of expandable graphite and ammonium polyphosphate on the flame-retardant and mechanical properties of rigid polyurethane foams. J. Appl. Polym. Sci. **114**(2), 853–863 (2009)
47. Shi, L., Li, Z.M., Yang, M.B., Yin, B., Zhou, Q.M., Tian, C.R., Wang, J.H.: Expandable graphite for halogen-free flame-retardant of high-density rigid polyurethane foams. Polym. Plast. Technol. **44**(7), 1323–1337 (2005)
48. Modesti, M., Lorenzetti, A., Simioni, F., Camino, G.: Expandable graphite as an intumescent flame retardant in polyisocyanurate-polyurethane foams. Polym. Degrad. Stabil. **77**(2), 195–202 (2002)
49. Modesti, M., Lorenzetti, A.: Improvement on fire behaviour of water blown PIR-PUR foams: use of an halogen-free flame retardant. Eur. Polym. J. **39**(2), 263–268 (2003)
50. Chalivendra, V.B., Shukla, A., Bose, A., Parameswaran, V.: Processing and mechanical characterization of lightweight polyurethane composites. J. Mater. Sci. **38**(8), 1631–1643 (2003)
51. Parameswaran, V., Shukla, A.: Processing and characterization of a model functionally gradient material. J. Mater. Sci. **35**(1), 21–29 (2000)
52. Bian, X.C., Tang, J.H., Li, Z.M.: Flame retardancy of hollow glass microsphere/rigid polyurethane foams in the presence of expandable graphite. J. Appl. Polym. Sci. **109**(3), 1935–1943 (2008)
53. Bian, X.C., Tang, J.H., Li, Z.M.: Flame retardancy of whisker silicon oxide/rigid polyurethane foam composites with expandable graphite. J. Appl. Polym. Sci. **110**(6), 3871–3879 (2008)
54. Zhang, A.Z., Zhang, Y.H., Lv, F.Z., Chu, P.K.: Synergistic effects of hydroxides and dimethyl methylphosphonate on rigid halogen-free and flame-retarding polyurethane foams. J. Appl. Polym. Sci. **128**(1), 347–353 (2013)
55. Feng, F.F., Qian, L.J.: The flame retardant behaviors and synergistic effect of expandable graphite and dimethyl methylphosphonate in rigid polyurethane foams. Polym. Compos. **35**(2), 301–309 (2014)
56. Thirumal, M., Khastgir, D., Singha, N.K., Manjunath, B.S., Naik, Y.P.: Mechanical, morphological and thermal properties of rigid polyurethane foam: effect of the fillers. Cell. Polym. **26**(4), 245–259 (2007)

57. Saint-Michel, F., Chazeau, L., Cavaille, J.Y.: Mechanical properties of high density polyurethane foams: II effect of the filler size. Compos. Sci. Technol. **66**(15), 2709–2718 (2006)
58. Javni, I., Zhang, W., Karajkov, V., Petrovic, Z.S., Divjakovic, V.: Effect of nano- and micro-silica fillers on polyurethane foam properties. J. Cell. Plast. **38**(3), 229–239 (2002)
59. Goods, S.H., Neuschwanger, C.L., Whinnery, L.L., Nix, W.D.: Mechanical properties of a particle-strengthened polyurethane foam. J. Appl. Polym. Sci. **74**(11), 2724–2736 (1999)
60. Chatterjee, A., Mishra, S.: Nano-calcium carbonate ($CaCO_3$)/polystyrene (PS) core-shell nanoparticle: it's effect on physical and mechanical properties of high impact polystyrene (HIPS). J. Polym. Res. **20**(10), 249 (2013)
61. Cao, X., Lee, L.J., Widya, T., Macosko, C.: Polyurethane/clay nanocomposites foams: processing, structure and properties. Polymer **46**(3), 775–783 (2005)
62. Widya, T., Macosko, C.W.: Nanoclay-modified rigid polyurethane foam. J. Macromol. Sci. Phys. **B44**(6), 897–908 (2005)
63. Weigand, E.: Properties and applications of recycled polyurethanes. In: Branderup, J., Bittner, M., Menges, G., Micheali, W. (eds.) Recycling and Recovery of Plastics. Hanser Publishers, Munich, Germany (1996)
64. Zia, K.M., Bhatti, H.N., Bhatti, I.A.: Methods for polyurethane and polyurethane composites, recycling and recovery: a review. React. Funct. Polym. **67**(8), 675–692 (2007)
65. Simon, D., Borreguero, A.M., de Lucas, A., Rodriguez, J.F.: Recycling of polyurethanes from laboratory to industry, a journey towards the sustainability. Waste Manage. **76**, 147–171 (2018)
66. Simon, D., Rodriguez, J.F., Carmona, M., Serrano, A., Borreguero, A.M.: Glycolysis of advanced polyurethanes composites containing thermoregulating microcapsules. Chem. Eng. J. **350**, 300–311 (2018)
67. Modesti, M., Simioni, F.: Chemical recycling of reinforced polyurethane from the automotive industry. Polym. Eng. Sci. **36**(17), 2173–2178 (1996)
68. Raßhofer, W., Weigand, E.: Advances in Plastics Recycling, Vol. 2: Automotive Polyurethanes. CRC Press, Boca Raton (2001)
69. Behrendt, G., Naber, B.W.: The recycling of polyurethanes (review). J. Univ. Chem. Technol. Metall. **44**(1), 3–23 (2009)
70. Sendijarevic, I., Harris, J.G., Hoffmann, S., Heric, S., Lathia, N., Sendijarevic, V.: Polyether polyols from scrap polyurethanes—for use in rigid and flexible foams. Polyurethanes Mag. Int. **8**(1), 55–60 (2010)

Chapter 3
PU Rigid Nanocomposite Foams Containing Plate-Like Nanofillers

3.1 Introduction

For plate-like nano-sized fillers, clays, graphene oxide and graphene can be given as the nano-filler examples that have been used in PURNCFs [1–13]. However, among PURF/plate-like nano-sized filler published works, the most commonly investigated filler is nanoclay due to its outstanding properties such as reduced gas permeability, flame retardancy, low price, environmentally friendliness, excellent expansibility, high temperature resistance, thermal insulation [1–10].

In the area of polymer nanocomposites, the most extensively studied nanoclay types can be named as montmorillonite (MMT) and vermiculite (VMT) since they can form favorable interactions with polymer chains due to their advantageous physicochemical properties. In terms of morphology, montmorillonite and vermiculite nanoclays belong to 2:1 layered silicates in which an octahedral plane consisting of magnesium hydroxide groups is situated between two tetrahedral planes of Si atoms. The typical layer thickness of montmorillonite clay is approximately 1 nm. Negative charges on MMT nanolayers are counterbalanced by the presence of Na^+, Li^+, K^+ and Ca^{+2} cations which are positioned between nanoclay layers [14–16]. Nanoclay layers can be decorated with functional polar groups due to the fact that intermolecular forces between clay nanoplatelets are not really that strong enough. Specifically, MMT nanoclay which is classified within the smectites family has considerable and characteristic values such as 800 m^2/g and 80–150 mEq/100 g in terms of specific surface area and cation exchange capacity (CEC), respectively [17]. Different types organic cations such as cationic surfactants and alkylammonium cations with specific physical and chemical properties can be utilized and exchanged with inorganic cations of clay nanolayers to enhance positive secondary bonding forces between nanoclay layers and lengthy polymer chains. Accordingly, the pristine hydrophilic surface property of nanoclay surface can be converted into an organophilic character which has a strong attraction to organic compounds. Surface modification of nanoclay surface with alkylammonium organic cations reduces the surface energy of nanoclay

E. Burgaz, *Polyurethane Insulation Foams for Energy and Sustainability*,
Advanced Structured Materials 111,
https://doi.org/10.1007/978-3-030-19558-8_3

silicate layers [14–18]. Thus, by this way, the physical interactions between polymer chains and nanoclay could be mainly enhanced. Besides the surface energy decrease of nanoclay, the interlayer distance between clay nanolayers was increased by the surface modification with organic cations. Also, this change in morphology of nanoclay drives the diffusion of polymer chains inside surface-modified clay galleries much more easier compared to that of unmodified nanoclay [14–18].

Among PURF/clay published works, the most extensively studied clays are montmorillonite [1–9] and vermiculite [6, 10] clays which have established an extremely important position both in academia and industry due to their outstanding and improved properties in terms of higher aspect ratio, heat deflection temperature, dimensional stability, reduced gas permeability, impact behaviour, compression and tensile strengths and flame retardant behavior. Nanoclay is a special and important nanomaterial which can be used to enhance mechanical properties of PURFs due to its very large lateral dimension to thickness ratio in the range of 100–1000, a maximum lateral dimension of 10 μm and a minimum thickness of 1 nm [19–25]. The effects of clay addition to PURFs have been investigated by several researchers in terms of morphology, density, thermal stability, thermal conductivity, flame retardant, mechanical and thermal properties [1–10]. Previously, several research groups prepared PURNCFs consisting of montmorillonite nanoclay by using unmodified original nanoclay and nanolayered silicates modified with organic functional groups via in situ polymerization method. Moreover, the effects of nanoclay addition to PURFs were systematically investigated [1–3]. In order to generate a well-dispersion of nanoclay in PURFs, usually clay was ultrasonicated either in polyol or isocyanate component [3]. In addition, based on gas chromatography results, it was shown that the blowing agent diffusion throughout the closed cell structure is suppressed by more than 82% with 1 wt% clay incorporation to PURFs. The reason behind permeability reduction of blowing agent was explained such that nanoclay behaves like a diffusion blockade against blowing agent due to its smaller particle size and homogenous dispersion within the PURNCF matrix [3].

PU foams are fabricated via a very complicated reaction mechanism in which reactive polymerization, foaming and blowing occur at the same time [2]. During the reactive foaming process, nanoclays can also serve as nucleation centers for the creation of many bubbles with smaller sizes. In addition to their major role on the formation of more bubbles, nanoclays can also obstruct the growth of extra more bubbles by increasing the viscosity of reaction medium [3]. Throughout the time of reactive foaming event, nanoclay particles perform as additional seeding cores in PURNCFs for the generation of uniform distribution of cells with smaller sizes and higher cell densities. Hence, this positive improvement in closed-cellular morphology paves the way for inducing better gas barrier properties in PURNCFs compared to the PURF system without any clay [2, 25]. In accordance with improved gas barrier properties of clay containing PURNCFs, the use of nanoclay in PURFs also decreases the thermal aging behavior of rigid foams. Thermal aging which is defined as the increase of thermal conductivity over time usually occurs in PURFs since CO_2 with low thermal conductivity leaves the closed cell structure and is exchanged with N_2 and O_2 which have higher thermal conductivities [26, 27]. It was previously reported that smaller cell sizes of PURFs lead the way to lower thermal conductivity values, and smaller cell sizes in PURFs can be obtained by using different types of catalysts and surfactants and nanoclay particles [28–31]. Thus, in order to reduce the thermal conductivity increase over time, in addition to different combinations of surfactants and catalysts, nanoclays should be used in PURFs since their nanoplate-like morphology is very suitable for acting as an obstacle against gas diffusion. Previously, it was shown that dispersion of vermiculite clay in the isocyanate component leads to the formation of PURNCFs with better barrier properties [8].

Generally, based on rheological measurements, higher loadings of clay above 1 wt% lead to significant increase of viscosity and yield stress such that reactive components such as polyol and isocyanates cannot be efficiently mixed [3, 4]. Generally, more than 1 wt% clay addition deteriorates both mechanical and physical properties of PURNCFs. Previously, an increase in the opening of closed cell structure was observed in PURFs with MMT clay addition [32]. In addition to the significant opening of closed cell structure, compressive strength of PURNCFs consisting of 2 wt% of clay (Cloisite® 25A) was found to be improved by 24%. However, at higher clay loadings above 2 wt%, the compressive strength of PURNCFs reduced accordingly due to the considerable amount of cell opening [32].

The morphological and mechanical properties of PURNCFs are strongly correlated with evenly distribution of nanoclay particles throughout PU matrix. As a consequence, people both in academia and industry have used many sample preparation methods such as using different mixing conditions, ultrasonication, solution mixing, using various intercalation agents and performing diverse surface modification procedures of nanoclay to maximize the degree of organoclay distribution within the PURF matrix [7]. Previously, Macosko et al. modified the surface of nanoclay with organotin and hydroxyl functional groups, and by this way, the exfoliation of organoclay nanolayers throughout PURNCF matrix was realized [2]. The degree of exfoliation/intercalation of nanoclays in isocyanates or polyols was found to be very strongly affected by rheological properties [8]. On the other hand, intercalation or exfoliation level of clays in PURF, polyol or isocyanate does not show a directly proportional relationship with the aspect ratio of filler. In addition, it was found out that aggregation behavior of nanofiller is strongly dependent on the yield strength of dispersion [8].

Morphology, physical and mechanical properties of PURNCFs consisting of clay nanoplatelets mainly depend on the functional organic groups that are on the surface of clays and also synthesis methods [1, 2, 5, 20, 32, 33]. In addition, the surface modification of clay nanolayers via effective synthetic methods, and the use of these organomodified clays in PURFs lead to significant improvements in terms of rigid foam properties compared to the case of using unmodified clays in PURFs. Cao et al. [2] investigated the consequences of using a modified clay (Cloisite® 30B) and a higher molecular weight polyol ($M_n = 540$ g/mol) on less rigid PURF properties and compared those results with those of pure PURF. PURNCF consisting of 5 wt% clay displayed larger values of compressive modulus and strength compared to those of pure PURF [2]. In addition, the absolute compressive strength value for PURNCF consisting of 5 wt% clay was measured as almost 6 times higher compared to that of pure PURF [2]. On the other hand, PURNCFs consisting of a much lower molecular weight polyol showed systematic reductions in relation to reduced compressive strength and modulus values with incorporation of 1 and 5 wt% nanoclay amounts [2, 3].

However, in the fabrication of ideal PURNCFs consisting of nanoclay particles, the good interfacial adhesion between polymer matrix and clay nanolayers is much more important than the specific conditions of synthesis methods [1–5]. Based on previous experiments of PURCNFs consisting of clays, the compressive modulus values for nanocomposite foams consisting of 4 wt% Cloisite® Na and Cloisite® 30B clays were found to be improved by 9 and 23% along the direction of foam rise, respectively [1]. Previously, PURNCFs were fabricated via the intercalation/exfoliation of nanoclay Cloisite® 30B by applying chemically neutral dimethylol butanoic acid. It was pointed out that values of thermal conductivity, compression strength and foam density reduce while decomposition and glass transition temperatures increase with the addition of higher amounts of nanoclay [9]. Here, the exfoliation or intercalation Cloisite® 30B nanolayers was made possible by functional groups that belong to both clay and dimethylol butanoic acid which provide good interfacial adhesion or interactions between nanoclay layers and PURF matrix.

There are a few studies about PURNCFs consisting of graphene oxide (GO) and graphene [11–13, 34, 35] in the literature. Graphene is a different form of carbon and it has a two-dimensional planar sheet geometry with a thickness of one carbon atom. Each graphene plate consists of carbon atoms with sp^2 bonds which are densely loaded in a mass of hexagonal crystal lattice [36]. Graphene has been considered to be the thinnest nanomaterial with substantial thermal, mechanical and electronic properties [12, 35, 36]. Graphene oxide is generally synthesized by using Hummers method through oxidation of graphite. Graphene oxide is a cheaper, more functional and effective nanofiller compared to graphene since it has surface functional groups consisting of oxygen atoms. Due to polar groups on its surface, the solubility of GO in solvents such as water and other polar solvents is pretty high. As a consequence, graphene oxide was utilized as an efficient plate-like nanoadditive in fabrication of polymer nanocomposites with enhanced properties and diverse applications [12, 13, 34, 37–39].

Previously, electromagnetic interference (EMI) shielding and electrical properties of low density PURNCFs consisting of graphene nanoparticles were studied. In accordance with experimental data, it was shown that average cell size of PURNCFs can be reduced to remarkable levels with the addition of very low amount of graphene in the range of 0.17–0.35 wt% [40]. However, using reduced graphene oxide in high density PURFs did not reduce the cell size that much. Furthermore, thermal conductivity increased with the addition of graphene to high density PURFs [41]. The consequences of using graphene on thermal insulation properties of PURFs was studied by fabrication and characterization of PURNCFs consisting of 0.3 and 0.5 wt% graphene [11]. It was found that the radiative contribution of thermal conductivity value at the beginning reduces with graphene incorporation via the mechanism of cell size reduction and elevation of extinction coefficient. In addition, PURNCFs consisting of 0.3 and 0.5 wt% graphene exhibited slower aging rates in comparison with that of pure PURF [11]. Previously, functionalized graphene oxide (FGO) particles were incorporated into PURFs to improve thermal and mechanical properties, and the optimum content of FGO for improving mechanical properties was found out to be 0.4 wt%. In addition, thermal properties was maximized with the addition of 1 wt% FGO to PURFs [42].

Very recently, graphite and GO nanoparticles were used as fillers in PURFs that were produced as main materials for the transport of liquefied natural gas. Specifically, the effects of fillers on the closed-cellular morphology, mechanical and thermal properties at room and cryogenic temperatures were examined [13]. Based on the results, it was found out that the addition of more than 1 wt% graphite and 0.1 wt% GO disrupts the closed-cellular morphology. The best mechanical properties were achieved with the addition of 0.05 wt% GO at room and cryogenic temperatures. Also, the thermal conductivity reduced with addition of less than 1 wt% graphite and 0.1 wt% GO due to the well preservation of cell structure [13]. Previously, the effects of using GO that was functionalized with polyol on compressive mechanical properties, closed-cellular morphology, foaming kinetics and thermal conductivity of water blown PURFs were studied. Based on experimental results, it was found out that both cell size and thermal conductivity reduce, also mechanical properties get deteriorated with inclusions of very small amounts (0.017, 0.033 and 0.088 wt%) of GO to water blown PURF system [12]. All of these studies showed that the addition of less than 1 wt% graphene or graphene oxide to PURFs is enough in order to obtain the best improvement in terms of mechanical and thermal properties.

3.2 Morphology

Cao et al. synthesized montmorillonite (MMT) based PURNCFs by using the batch process and the MMTs that were used in PU rigid nanocomposite foams were surface modified with organic functional groups by using the in-situ polymerization [2]. Authors used transmission electron microscopy (TEM) and X-ray diffraction (XRD) to perceive extent of organoclay distribution within PURNCF matrix and the overall morphology of the system. Based on the morphological results, it was found out that physical and mechanical properties of PURNCFs consisting of organoclays extensively depend on the organic functional groups that are present on the surface of organoclays, various important steps in the synthesis methods, and molecular weight of polyols that were used as the major component for the reactive foaming process [2]. Thus, the nature of chemical reactions and physical interactions between system components such as polyols, polyisocyanates, organic functional groups of organoclay, catalysts, blowing agents, etc. affect the overall performance of PURNCFs such as physical and mechanical properties.

Authors surface modified the surface of MMTs with hydroxyl and organotin functional groups, and showed that nanolayers of organoclay can be exfoliated throughout the PURNCF matrix [2]. It was also shown that the cell size diminished and cell density of foams elevated with the incorporation of organoclay into PURNCFs in comparison with the case of pure PURF. Based on mechanical and thermomechanical test results, it was also discovered that glass transition temperature increased about 6 °C, reduced compressive modulus increased about 650% and reduced modulus increased about 780% with the addition of 5% organoclays to the PURNCF system consisting of a high molecular weight polyol. On the other hand, authors found out that the system which had a highly-crosslinked structure exhibited the opposite behaviors. The general difference between the two systems was explained such that the H-bonding between the organoclay and system components led to property improvements in PURNCF system in which a high molecular weight polyol was used as the major reaction component [2].

Figure 3.1 shows the SEM images of specimens that were prepared from the freeze-fractured surface of pure PURF and PURNCFs consisting of organoclays. In addition, SEM micrographs show the morphological features within cross-sections which are parallel to foam rising direction [2]. Based on the SEM images, it was shown that both pure PURF and PURNCFs consisting of organoclays such as MMT-Tin and MMT-OH displayed closed-cellular morphologies which have

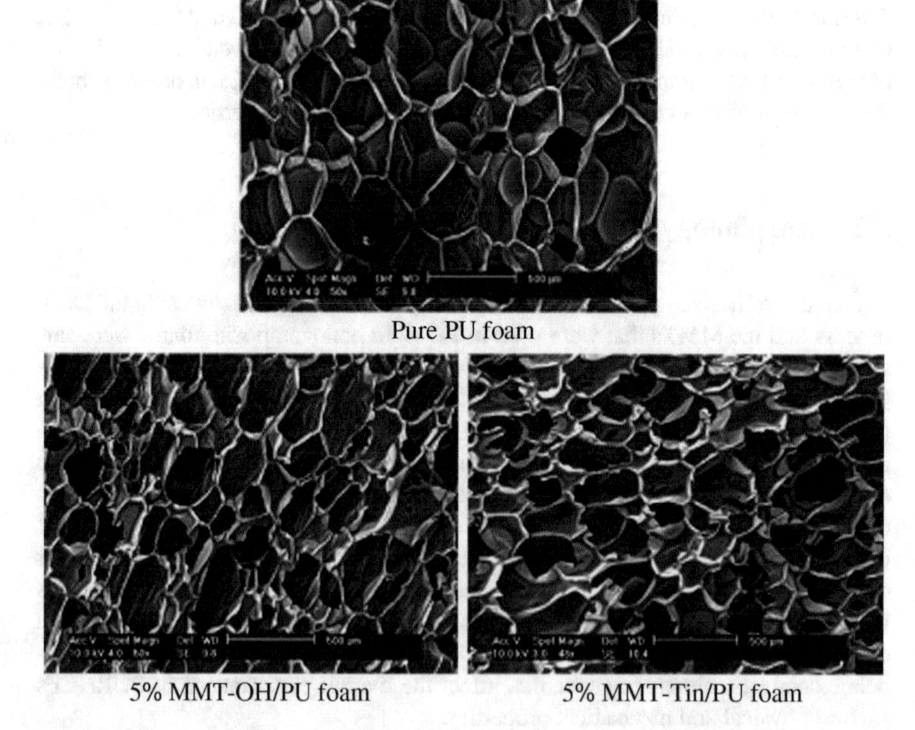

Fig. 3.1 SEM micrographs of pure PURF and PURNCFs consisting of different types of organoclays from sections parallel to foam growth direction. Reproduced with permission from Ref. [2]. Copyright 2004 Elsevier Ltd.

thermodynamically-stable cellular structures with pentagonal and hexagonal geometries. In addition, it was reported that the cell density is much smaller and the cell sizes are much larger in pure PURF compared to the case of PURNCFs consisting of 5 wt% organoclays. Furthermore based on cell density and cell size measurements from actual SEM images, it was also stated that PURNCFs consisting of 5 wt% organoclays has a lower bulk density, higher cell density (higher number of cells per unit sample volume) and much more smaller cell sizes compared to those of pure PURF sample. Thus, this result showed that the presence of organoclay layers within the PURNCF matrix act as extra nucleation sites, especially behave like heterogenous nucleation centers during the closed-cellular morphology formation [2, 43, 44], and this mechanism led to the difference in terms of cellular morphology between pure PURF and PURNCFs consisting of organoclay MMT layers.

In terms of the characteristic details such as cell size and cell density of closed-cellular morphology, it was shown that there is a very little difference between PURNCFs consisting of MMT-Tin and MMT-OH organoclays [2]. To understand the detailed morphology of PURNCFs consisting these various organofunctional clays, authors also performed XRD experiments. Based on XRD experiments, authors fount out very little difference in terms of characteristic peaks. Based on the information that was collected based on SEM and XRD experiments, authors concluded that more detailed investigation should be carried out in order to understand the morphological details of PURNCFs consisting of MMT-Tin and MMT-OH organoclays [2].

Seo et al. fabricated PURNCFs consisting of nanoclay which was surface-altered by PMDI via application of ultrasound technique [20]. Based on TEM results, authors showed that the interlayer spacing between PMDI surface modified clays in PURNCFs was increased with the application of ultrasound. In addition, based on XRD and TEM characterization results, it was also reported that using the ultrasound method during surface modification of nanoclay with PMDI chains enhanced the yield of surface modification process by separating clay aggregates and eased the intercalation of polymer chains within the silicate nanolayers [20]. Based on mechanical test results, it was shown that the maximum values in terms of tensile and flexural strengths of PURNCFs were obtained with the incorporation of 3 wt% nanoclay based on PMDI amount [20]. Thus, the reason behind the increase of tensile and flexural strengths of PURNCFs was explained such that the incorporation of ultrasound during surface modification helped to disperse clay layers homogenously within the PURNCF matrix, and increased the mechanical properties accordingly due to better interfacial bondings between PMDI groups on the surface of nanoclay and PU chains of PURNCF matrix. Moreover, it was also reported that the fire resistance properties of PURNCFs increased with the integration of ultrasound method in comparison with PURNCFs in which no ultrasound was applied during the surface treatment of clay nanolayers. In addition, authors also showed that using ultrasound technique during the surface modification of nanoclay led to reduction of cell sizes and thermal conductivity values in PURNCFs in comparison with system in which no ultrasound was applied. Based on these results, authors concluded that both smaller cell sizes and reduction of thermal conductivity values in PURNCFs were attributed to homogenous distribution of PMDI-modified nanoclay throughout PURNCF matrix due to the integration of ultrasound method during the surface modification process of clay nanolayers [20].

Figure 3.2 shows SEM micrographs of (a) pure PURF, (b) PURNCF consisting of clay without performing any ultrasound, and (c) PURNCF consisting of clay with the application of ultrasound method. In accordance with SEM data, it was pointed out that cell sizes in pure PURF, PURNCFs that were prepared without and with ultrasound were found to be around 400, 350 and 280 μm, respectively [20]. Thus, based on SEM data, it is plainly noticed that using PMDI-clay prepared without ultrasound in PURFs decreased the cell size efficiently due to the act of clay as extra nucleation sites [2, 45–47] during the foaming process in comparison with that of pure PURF. Secondly, the addition of PMDI-clay prepared with ultrasound to PURFs decreased the cell size efficiently due to effective dissolution of clay aggregations and increase of intermolecular interactions between clay and PURFs with the integration of ultrasound method compared to PURNCFs consisting of PMDI-clay prepared without ultrasound [20].

Han et al. synthesized PURNCFs consisting of organoclay particles with improved thermal insulation properties. In their PURNCFs, authors used organoclay as the plate-like nanofiller which was surface modified with polymeric diphenylmethane diisocyanate (PMDI) by using a silane coupling agent [7]. Authors used FT-IR, nuclear magnetic resonance (NMR) spectroscopy to comprehend the chemical structure of organoclay-modified PMDI in detail. In accordance with TEM and wide-angle X-ray diffraction (WAXRD) data, authors pointed out interlayer spacing between clay nanolayers in PURNCFs increased due to the use of silane coupling agent during surface modification process. The reason behind increase of interlayer clay spacing was explained such that chemical bond generation between silane molecule and organoclay nanolayers basically leads to elevation of distance between organoclay layers and also well-dispersion of clay nanolayers within the PURNCF matrix [7]. Based on mechanical testing results of PURNCFs, it was shown that there is no big difference in terms of compressive and flexural strengths between PURNCFs consisting of organoclay layers that were prepared with and without silane coupling agent. On the other hand, in accordance with SEM and thermal conductivity data, it was pointed out that in terms of thermal conductivity and cell size, PURNCFs consisting of organoclay-silane coupling agent system had much lower values compared to PURNCFs consisting of organoclay without the silane coupling agent. In conclusion, authors interpreted these findings such that using silane coupling agents during surface modification of organoclay led to the improved exfoliation of organoclay nanolayers within the PURNCF matrix, and thus decreased the values of thermal conductivity and cell size in comparison with the system that did not contain any silane coupling agent [7].

Figure 3.3 depicts SEM micrographs of (a) PURNCF/organoclay system (b) PURNCF/organoclay system containing silane molecule prepared with no ultrasound and (c) PURNCF/organoclay system containing silane molecule prepared with application of ultrasound. According to experimental data, it is obviously noticed that cell size of PURNCF consisting of silane molecule and organoclay system is the smallest among SEM micrographs. The reason behind this difference in terms of cell

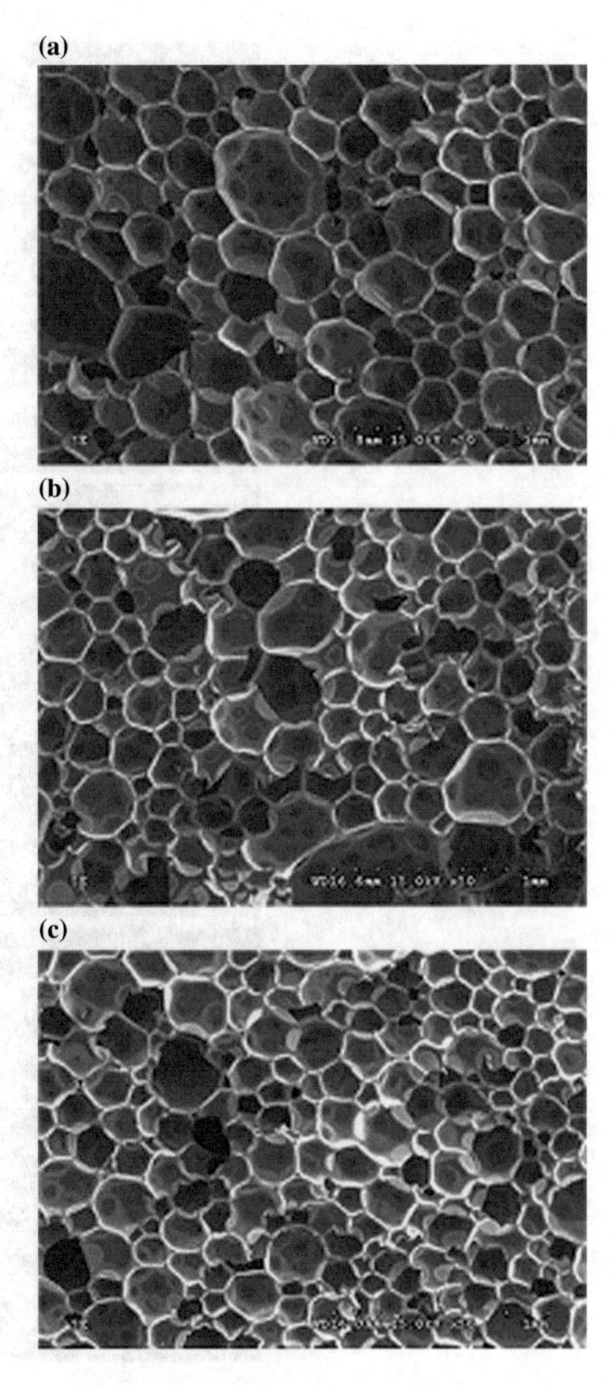

Fig. 3.2 SEM images of **a** pure PURF without clay, **b** PURNCF consisting of clay without ultrasound, and **c** PURNCF consisting of clay with ultrasound. Reproduced with permission from Ref. [20]. Copyright 2006 Wiley Periodicals, Inc.

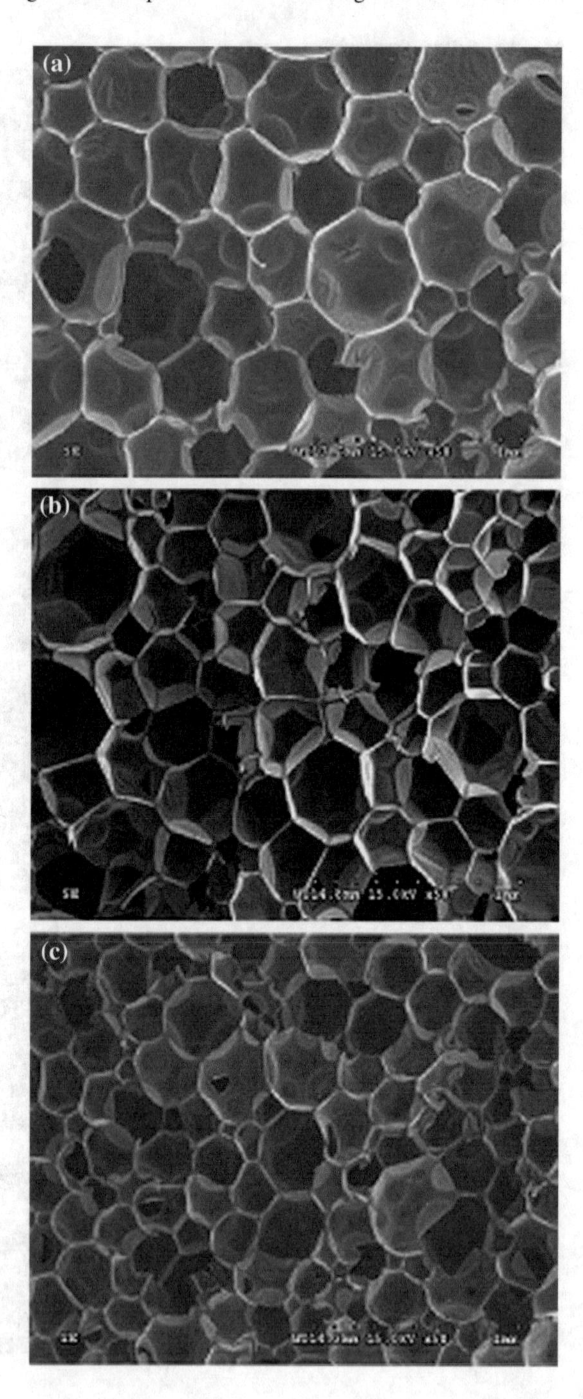

Fig. 3.3 SEM micrographs of **a** PURNCF/organoclay system, **b** PURNCF/organoclay system containing silane molecule prepared with no ultrasound and **c** PURNCF/organoclay system containing silane molecule prepared with application of ultrasound. Reproduced with permission from Ref. [7]. Copyright 2008 Wiley Periodicals, Inc.

size was explained due to the enhanced exfoliated morphology of organoclay in PURNCF consisting of silane molecule and organoclay system with the application of ultrasound method [7]. Furthermore, it was pointed out that exfoliated morphology of organoclay nanolayers which were surface-modified via reactions of silane molecules in PURNCF matrix could function as seeding media and locations for development of bubbles during the reactive foaming process [1, 7, 46, 48]. Consequently, it was reasoned that nucleation behaviour of exfoliated organoclay within PURNCF consisting of silane molecule and organoclay system could be responsible for cell size reduction [7].

Widya et al. incorporated montmorillonite type of organoclay into PURFs by ultrasonically dispersing clay nanolayers in the isocyanate component [3]. Authors used rheology and small angle X-ray scattering (SAXS) to understand extent of clay dispersion within PURNCFs. In accordance with experimental data, it was found out that distribution of clay layers extensively improved due to the use of toluene as the dispersion solvent during the process. Based on SEM results, authors found out that the cell size decreased and also the cell density increased with the incorporation of 1 wt% clay into 300-index foams. In addition, gas chromatography (GC) results revealed that the clay addition suppressed the outflow of blowing agent from the closed-cellular morphology of PURNCFs by more than 82%. The reason behind this permeability reduction was explained due to reduction of cell size and homogenous distribution of nanoclay that acts as a diffusion blockade against discharge of blowing agents from the PURNCF system. In addition, based on mechanical testing results, no improvement was noticed in mechanical strength of foams with incorporation of 1 wt% clay to the system. Moreover, it was pointed out that larger values in relation to yield stress and viscosity were obtained with inclusion of very high amounts of nanoclay to PURNCF system [3].

Figure 3.4 shows SEM micrographs of pure PURF and PURNCFs consisting of 1 wt% nanoclay. Based on SEM results, it was shown that the cell sizes in 200- and 250-index foams increase with the addition of nanoclay. On the other hand, it is clearly seen that incorporation of nanoclay leads to reduction of cell size and cell density elevation in 300-index based PURNCFs [3]. Based on cell size measurements, it was revealed that more broken cells were observed in 200-index compared to 300-index pure PURFs. Based on these findings and predictions, it was concluded that the thermal conductivity of PURNCFs consisting of 1 wt% nanoclay could decrease about 13% due to this degree of reduction in cell size [3].

Harikrishnan et al. fabricated nanodispersions of clays in polyurethane rigid foams. During sample preparation, authors used both pristine nanoclay and nanoclays containing various aspect ratios that were surface modified with organic functional groups. In addition, nanofillers were dispersed in polyol and polyisocyanate components, separately [8]. Authors characterized these nanodispersions by using FT-IR, rheology, XRC and cryo-TEM. The effect of foaming reaction on exfoliation was investigated by comparing the conditions of filler dispersions in components and in PURFs. Based on these experimental findings, it was pointed out that properties of PURNCFs are directly correlated with physicochemical and rheological conditions of nanodispersions [8].

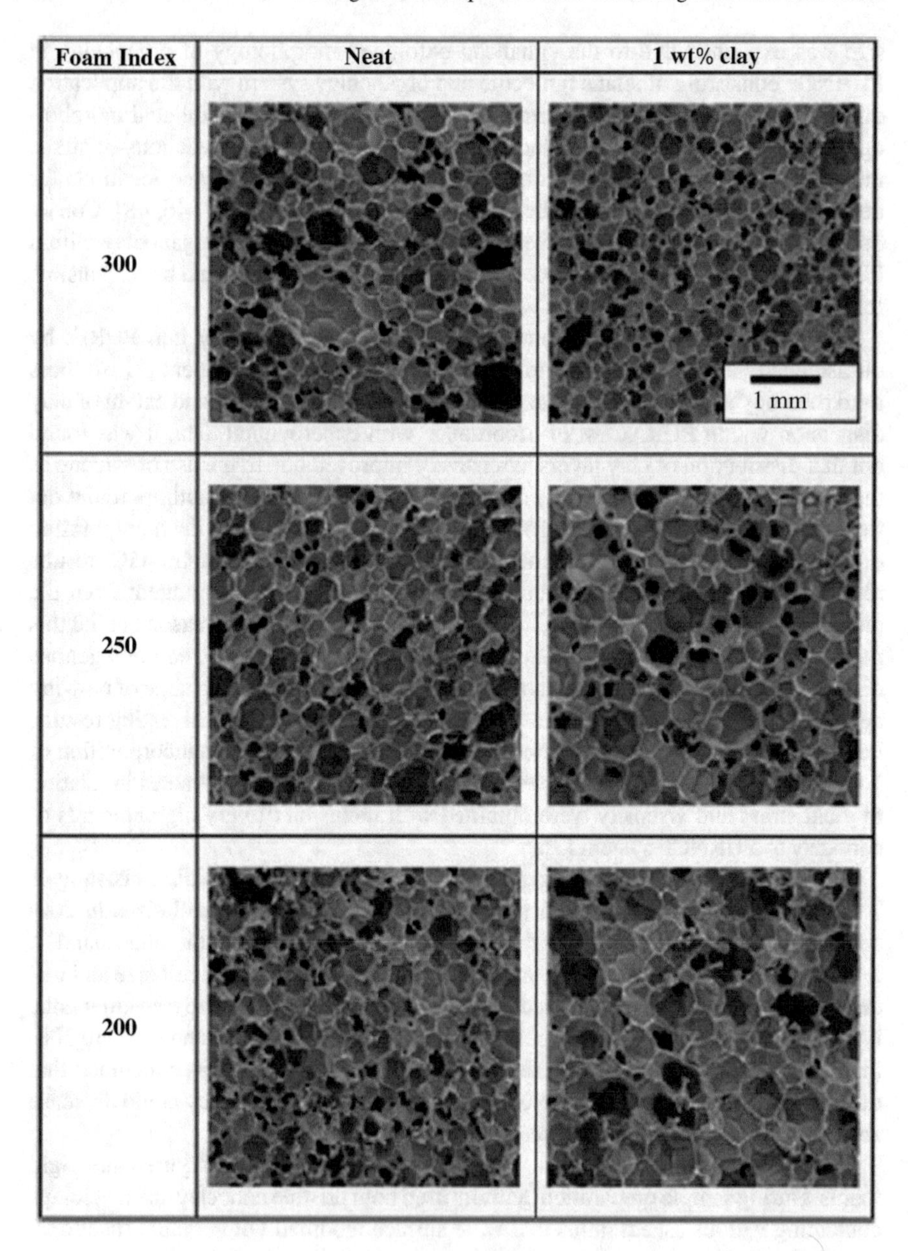

Fig. 3.4 SEM micrographs of pure PURF and PURNCFs consisting of 1 wt% nanoclay. Each micrograph has the same amount of magnification. Reproduced with permission from Ref. [3]. Copyright 2005 Taylor & Francis LLC

Figure 3.5 shows SEM micrographs of (a, b) pure PURFs, (c) PURF prepared from a laponite-MDI dispersion, (d) PURF prepared from a PEtP-laponite-MDI dispersion, (e) PURF prepared from a MMT-MDI dispersion, (f) PURF prepared from a MMT-PEtP dispersion, (g) PURF prepared from a vermiculite-MDI (VMT) dispersion, and (h) PURF from a VMT-PEtP dispersion. Based on these SEM results, it was shown that the inclusion of modified laponite to either in polyol or polyisocyanate component did not lead to a considerable difference in the closed-cellular morphology. In addition, it was pointed out that cell sizes of PURNCFs containing VMT are much smaller and homogenous compared to those of pure PURFs [8].

Semenzato et al. prepared and characterized diphosphonium based montmorillonite which was obtained via insertion of quaternary diphosphonium salt molecules within nanoclay layers. This surface modified nanoclay as a plate-like nanofiller was used in the preparation and thermal behavior analysis of PURNCFs [5]. Based on XRD results, it was shown that the synthesized diphosphonium based montmorillonite had an interlayer d-spacing of 1.90 nm, and this length scale was compatible with packing of intercalated chains within two nanolayers of clay. Based on TGA data, it was revealed that weight loss in diphosphonium based MMT was observed to be around 18% at about 400 °C. In general, during experiments, 5 wt% of diphosphonium based MMT was used as the nano-plate filler in the fabrication of PURNCFs. In addition to incorporation of diphosphonium based MMT into PURFs, commercial MMTs such as ammonium-modified MMT and unmodified pristine MMT were applied as plate-like nanofillers with Al-hypophosphite in PURNCFs [5].

Figure 3.6 shows SEM images of (a) pure PURF, (b) PURNCF consisting of 10% phosphinate and diphosphonium based MMT. According to SEM data, it was disclosed that cell size moderately got larger when both phosphinate and diphosphonium based MMT were used as plate-like nanofillers in PURNCF in comparison with that of pure PURF [5].

Kim et al. fabricated PURNCFs consisting of nanoclay by using a specific method in which clay nanolayers were intercalated and exfoliated with the help of neutralized dimethylol butanoic acid (DMBA) [9]. In accordance with experimental data, it was shown that values of foam density and compression strength decreased while gel, tack-free and cream time increased with increasing amounts of nanoclay into PURFs. On the other hand, it was also reported that the thermal conductivity, cell size, the amount of closed cells and the volume change as a function of temperature decreased with the incorporation of nanoclay into PURF system. Based on thermomechanical and thermal stability results, it was pointed out that nanoclay incorporation elevated temperatures such as glass transition and thermal decomposition due to reduced mobility of PU chains by barrier effects of nanoclay platelets [9].

Figure 3.7 displays SEM micrographs of pure PURF and PURNCFs consisting of different amounts of nanoclay. According to SEM data, it is evidently seen that mean size of closed cells in PURNCF matrix decreases with the addition of nanoclay to PURFs. It was pointed out that the closed-cellular morphology in these foams have spherical cellular structure with polyhedral geometry and the cell size in these foams reaches to its minimum value with the addition of 2 pphp nanoclay. Thus based on these results, in agreement with other studies [2, 3, 5, 7, 20], it was confirmed that

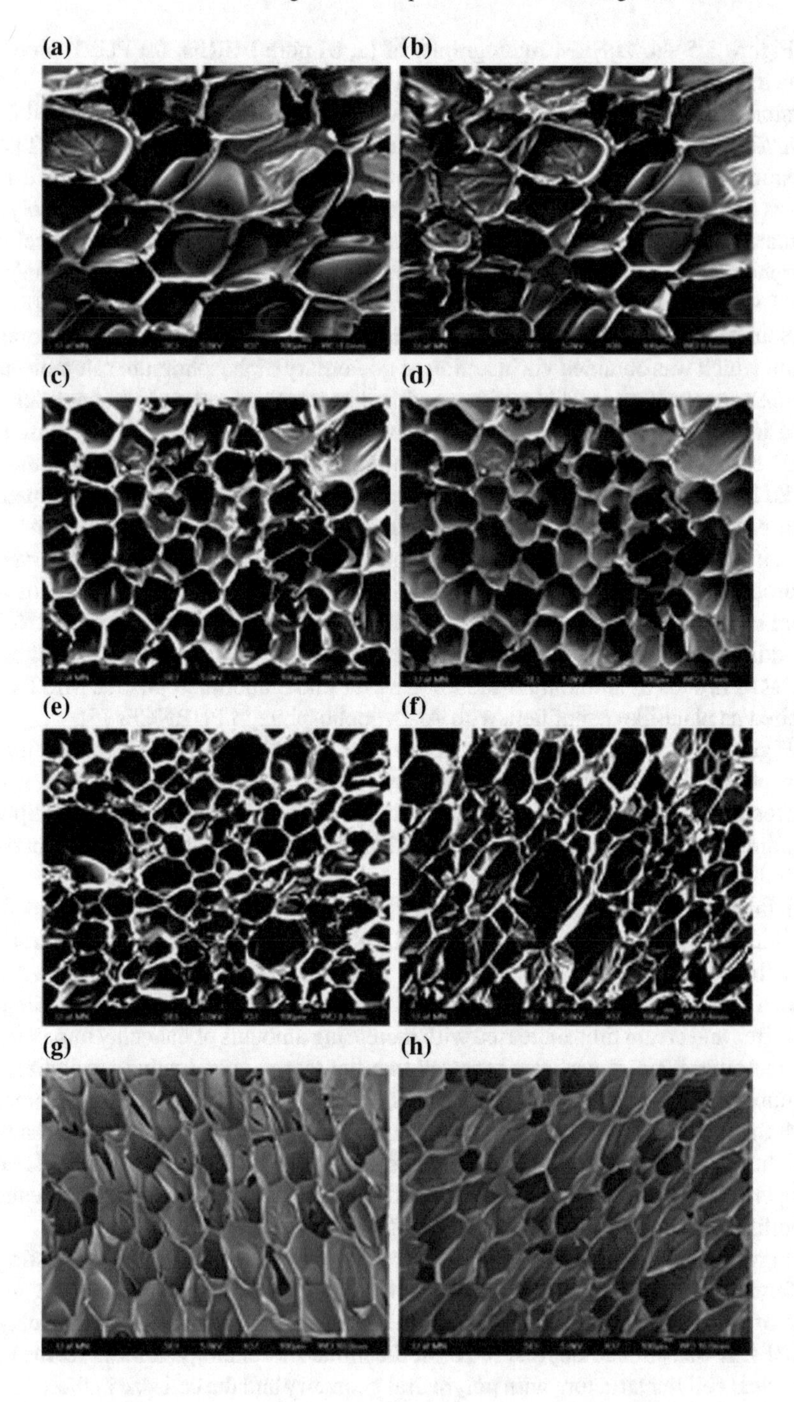
(a)
(b)
(c)
(d)
(e)
(f)
(g)
(h)

◂**Fig. 3.5** SEM micrographs of **a**, **b** pure PURFs, **c** PURF prepared from a laponite-MDI dispersion, **d** PURF prepared from a PEtP-laponite-MDI dispersion, **e** PURF prepared from a MMT-MDI dispersion, **f** PURF prepared from a MMT-PEtP dispersion, **g** PURF prepared from a vermiculite-MDI (VMT) dispersion, and **h** PURF from a VMT-PEtP dispersion. Reproduced with permission from Ref. [8]. Copyright 2009 American Chemical Society

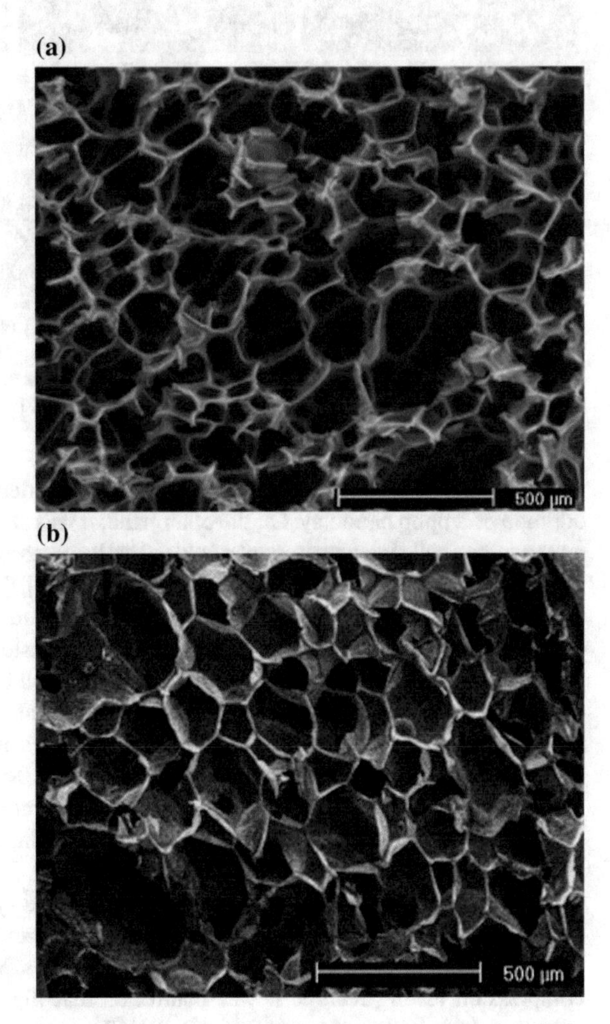

Fig. 3.6 SEM micrographs of **a** pure PURF and **b** PURNCF consisting of 10% phosphinate and diphosphonium based MMT. Reproduced with permission from Ref. [5]. Copyright 2009 Elsevier Ltd.

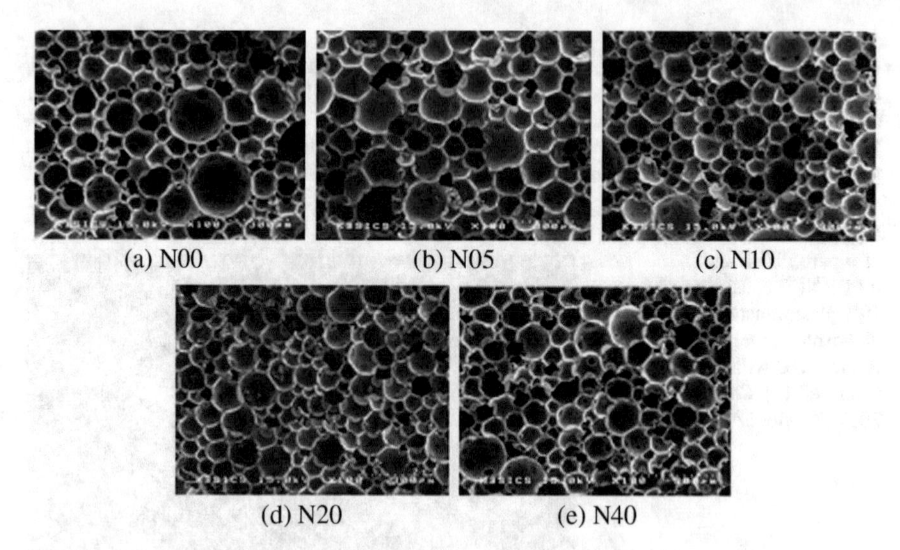

Fig. 3.7 SEM micrographs of pure PURF and PURNCFs consisting of different amounts of nanoclay. Reproduced with permission from Ref. [9]. Copyright 2010 Wiley Periodicals, Inc.

nanoclay acts as a nucleation agent and leads to the decrease of cell size with the addition of 2 pphp nanoclay. On the other hand, it was also revealed that at high clay amounts, the cell size and amount of closed cells increase due to the poor dispersion and interactions of nanoclay with the polymer chains [9].

Palanisamy fabricated PU rigid nanocomposite foams consisting of nanoclay by substituting some of synthetic polyol with a type of castor oil [4]. Basically, authors produced diethanol amide from hydroxylated castor oil by using the transamidation process, and the synthesized polyol was used in the fabrication of water-blown PU foams. Surface-modified MMT clay was incorporated in the amounts of 0.5, 1.0, 2.0, and 5.0% as nanofillers into the synthesized PURNCFs. According to XRD and rheology data, it was pointed out that no substantial change in viscosity was observed with shear rate in polyol–clay mixtures consisting of 1% clay loading. However, with the addition of more than 2 wt% clay, it was pointed out that there was shear thinning in PURNCFs. Authors investigated systematically the effects of modified clay on mechanical properties such as compression strength, compression modulus, foam density, and morphology of PURNCFs. Based on foam density and compression testing results, it was confirmed that the incorporation of nanoclay enhanced the mechanical properties of PURNCFs. In accordance with DSC and DMA experimental data, it was revealed that exfoliation and plasticizing effect occurred with the addition of 1 wt% of nanoclay. Based on optical microscopy (OM) and SEM results, it was confirmed that the exfoliation mechanism in PURNCFs consisting of 1 wt% or more nanoclay leads to much more reduced cell size and higher nucleation rates in comparison with intercalated PURNCF systems [4].

Figure 3.8 shows SEM micrographs of pure PURF and PURNCFs consisting of different amounts of nanoclay. Based on SEM results, it is clearly observed that there

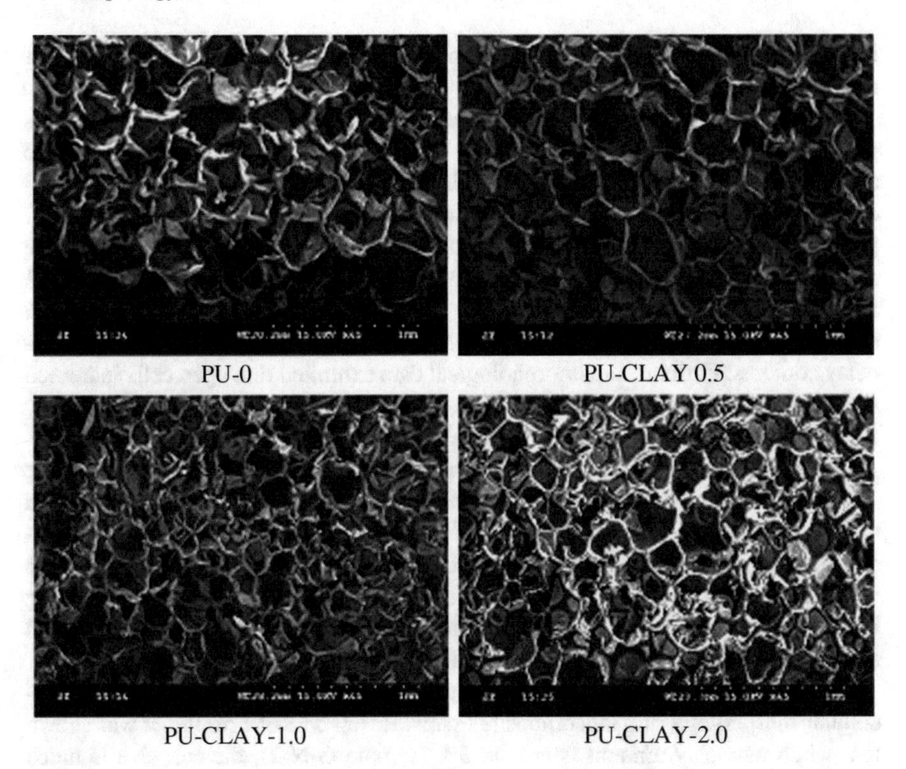

Fig. 3.8 SEM micrographs of pure PURF and PURNCFs consisting of various amounts of nanoclay. Reproduced with permission from Ref. [4]. Copyright 2013 Society of Plastics Engineers

is a clear reduction in cell size with incorporation of nanoclay to PURF system. The decrease of cell size with increasing amount of clay addition was explained such that the cell growth was hindered due to the viscosity elevation in the company of nanoclay, and this mechanism led to the decrease of cell size in PURNCFs consisting of nanoclay [4].

Another reason for the decrease of cell size in PURNCFs consisting of nanoclay was explained such that the presence of clay led to the nucleation of more bubbles and decreased the growth of bubbles. Eventually, this caused the decrease of cell size due to the less amount of gas available in the system. The enhanced nucleation in exfoliated PURNCFs containing nanoclay was also explained such that the smaller dimensions and very high surface areas of nanoclay with very widely dispersed morphology in PURNCFs increased the number of interactions among polymer matrix, gas and nanoclay particles. This behavior also led to the higher nucleation rates in exfoliated PURNCFs compared to intercalated ones [4]. Based on these results, it is well noted that the optimum concentration of nanoclay in exfoliated PURNCFs should be kept around 1 wt% in order to have a minimum cell size and thermal conductivity behavior in PURNCFs. However, more than 1 wt% addition of nanoclay to PURF system leads to the deterioration of PURNCF properties due to reduction in number of interactions between PU chains and nanoclay particles as a result of nanoclay aggregation in the system.

Hu et al. fabricated phenol-urea-formaldehyde based glass fiber/nanoclay composites to improve flame retardancy, toughness and compression strength of phenolic foams. Authors investigated closed-cellular morphology, flame retardant, mechanical and thermal decomposition properties of phenolic foam nanocomposites in detail [49]. Based on these results, it was shown that pulverization rate of phenolic foams reduced extensively with incorporation of nanoclay and glass fibers to pristine PURF. Based on mechanical property results, it was shown that impact and compression strength values of PURNCFs enhanced with inclusion of higher glass fiber and nanoclay contents. Furthermore, morphological data exhibited that open cells increased in number with the addition of glass fibers, however the smaller amounts of nanoclay leads to the control of cell size in the nanocomposite foams consisting of nanoclay. Furthermore, it was also reported that higher amounts of nanoclay led to the increase of cell wall thickness and number of open cells in the nanocomposite foams. Based on thermal stability results, inclusion of nanoclay enhanced thermal stability but diminished flame retardant properties such as heat release rate and smoke release values of PURNCFs. So overall, based on these results, it was confirmed that using both nanoclay and glass fiber has synergetic effects in terms of improving mechanical and flame retardant properties and thermal stability of nanocomposite foams [49].

Figure 3.9 shows the dependence of nanoclay and glass fiber addition on closed-cellular morphology of nanocomposite foams. Based on SEM results, it was shown that when nanoclay amount is around 3% (sample G-N-2), the cell size is much smaller compared to the sample G-4 filled with only glass fiber. The major difference in terms of cell size between samples G-4 and G-N-2 was explained such that nanoclay incorporation into foams operates as an appropriate reinforcing and efficient geometrical obstacle and leads to smaller cell sizes and high compression strength values in nanocomposite foams. However, on the other hand, it was also revealed that when nanoclay was used in very high amounts in the presence of glass fibers, some of the cellular structure was observed to be destroyed and collapsed in sample G-N-3 foam in comparison with other two foams. Based on these observations, and also in agreement with other studies [4], it was concluded that the optimum quantity

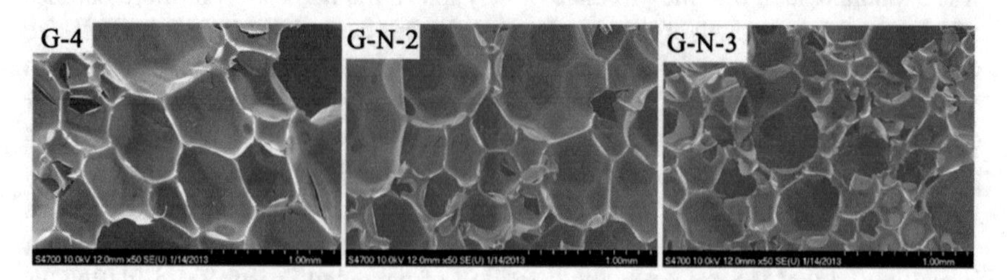

Fig. 3.9 Effect of nanoclay and glass fiber on closed-cellular morphology of nanocomposite foams. Reproduced with permission from Ref. [49]. Copyright 2015 Society of Plastics Engineers

for the nanoclay should not be exceeded in order to produce nanocomposite foams with improved physical, thermal and mechanical properties [49].

Danowska et al. fabricated PURNCFs consisting of nanoclay with the aim of investigating the effect of three different nanofillers such as MMT surface-modified with quaternary ammonium salt, a synthetic layered silicate and Bentonite clay on the properties of PURNCFs [50]. Nanoclay based PURNCFs were prepared by performing a single shot laboratory based process based on a foaming reaction involving 2:1 ratio of NCO to OH groups. Also, it was reported that required contents of catalysts, water, nanoclays, oligoether polyol and PMDI were present in the reaction mixture during the reactive foaming process. Authors investigated the thermal and thermomechanical properties of PURNCFs by using DMA, TGA, LOI and thermal conductivity measurements. Based on the experimental results, it was shown that PURNCFs that were prepared with selected nanofillers had higher mechanical strength and improved fire resistance properties [50].

Figure 3.10 shows SEM surface morphologies of (a) pure PURF and PURNCFs containing 3 wt% of different nanoclays such as (b) MMT surface-modified with quaternary ammonium salt, (c) a synthetic layered silicate, and (d) Bentonite. According to SEM data, it was found out that incorporation of 3 wt% MMT surface-modified

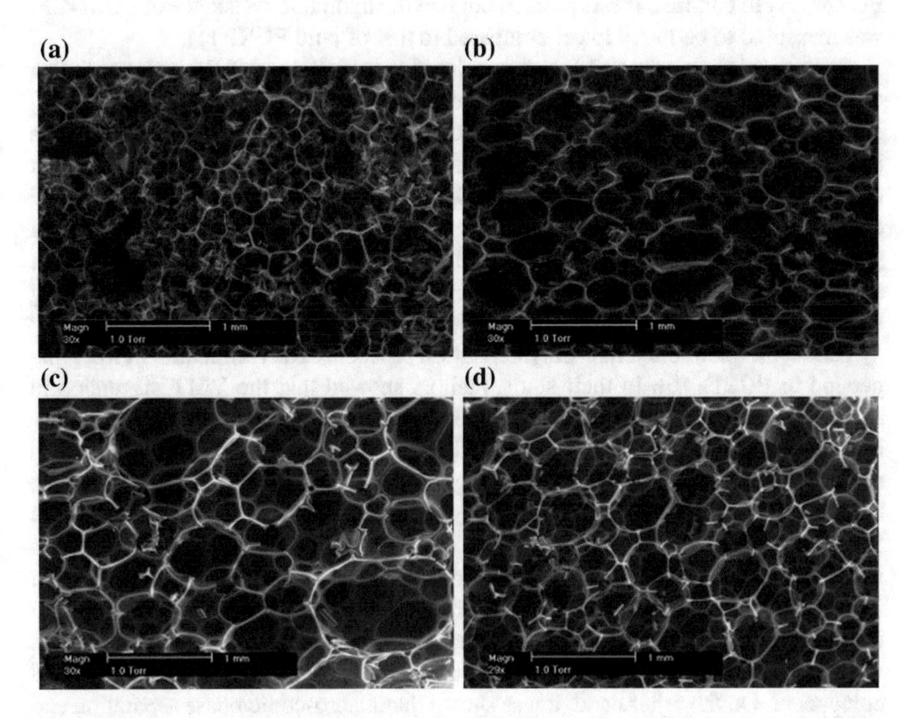

Fig. 3.10 Surface morphologies of **a** pure PURF and PURNCFs containing 3 wt% of different nanoclays such as **b** MMT surface- modified with quaternary ammonium salt, **c** a synthetic layered silicate, and **d** Bentonite. Reproduced with permission from Ref. [50]. Copyright 2013 Wiley Periodicals, Inc.

with quaternary ammonium salt leads to cell density elevation and reduction of cell size (smaller than 300 μm) with a much more ordered closed-cellular structure compared to pure PURF. In addition, it was also reported that with the addition of synthetic layered silicate, mechanical properties of PURNCFs decreased and closed-cellular morphology was disrupted due to the increase of cell size to the range of 400–900 μm in these nanocomposite foams. However, among the different types of nanoclays used in the study, and also in agreement with the results from other works [2–5, 7, 9, 20, 49], it was shown that the addition of 3% MMT surface-modified with quaternary ammonium salt resulted in the decrease of cell size due to the beneficial effects of using lower amounts of nanoclay that act as nucleation centers in the reduction of cell sizes in nanoclay based PURNCFs [50].

Mondal et al. synthesized PURNCFs consisting of different types of nanoclay with different concentrations [1]. The densities of PURNCFs were adjusted to be 140–160 kg/m^3 so that they can be used in structural applications such as construction and underwater buoyancy. Authors used wide-angle XRD and TEM in order to confirm the nanocomposite structures of PURNCFs. Based on the mechanical testing results, it was confirmed that inclusion of nanoclay fillers to PURF system enhanced properties such as storage and compressive modulus but reduced mean cell size in PURNCFs. In contrast, it was pointed out that the hydraulic resistance of PURNCFs was measured to be much lower compared to that of pure PURF [1].

Figure 3.11 shows the SEM micrographs of pure PURF and PURNCFs consisting of different types of nanoclay with different concentrations. Based on SEM results, it is apparently noticed that the addition of clay to pure PURF considerably decreases the cell size in PURNCFs consisting of different types of clay. In agreement with other similar studies [2–5, 7, 9, 20, 49, 50], authors explained the reason behind the cell size reduction such that nanoclay acts as nucleation centers for the production of gas bubbles. In addition, the viscosity of the reaction medium becomes much higher and the combination of smaller cells is prevented by the presence of nanoclay during the reactive foaming process of PURNCFs [1].

Park et al. found out a new polymerization procedure for vermiculite (VMT) dispersion in PURFs [6]. In their study, authors showed that the VMT dispersion in MDI was mainly enhanced when VMT was surface-modified with cetyltrimethylammonium bromide (CTAB) via the cation exchange reaction. In addition, it was also reported that the application of a high intensity dispersive mixing method the onto polyol-clay blend and combining this polyol blend with the VMT dispersion in MDI increased the dispersion degree of VMT in the nanocomposite foams. Authors used XRD, rheology and microscopy to understand the degree of VMT dispersion in nanocomposite foams. Based on electron microscopy results, it was shown that clay nanoparticles were substantially intercalated and exfoliated in PURNCFs that were prepared with the sample preparation method VMT dispersion in the polyol component. On the other hand, it was shown that macroscopic phase separation and poor distribution of clay nanoparticles in PURFs were observed in PURNCFs that were prepared with simple mixing of organoclay in MDI. Based on mechanical and permeability test results, it was further revealed that the PURNCFs that were prepared with polyol-assisted VMT dispersion or master batch (MB) method exhibited

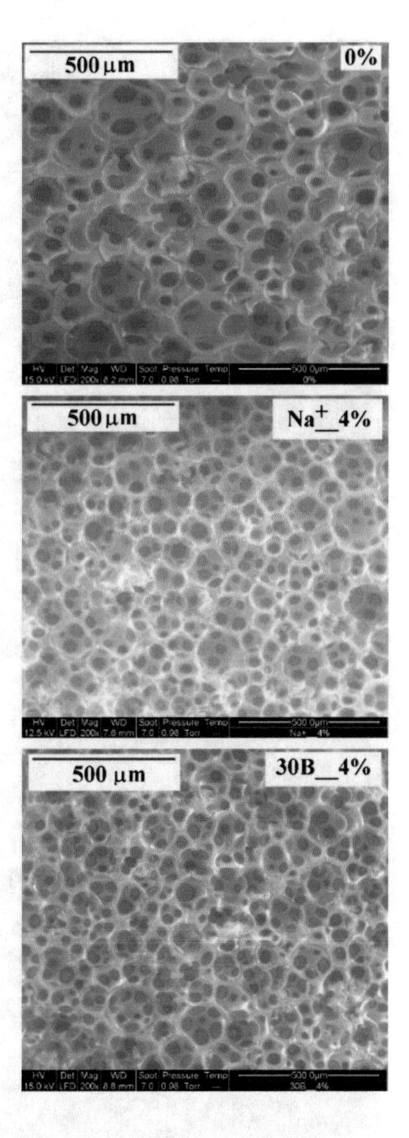

Fig. 3.11 SEM micrographs of pure PURF and PURNCFs consisting of different types of nanoclay with different concentrations. Reproduced with permission from Ref. [1]. Copyright 2006 Wiley Periodicals, Inc.

about 85% improvement in terms of elastic modulus and 40% reduction in terms of CO_2 permeability at 3.3 wt% nanoclay loadings [6].

Figure 3.12 shows the TEM images of PURNCFs such as (a, b) PURF-MB system consisting of 2.2 wt% clay, (c, d) PURF-MB system consisting of 4.4 wt% clay, and (e, f) PURF-CTAB-VMT system consisting of 1.9 wt% clay that was prepared with direct mixing method. In the TEM micrographs, the dark areas correspond to CTAB-VMT-rich areas whereas light gray areas refer to PU-rich regions of PURNCFs. Based on TEM results, it was shown that single sheets of CTAB-VMT clays with almost exfoliated structure were found out in PURNCFs that were prepared with the MB method. In addition, also in agreement with XRD results, it was confirmed that the

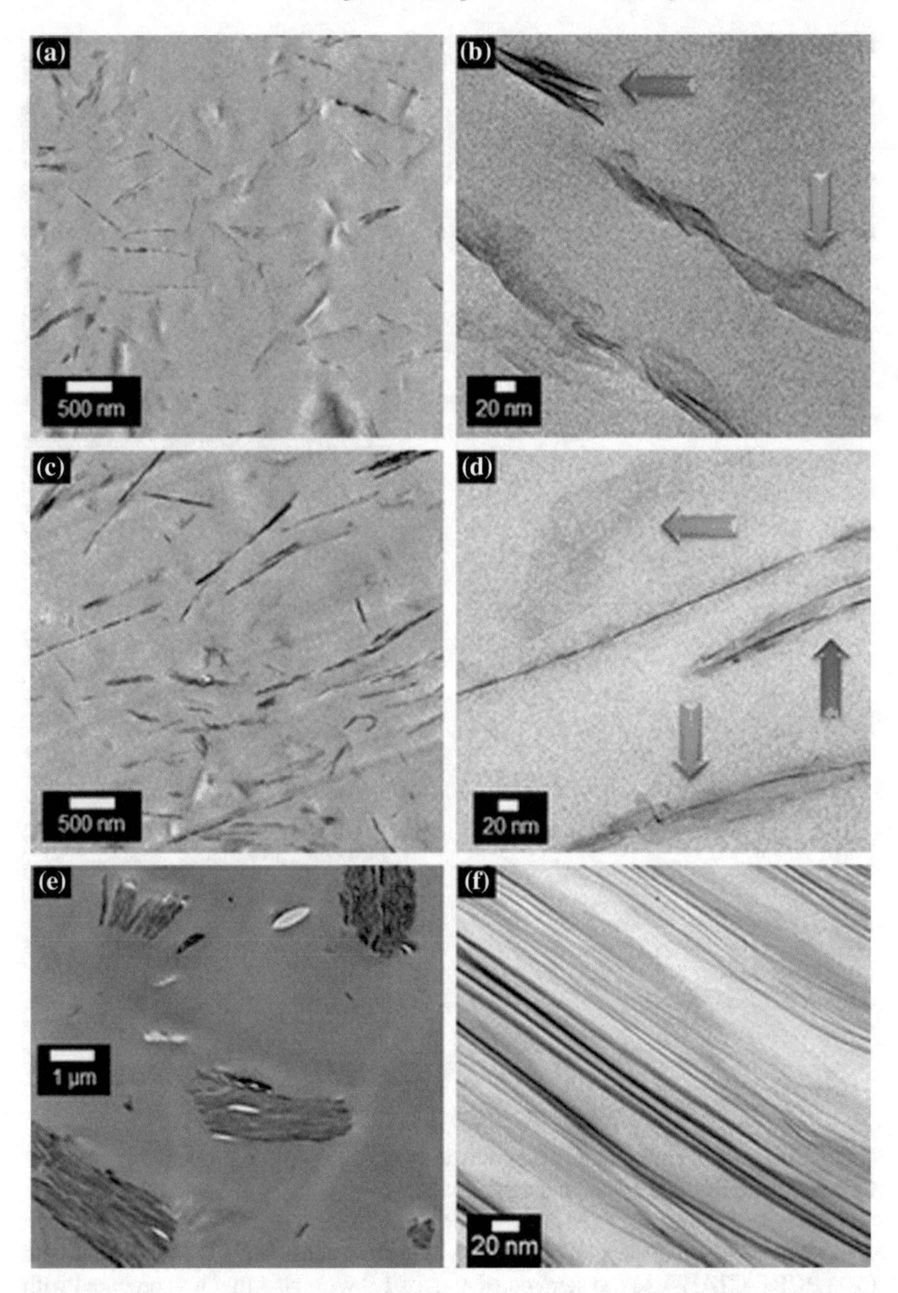

Fig. 3.12 TEM images of PURNCFs such as **a**, **b** PURF-MB system consisting of 2.2 wt% clay, **c**, **d** PURF-MB system consisting of 4.4 wt% clay, and **e**, **f** PURF-CTAB-VMT system consisting of 1.9 wt% clay that was prepared with direct mixing method. Reproduced with permission from Ref. [6]. Copyright 2013 American Chemical Society

diffusion of polyol into the interlayers of organoclay and high shear mixing led to the formation of both intercalation and exfoliation in PURNCFs [6].

Thus, it was also reported that by firstly dispersing CTAB-VMT in polyol and then mixing this solution with MDI led to the homogeneous dispersion of clay nanolayers due to better intermolecular interactions between CTAB-VMT and polyol in comparison with using only MDI as the dispersion medium. Thus, better intermolecular interactions between CTAB-VMT and polyol were shown to be the major driving force for the observation of both intercalation and exfoliation regions in the TEM micrographs [6].

On the contrary, in Fig. 3.12e, f, aggregates of several clay layers and no exfoliation mechanism were observed in PURF-CTAB-VMT system consisting of 1.9 wt% clay which was prepared by the direct mixing method. According to these results, it was also pointed out that it was really hard to break apart the smaller sized clay aggregates in PURNCFs prepared by using the direct mixing method. Furthermore, these systems consisted of aggregates of several clay layers that were distributed within the PURNCF matrix [6]. Thus, based on these results, it is well-noted that using suitable PURF components such as polyols or isocyanates with appropriate physicochemical properties that are compatible with surface functional groups of clays is the major driving force for the production of advanced PURNCFs with exfoliated morphologies. Consequently, this specific property or behavior improved physical and mechanical properties.

Patro et al. fabricated PURNCFs consisted of unmodified VMT clay by distribution of nanoclay fillers either in polyol or isocyanate component before the blending process [10]. In accordance with experimental data, it was shown that the incorporation of 5 pphp clay increases the viscosity of polyol and decreases the gel and rise times during the reactive foaming process. The reasons behind these observations were explained such that nanoclay functioned as a diverse kind of catalyst for polymerization and foaming reactions during the free rising of foams [10]. According to XRD and TEM experimental data, it was confirmed that nanoclay is partially exfoliated throughout PURNCF matrix. In addition, it was found out that the presence of clay in PURNCFs leads to the generation of gas bubbles which also drive the formation of smaller cells with almost identical sizes in final PURNCFs. Moreover, based on SEM results, it was shown that the closed cell contents in PURNCFs increase slightly with increasing amount of nanoclay. Based on the mechanical testing results, it was reported that the best mechanical properties were obtained when 2.3 wt% of nanoclay was dispersed in only isocyanate component before the reactive foaming process. Based on thermal conductivity results, it was revealed that the thermal conductivity of PURNCFs consisted of nanoclay was observed to be 10% lower than that of pure PURF [10].

Figure 3.13 displays environmental scanning electron microscopy (ESEM) micrographs revealing the closed-cellular morphology of (a) pure PURF and PURNCFs consisting of different amounts of VMT such as (b) 3, (c) 5 and (d) 8 pphp. Based on ESEM results, it is clearly noticed that cells become smaller and more uniform with incorporation of VMT. The reason behind the cell size reduction in PURNCFs with VMT addition was explained such that the clay addition leads to gas-bubble

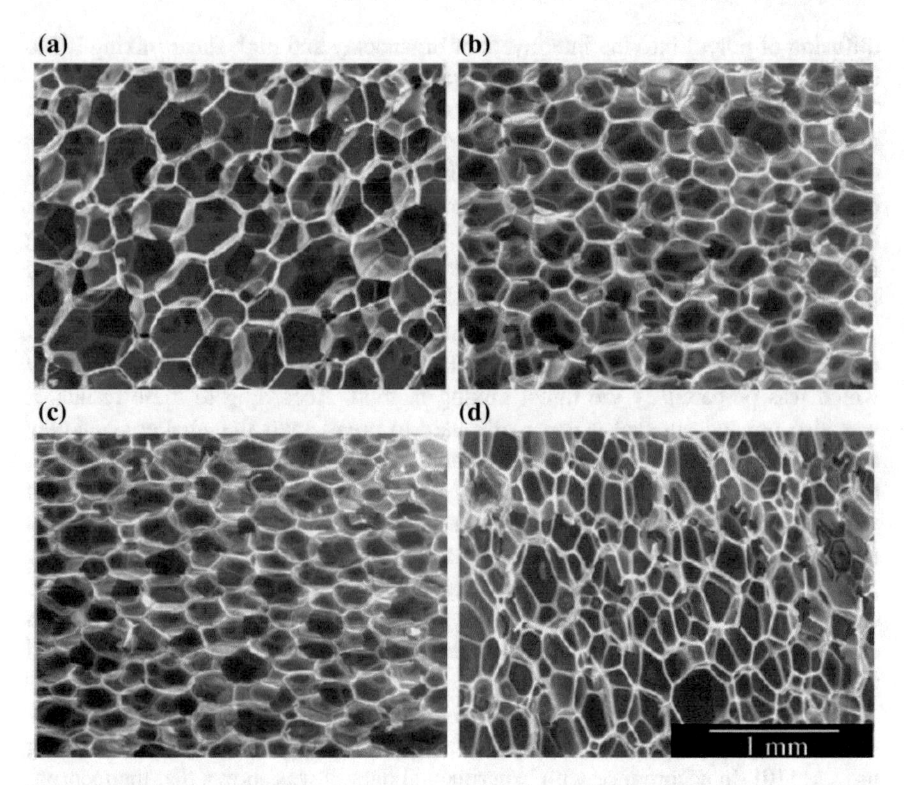

Fig. 3.13 ESEM micrographs showing the closed-cellular structure of **a** pure PURF and PURNCFs consisting of different amounts of VMT. Reproduced with permission from Ref. [10]. Copyright 2008 Society of Plastics Engineers

nucleations during the reactive foaming process, thus producing higher number of cells with smaller sizes per unit volume and uniform shapes in PURNCFs [10]. Also, based on SEM cell size measurements, it was pointed out that closed cells in PURNCFs increased slightly with inclusion of VMT. As a consequence, based on this finding, it was concluded that unlike MMT, the breaking up of cells and also increasing the cell-opening process were not observed with the addition of VMT to PURFs [10, 32].

Lorenzetti et al. examined the influences of using graphene on the improvement of thermal insulation properties of PURFs by preparing and characterizing pure PURF and PURNCFs consisting of 0.3 and 0.5 wt% graphene [11]. In accordance with on experimental data, radiative component related to thermal conductivity at the beginning diminished with incorporation of 0.3 wt% graphene via mechanisms of cell size reduction and elevation of extinction coefficient. Based on thermal conductivity measurements as a function of time, it was also revealed that slower aging rates were observed in graphene-based PURNCFs compared to that of pure PURF [11].

Figure 3.14 shows SEM micrographs at low magnifications of (a) pure PURF and PURNCFs consisting of (b) 0.3 and (c) 0.5 wt% graphene. Based on SEM

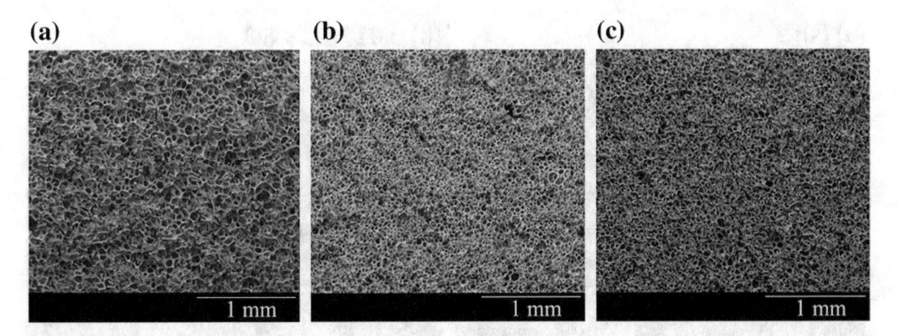

Fig. 3.14 SEM images at low magnifications of **a** pure PURF and PURNCFs consisting of **b** 0.3 and **c** 0.5 wt% graphene. Reproduced with permission from Ref. [11]. Copyright 2015 John Wiley & Sons, Ltd.

results, it was revealed that cell sizes in pure PURF, PURCNFs consisting of 0.3 and 0.5 wt% graphene nanoparticles were found to be 210, 152 and 220 μm, respectively [11]. Based on these results, the observation of cell sizes was explained such that in PURNCF consisting of 0.3 wt% graphene, a smaller cell size was observed in comparison with pure foam due to the nucleating agent behavior of graphene at this small concentration. However, it was also discussed that at higher concentrations of graphene up to 0.5 wt%, the increase of viscosity of the reaction medium leads to larger aggregates of graphene and this mechanism hindered the well dispersion of graphene nanoparticles within PURNCFs and led to the formation of larger cell sizes and consequently the reduction of physical and mechanical properties. Also, based on extra high magnification ESEM images, authors proved that no large agglomerates were present in PURNCF consisting of 0.3 wt% graphene, but conversely, PURNCF filled with 0.5 wt% graphene revealed the existence of a few big agglomerates [11].

Santiago-Calvo et al. produced water blown PURFs by using in situ polymerization method via functionalization of polyols with graphene oxide (GO) [12]. Authors investigated the effects of using GO functionalized with polyol on the closed-cellular morphology, foaming kinetics, thermal conductivity and compressive mechanical properties of PURFs. Based on experimental results, it was shown that about 33% cell reduction occurred with the additions of small amounts (0.017, 0.033 and 0.088 wt%) of GO into the PURF system [12]. In addition, it was revealed that thermal conductivity of nanocomposite foams reduced with incorporation of GO. In contrast, based on mechanical testing data, it was pointed out that inclusion of GO particles basically reduced mechanical properties of PURNCFs. Based on infrared expandometry, FT-IR spectroscopy and reaction temperature measurements, it was shown that the presence of GO particles leads to the formation of high amounts of urea groups during reactive foaming process, and thus because of this, the final nanocomposite foams showed higher expansion parameters [12].

Figure 3.15 shows SEM micrographs revealing the foam growth planes of (a) pure PURF and PURNCFs filled with (b) 0.017, (c) 0.033 and (d) 0.083 wt% GO. Based on SEM results, it was shown that the cell size was reduced up to 33% in

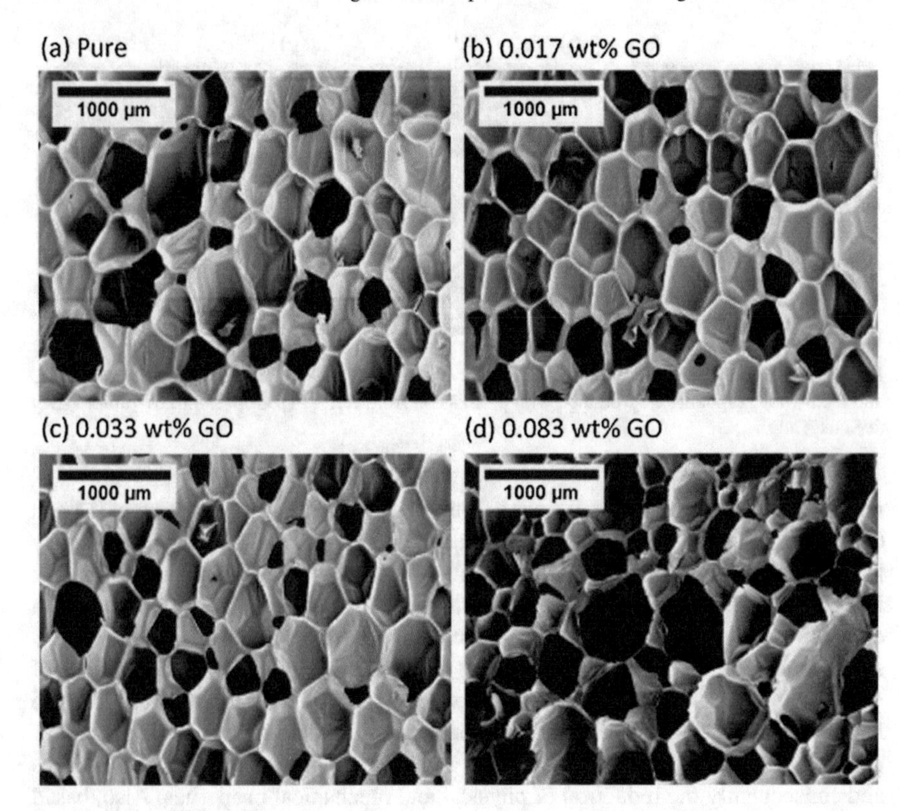

Fig. 3.15 SEM micrographs revealing the foam growth planes of **a** pure PURF and PURNCFs filled with diverse GO contents. Reproduced with permission from Ref. [12]. Copyright 2017 Elsevier Ltd.

PURNCF consisting of 0.083 wt% GO, and this behavior proved that GO is a very effective cell nucleating agent during the reactive foaming process [12]. In addition, it was also revealed that PURNCFs consisting of lower amounts of GO (equal or below 0.033 wt%) exhibited cell anisotropy compared to pure PURF. However, the anisotropy ratio of cells was observed to be much more reduced for PURNCFs consisting of higher concentrations of GO (0.083 wt% GO) and the reduction of cell anisotropy was attributed to higher contents of open cells due to increasing amount of GO in PURNCFs [12].

PURFs with advanced thermal insulation properties both in very cold temperatures and highly corrosive environments have made possible the use of these fascinating materials in applications such as liquefied natural gas (LNG) carrier cargo systems. Due to the harsh conditions that were experienced by these systems, the mechanical and thermal insulation properties of PURFs that are specifically used in these applications should be improved in order to meet the demands coming from this particular industry.

Recently, Kim et al. fabricated nanoparticle enhanced PURFs consisting of different amounts of graphite and nano-sized graphene oxide particles to solve the industrial problems that were faced by cargo systems in which liquefied natural gas is transported [13]. Authors used SEM analysis to comprehend the effects of using nanoparticles on the closed-cellular morphology of developed PURNCFs. In addition, the mechanical properties of PURNCFs were investigated at both room and at 110 K which correspond to boiling temperatures of LNG. Furthermore, based on morphology, thermal conductivity and macroscopic failure tests, it was found out that the morphology and mechanical strength of PURNCFs were improved and highly dependent on weight percent of nano-graphene oxide particles compared to pure PURFs [13].

Figure 3.16 displays SEM micrographs of PURNCFs consisting of different GO nanoparticle contents. Based on SEM results, it was shown that for instance in pure PURF, the closed-cell morphology and non-uniform cell sizes were typically observed without the presence of any GO nanoparticles [13]. In addition, it was also reported that cells were basically damaged with the addition of GO more than 0.05 wt%. According to SEM results in Fig. 3.16, it was pointed out that the highest amount of damaged closed cells was observed in PURNCF consisting of 0.4 wt% GO nanoparticles. Based on this observation, it can be said that increasing the amount of GO content from 0.05 to 0.4 wt% (almost eight times) leads to

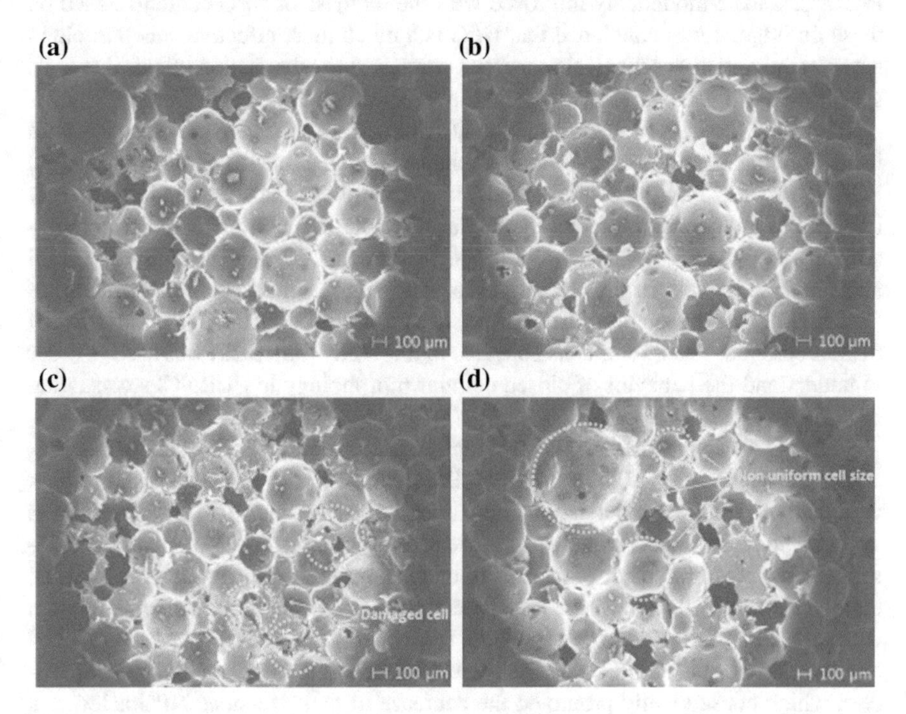

Fig. 3.16 SEM images of PURNCFs consisting of different GO nanoparticle contents. Reproduced with permission from Ref. [13]. Copyright 2017 Elsevier Ltd.

GO aggregation in PURNCFs thus causing the decrease of intermolecular interaction between nanoparticles and polymer chains. This outcome basically disturbs the closed-cellular morphology in such a way that closed cells become more non-uniform and their shapes are destructured more compared to PURNCFs in which very small amounts of GO are used during the reactive foaming process. Moreover, if the GO content in PURNCFs reached to 0.4 and 0.7 wt%, it was revealed that in the PURNCF consisting of 0.4 wt% GO, the cell size became the most non-uniform but the closed cellular structure was better compared to that containing 0.7 wt% GO. Thus, based on this finding, in agreement with other studies [2–5, 7, 9, 20, 49, 50], it was concluded that nucleation in PURNCFs is affected by the amount and distribution of GO which acts in the system as the nucleation agent [13].

Hoseinabadi et al. fabricated PURNCFs consisting of nanoporous graphene (NPG), and they investigated the density, mechanical, morphological, and thermal-resistant characteristics in relation to NPG amount [35]. In their study, authors basically synthesized the polyols for the fabrication of PURNCFs and at the same time, they varied the NPG content from 0.1 to 0.5 wt% in PURNCFs. SEM experiments were performed in order to examine NPG distribution and cell size characteristics in PURNCFs. According to experimental data, it was pointed out that inclusion of only 0.25 wt% NPG enhanced compressive strength and modulus values of PURNCFs [35]. In accordance with TGA data, it was shown that degradation temperatures of PURNCFs were moderately improved with the increase of NPG content. Based on these findings, it was concluded that NPG is a much more effective nanoparticle in terms of enhancing mechanical properties compared to other nanoparticles due to its special two-dimensional geometry and higher specific surface area [35].

In most of the PURNCF works in the literature, the ultimate goal of each research is to find the best structure-property relationship in order to develop advanced PURNCFs with improved properties. Thus, for this reason, morphology is very important to understand the closed cellular structure and establish the correct relationships between morphology and properties. In this case, SEM is a very useful and suitable characterization technique for studying the closed cellular structure of PURNCFs since the length-scale of these structures are within the operation limits of SEM technique. Thus, for this reason, also in this study [35], SEM was used in order to understand the behavior of closed cellular morphology in PURNCFs consisting of different amounts of GNPs.

Figure 3.17 shows the SEM micrographs of (a) pure PURF and PURNCFs consisting of (b) 0.1 and (c) 0.25 wt% NPG. Based on SEM experimental results which were obtained at the same magnification along the free-rising foam direction, it was shown that the cells were found to be in the form of spherical and polyhedral geometries [35]. According to cell size measurements based on SEM data, it was reported that the average cell size in pure PURF decreased from 334 to 232 μm and 180 μm with inclusion of 0.1 and 0.25 wt% NPG, respectively. As a result, it was concluded that the number of bubbles increased with increasing amounts of GNP nanoparticles, which consequently promoted the decrease of cell size since NPG acted as a nucleation agent during the reactive foaming process in PURNCFs [35].

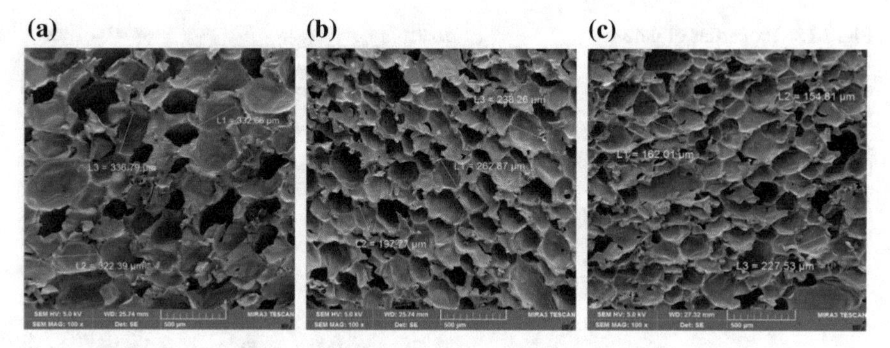

Fig. 3.17 SEM micrographs of **a** pure PURF and PURNCFs consisting of **b** 0.1 and **c** 0.25 wt% NPG. Reproduced with permission from Ref. [35]. Copyright 2017 Wiley Periodicals, Inc.

3.3 Mechanical Properties

Montmorillonite (MMT) based PURNCFs were synthesized by using the batch process and the MMTs that were used in PU rigid nanocomposite foams were surface modified with organic functional groups via in-situ polymerization [2]. Based on morphological results, it was found out that physical and mechanical properties of PURNCFs consisting of organoclays extensively depend on organic functional groups on the surface of organoclays, various important steps of synthesis methods, and molecular weights of polyols that were used as the major components for the reactive foaming process [2]. Thus, the nature of chemical reactions and physical interactions between system components such as polyols, polyisocyanates, organic functional groups of organoclay, catalysts, blowing agents, etc. affect the overall performance of PURNCFs such as physical and mechanical properties.

Figure 3.18 shows the properties of PURNCF consisting of 5 wt% MMT-OH and MMT-Tin (a) reduced compressive modulus and strength (b) glass transition temperature. According to mechanical properties data in Fig. 3.18a, PURNCFs consisting of MMT-OH and MMT-Tin exhibited considerably enhanced reduced compressive strength and modulus values in comparison with that of pure PURF [2]. Specifically, it was revealed that in comparison with pure PURF, the reduced strength and modulus of PURNCF consisting of MMT-OH increased about 650 and 610%, respectively. On the other hand, it was also shown that the reduced strength and modulus of PURNCF consisting of MMT-Tin increased about 610 and 760%, respectively compared to pure PURF. Based on glass transition temperature results of PURNCFs in Fig. 3.18b, it was also reported that PURNCF consisting of MMT-Tin has the maximum T_g which is 6 and 2 °C higher compared to pure PURF and PURNCF consisting of MMT-OH, respectively [2]. The reason behind the increase of T_g in PURNCF with MMT-Tin was explained such that homogenous dispersion of MMT within the polymer matrix leads to mobility reduction of PU chains due to the barrier effect imposed by MMT nanolayers. However, this effect was much more pronounced in PURNCF consisting

Fig. 3.18 Properties of pure PURF and PURNCFs consisting of 5 wt% MMT-Tin and MMT-OH. Reproduced with permission from Ref. [2]. Copyright 2004 Elsevier Ltd.

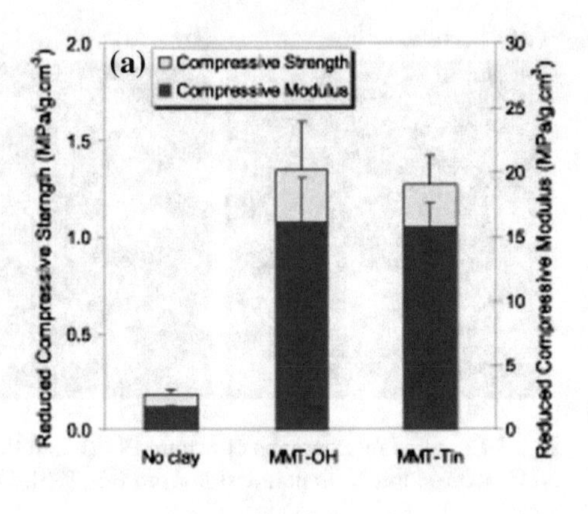

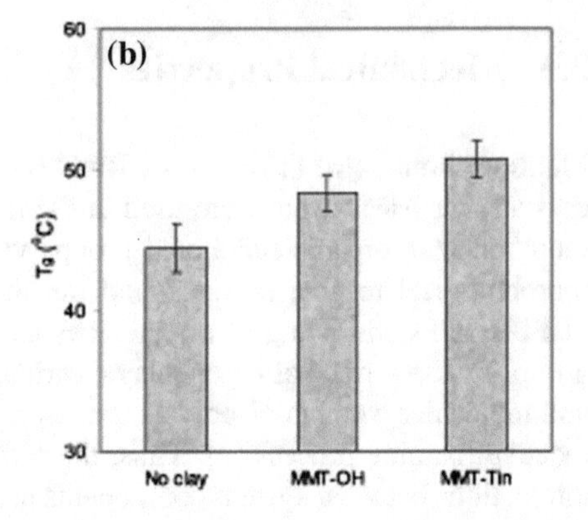

of MMT-Tin since organotin with catalytic property could increase the number of favorable intra-gallery reactions during the reactive foaming process [2].

PURNCFs containing nanoclay which was modified by polymeric diphenylmethane diisocyanate (PMDI) were fabricated by using the ultrasound technique [20]. Based on mechanical test results, it was shown that the maximum values in terms of tensile and flexural strengths of PURNCFs were obtained with the incorporation of 3 wt% nanoclay based on PMDI amount [20]. Thus, the reason behind increase of tensile and flexural strengths of PURNCFs was explained such that incorporation of ultrasound during surface modification helped to disperse clay nanolayers homogenously within PURNCF matrix and increased mechanical properties accordingly due to better interfacial bonding between PMDI functional groups of nanoclay and PU chains of PURNCF matrix. Moreover, it was also reported that the fire resistance properties of PURNCFs increased with the integration of ultrasound method in comparison with PURNCFs in which no ultrasound was applied during the surface treatment of clay nanolayers [20].

Figure 3.19 shows the effect of using different amounts of modified clay on the tensile strength of PURNCFs which were prepared with and without application of ultrasound. Based on results, it was shown that when ultrasound was applied, the

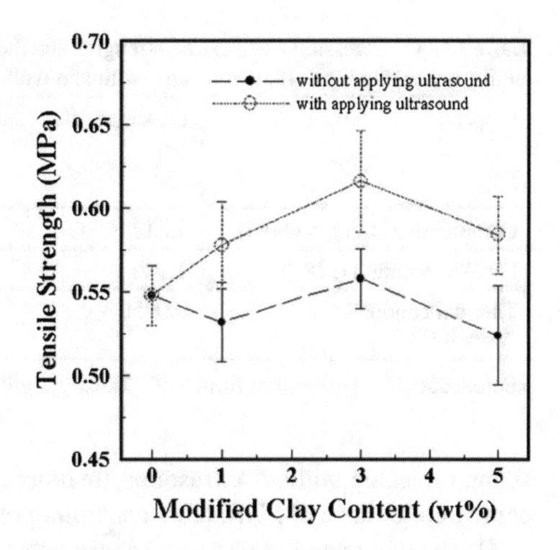

Fig. 3.19 Effect of the modified clay content on the tensile strength of PURNCFs which were prepared with and without application of ultrasound. Reproduced with permission from Ref. [20]. Copyright 2006 Wiley Periodicals, Inc.

maximum tensile strength of PURNCFs was observed as 0.62 MPa with incorporation of 3.0 wt% modified clay [20]. However, on the other hand, it is apparently noticed from Fig. 3.19 that tensile strengths of PURNCFs consisting of modified clays that were not treated with the ultrasound method displayed much lower tensile strength values compared to those which were treated with ultrasound. According to these results, it was revealed that tensile strength enhancement of PURNCFs consisting of modified clays with ultrasound was explained due to homogenous distribution of clay nanoparticles via ultrasound during sample preparation [20]. Generally, it is very well-known that the application of ultrasound is one of the most preferred and easiest method in order to break apart the aggregates of nanoparticles such as nanoclay, nanosilica and carbon nanotubes, etc. and to form a homogenous dispersion of nanoparticles during the sample preparation of polymer nanocomposites from the solution state.

Previously, PURNCFs consisting of organoclay particles with improved thermal insulation properties were synthesized. In their PURNCFs, authors used organoclay as the plate-like nanofiller which was surface modified with polymeric diphenylmethane diisocyanate (PMDI) by using a silane coupling agent [7]. Based on mechanical testing results of PURNCFs, it was shown that there is no big difference in terms of compressive and flexural strengths between PURNCFs consisting of organoclay layers that were prepared with and without silane coupling agent. In conclusion, authors interpreted these findings such that using silane coupling agents during surface modification of organoclay led to the improved exfoliation of organoclay nanolayers within the PURNCF matrix, and thus decreased the values of thermal conductivity and cell size in comparison with the system that did not contain any silane coupling agent [7].

Table 3.1 shows the mechanical properties such as compression and flexural strengths and thermal conductivity values of PURNCFs consisting of organoclay and silane molecule with and without ultrasound treatment. According to experimental data, mechanical properties of PURNCFs containing both organoclay and

Table 3.1 Compression and flexural strengths and thermal conductivity values of PURNCFs containing organoclay and silane molecule with and without ultrasound treatment

Property	PU/organoclay foam	PU/silane coupling agent/organoclay foam	
		Without ultrasound	With ultrasound
Compressive strength (MPa)	1.512	1.522	1.523
Flexural strength (MPa)	1.980	2.041	2.256
Thermal conductivity (W/m h °C)	0.0250	0.0234	0.0230

silane molecule without ultrasound treatment were observed to moderately higher compared to those of PURNCFs containing only organoclay [7].

On the other hand, it was also reported that compressive and flexural strengths of PURNCFs containing both organoclay and silane molecule with ultrasound treatment were observed to be higher compared to those of PURNCFs containing organoclay and silane coupling agent without ultrasound treatment. Thus, in accordance with experimental data, it was suggested that silane molecule is typically responsible for the exfoliation of organoclay nanolayers within the PURNCF matrix [7]. Thus, based on these results, PURNCFs containing both organoclay and silane molecule which were fabricated with ultrasound treatment gave the best results in terms of higher compressive and flexural strengths and lower thermal conductivity value. In conclusion, it is well-noted that using only organoclay in PURFs is not enough for obtaining improved mechanical properties. Besides the presence of organoclay, a surfactant with compatible functional groups should be used to create favorable interactions between organoclay and PU chains. At the same time, the aggregation behavior of organoclay could be a serious problem although nanoclay has compatible surface functional groups and the PURF system contains an additional component such as a surfactant. Thus, in this case, the aggregation behaviour of organoclay should be minimized by applying ultrasound treatment during sample preparation to maximize the number of favorable interactions between PU chains and organoclay nanolayers [7].

Montmorillonite based organoclay were added into PURFs by ultrasonically dispersing clay nanolayers in the isocyanate component [3]. Authors used small angle X-ray scattering (SAXS), SEM, compression testing, gas chromatography (GC) and rheology to understand the degree of clay dispersion, closed-cellular morphology and mechanical properties of PURNCF system consisting of organoclay. Furthermore, based on mechanical testing results, no improvement was noticed in mechanical strength of foams with incorporation of 1 wt% clay to the system. Moreover, it was pointed out that higher levels of viscosity and yield stress were obtained with the addition of very high amounts of nanoclay to the PURNCF system [3].

Figure 3.20 depicts effects of using nanoclay on compression (a) strength and (b) modulus of pure PURF and PURNCFs consisting of different foam index values.

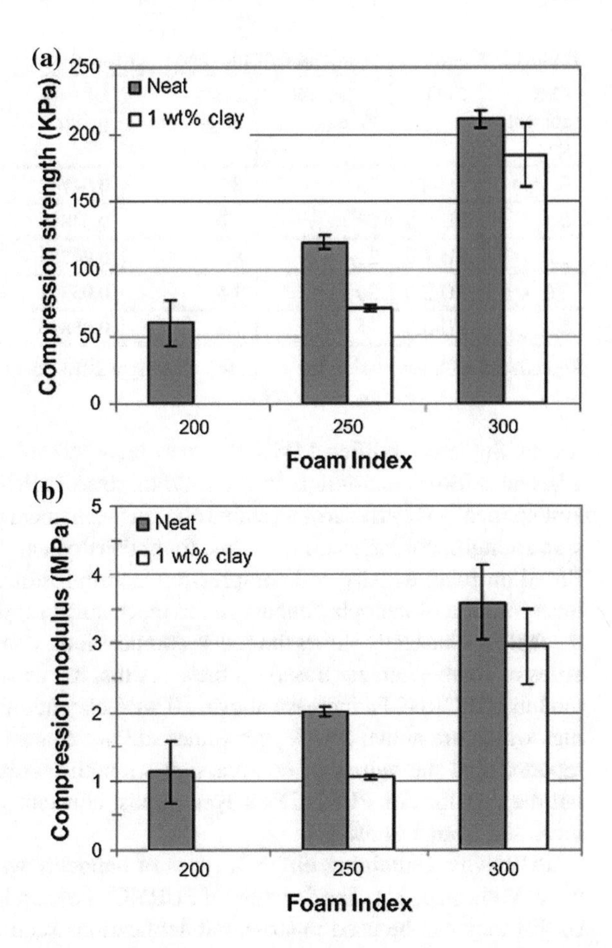

Fig. 3.20 Effect of using nanoclay on compression mechanical properties for pure PURF and PURNCFs consisting of various foam index values. Reproduced with permission from Ref. [3]. Copyright 2005 Taylor & Francis LLC

Based on the mechanical property testing results, it was shown that PURFs consisting of higher index values have much higher values of mechanical properties due to generation of more hard segments compared to lower index foams [3]. However, on the other hand, it was that the strength and modulus for the 250-index foams decreased with incorporation of nanoclay. The mechanical strength reduction with inclusion of nanoclay was explained such that the presence of clay might affect the formation of isocyanurate hard segments negatively, thus leading to a weaker PURNCF. On the contrary, it is apparently noticed from Fig. 3.20 that in the case of PURNCFs having 300-index, with inclusion of 1 wt% clay, the reduction of compression mechanical properties compared to that of pure PURF was not that high as in 250-index system [3].

PU rigid nanocomposite foams consisting of nanoclay were fabricated by substituting some of synthetic polyol with a type of castor oil [4]. Basically, authors produced diethanol amide from hydroxylated castor oil by using the transamidation process, and the synthesized polyol was used in the fabrication of water-blown PU

Table 3.2 Codes, compositions, densities and mechanical properties of foam samples

Clay content (%)	Code	Foam rise time (s)	Foam height (cm)	Density (g/cc)	Compression strength (kPa)	Compression modulus
0	PU-0	33	8.1	0.04742	52	1.50
0.5	PU-0.5	23	8.0	0.04818	79	1.52
1.0	PU-1.0	21	8.1	0.04785	93	1.60
2.0	PU-2.0	20	7.5	0.05558	110	1.65
5.0	PU-5.0	15	7.2	0.05896	113	1.66

foams. Surface-modified MMT clay was incorporated in the amounts of 0.5, 1.0, 2.0, and 5.0% as nanofillers into the synthesized PURNCFs. Authors investigated systematically the effects of modified clay on mechanical properties such as compression strength, compression modulus, foam density, and morphology of PURNCFs. Based on foam density and compression testing results, it was confirmed that the incorporation of nanoclay enhanced the mechanical properties of PURNCFs [4].

Table 3.2 basically shows the codes, compositions, densities and mechanical properties of foam samples. Based on these results, it was shown that the compression moduli of PURNCFs increase above 1.0 wt% clay addition but at higher clay loadings which are above 2 wt%, the values did not change that much [4]. It was also reported that the values of compression strength increased from 52 to 113 MPa, but the densities of PURNCFs stayed almost constant when the nanoclay content increased from 0 to 5% [4].

PURNCFs containing different types of nanoclay with different concentrations were synthesized [1]. The densities of PURNCFs were adjusted to be 140–160 kg/m^3 so that they can be used in structural applications such as construction and underwater buoyancy. Authors used wide-angle XRD and TEM in order to confirm the nanocomposite structures of PURNCFs. Based on mechanical testing results, it was confirmed that inclusion of nanoclay fillers to PURF system enhanced properties such as storage and compressive modulus but reduced mean cell size in PURNCFs. In contrast, it was pointed out that the hydraulic resistance of PURNCFs was measured to be much lower compared to that of pure PURF [1].

Table 3.3 displays compressive modulus values of pure PURF and PURNCFs containing different types of nanoclays with different concentrations along parallel and perpendicular to foam growth direction. Based on these results, it was shown that nanoclay incorporation clearly increases modulus parallel to foam rise direction, but it does not change the value of modulus perpendicular to foam growth direction [1]. The reason behind this finding was explained such that elongation of PURNCFs during mechanical testing causes the orientation of cell struts along the foam rise direction, thus leading to larger compressive strength value parallel to foam growth direction. In accordance with experimental data, it is also clearly observed that compressive

Table 3.3 Compressive modulus of pure PURF and PURNCFs containing different types of nanoclays with different concentrations along directions parallel and perpendicular to foam rise

Foam	PL (MPa)	PR (MPa)
Without clay	17.71 ± 0.61	15.28 ± 0.53
Na^+_1%	18.36 ± 0.85	17.19 ± 0.71
Na^+_4%	19.39 ± 0.62	16.22 ± 0.69
30B_1%	20.03 ± 0.79	16.67 ± 0.40
30B_4%	21.79 ± 0.85	14.82 ± 0.52

Reproduced with permission from Ref. [1]. Copyright 2006 Wiley Periodicals, Inc.

modulus values of PURNCFs consisting of Na^+ clay are almost close to the values that are observed in PURNCFs consisting of 30B clay [1].

PURNCFs containing unmodified VMT clay were synthesized by distribution of nanoclay fillers either in polyol or isocyanate component before the blending process [10]. In accordance with experimental data, it was shown that the incorporation of 5 pphp clay increases the viscosity of polyol and decreases the gel and rise times during the reactive foaming process. The reasons behind these observations were explained such that nanoclay functioned as a diverse kind of catalyst for polymerization and foaming reactions during the free rising of foams [10]. Based on the mechanical testing results, it was reported that the best mechanical properties were obtained when 2.3 wt% of nanoclay that was dispersed in only isocyanate component before the reactive foaming process. Based on thermal conductivity results, it was revealed that the thermal conductivity of PURNCFs consisted of nanoclay was observed to be 10% lower than that of pure PURF [10].

Figure 3.21 shows the compressive stress vs. strain data of pure PURF and PURNCFs consisting of VMT with different concentrations. In Fig. 3.21, VMT was spread out in (a) isocyanate and (b) polyol components, respectively before reactive foaming event. According to results, it was shown that incorporation of nanoclay extensively enhances the compressive strength values. Moreover, it was also pointed out that the maximum of yield point turns out to be much more apparent with nanoclay inclusion, showing that foams become much more brittle with increasing amount of nanoclay [10].

Based on the numerical data from compression results, it was shown that PURNCFs prepared with nanoclay dispersion in the isocyanate component exhibited enhanced mechanical properties than those in which nanoclay was distributed in the polyol component [10]. Furthermore, authors showed that the foam containing 5 pphi VMT prepared in isocyanate showed much higher compressive strength and modulus values than PURNCFs consisting of 5 pphp VMT clay that was dispersed in the polyol component [10]. The reason behind this finding was explained such that this behavior occurred due to generation of primary bonding forces between hydroxyl functionalities of VMT and isocyanates of PURNCF system [2, 10].

It was also revealed that bulk density of PURNCFs increase a little with the addition of nanoclay. Based on bulk density values of PURNCFs, reduced compressive strength and modulus values were evaluated and based on this numerical data, it was

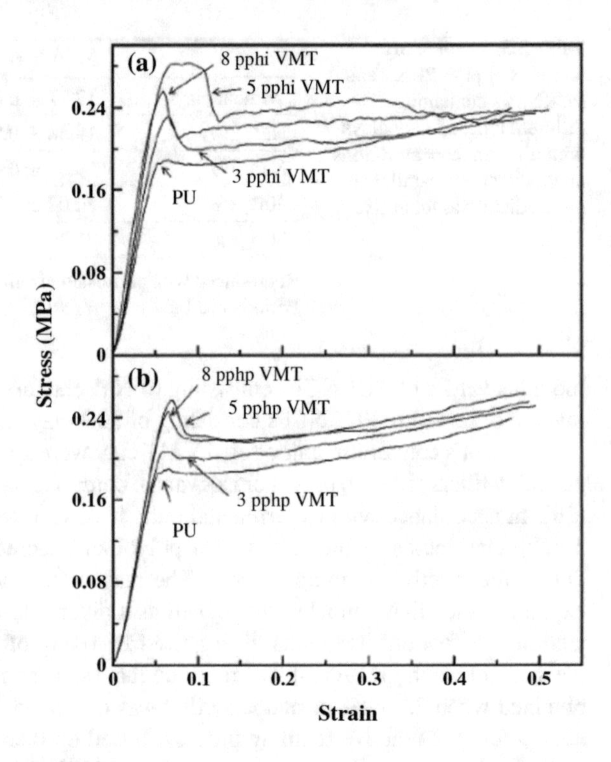

Fig. 3.21 Compressive mechanical properties of pure PURF and PURNCFs consisting of VMT with different concentrations. Reproduced with permission from Ref. [10]. Copyright 2008 Society of Plastics Engineers

revealed that reduced compressive strength and modulus values also increase with incorporation of nanoclay [10].

Thus, based on these findings, it is well-noted that the dispersion of nanoclay either in the polyol or isocyanate component determines the final properties of PURNCFs. Here, in PURNCFs consisting of VMT clay, the mechanical properties are directly affected by the preparation method [10]. Also, the type and surface functionality of nanoclay dictates whether the mechanical properties of final PURNCF product could be improved or not. Basically, physical intermolecular interactions or chemical bonds between surface functional groups of nanoclay and PURF components (polyol or isocyanate) determine if mechanical properties of PURNCFs would be improved or not.

Water blown PURFs were fabricated by utilizing in situ polymerization method via polyol functionalization with graphene oxide (GO) [12]. Authors investigated the effects of using GO functionalized with polyol on the closed-cellular morphology, foaming kinetics, thermal conductivity and compressive mechanical properties of PURFs. Moreover, it was found out that incorporation of GO reduced thermal conductivity values in PURNCFs. In contrast, based on mechanical testing data, it was pointed out that inclusion of GO particles basically reduced mechanical properties of PURNCFs. Based on infrared expandometry, FT-IR spectroscopy and reaction temperature measurements, it was shown that the presence of GO particles leads to the formation of high amounts of urea groups during reactive foaming process,

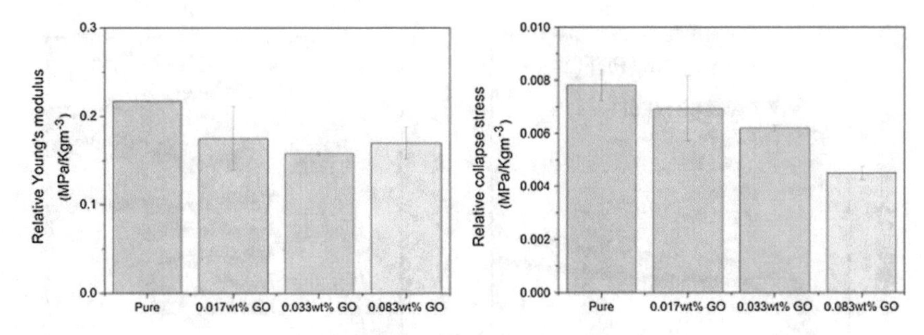

Fig. 3.22 Relative elastic modulus and collapse stress values of pure PURF and PURNCFs consisting of different GO contents. Reproduced with permission from Ref. [12]. Copyright 2017 Elsevier Ltd.

and thus because of this, the final nanocomposite foams showed higher expansion parameters [12].

Figure 3.22 shows relative elastic modulus and collapse stress values of pure PURF and PURNCFs consisting of different GO contents. According to results, it was revealed that increasing content of GO in PURNCFs basically decreases the mechanical properties and compressive test values of PURNCFs are much lower compared to those of pure PURF [12]. In addition, it was also pointed out that the reduction of mechanical properties in PURNCFs compared to pure PURF becomes much smaller at much lower amounts of GO particles, and the difference in terms of mechanical properties become much larger at very high GO contents [12]. The reason behind the large difference in mechanical properties between pure PURF and PURNCFs consisting of very high GO contents could be explained such that the closed cellular morphology at very high contents of GO (for instance 0.083 wt%) is disrupted or damaged due to the increase of viscosity and the aggregation behavior of GOs at these concentrations [12].

Nanoparticle enhanced PURNCFs consisting of different amounts of graphite and nano-sized graphene oxide particles were fabricated to solve industrial problems that were faced by cargo systems in which liquefied natural gas is transported [13]. In addition, the mechanical properties of PURNCFs were investigated at both room and at 110 K which correspond to boiling temperatures of LNG. Furthermore, based on morphology, thermal conductivity and macroscopic failure tests, it was found out that the morphology and mechanical strength of PURNCFs were improved and highly dependent on weight percent of nano-graphene oxide particles compared to pure PURFs [13].

Figure 3.23 shows the compression test results of pure PURF and PURNCFs consisting of GO with different concentrations at room and cryogenic temperatures. According to results, it was pointed out that PURNCFs with GO contents have much higher compressive strength values compared to pure PURF at room temperature, except the PURNCF sample consisting of 0.7 wt% GO [13]. In addition, it was also shown that the compressive strength of PURNCFs improved with incorporation of 0.05 wt% GO compared to that of pure PURF at room temperature. However, on

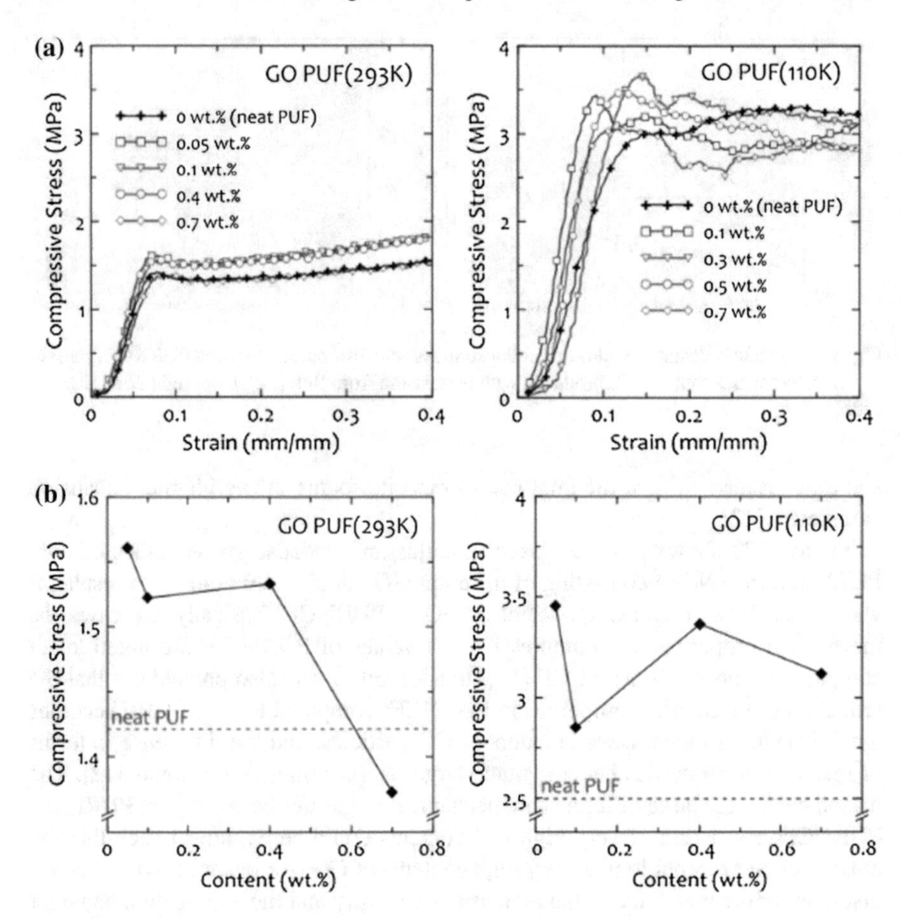

Fig. 3.23 Compression test results of pure PURF and PURNCFs consisting of GO with different concentrations at 293 and 110 K: **a** stress–strain curves and **b** compressive stress data versus various amounts of GO. Reproduced with permission from Ref. [13]. Copyright 2017 Elsevier Ltd.

the other hand, the compressive strength of PURNCF diminished with inclusion of 0.7 wt% GO compared to that of pure PURF at room temperature. In contrast, according to data in Fig. 3.21b, compressive stress values of all PURNCFs consisting of different amounts of GO are much higher compared to that of pure PURF at 110 K [13].

These results showed that mechanical properties of PURNCFs did not increase directly proportional with the amount of GO nanoparticles [13]. Based on results at 110 K, it was also shown that PURNCFs consisting of different amounts of GO nanoparticles did not exhibit any yield point before reaching strain levels of 0.1 [13]. However, on the contrary, it is apparently understood from the data in Fig. 3.23 that at room temperature, PURNCFs consisting of different amounts of GO nanoparticles exhibited yield points before reaching strain levels of 0.1 [13].

PURNCFs containing nanoporous graphene (NPG) were fabricated, and authors investigated foam density, closed-cellular morphology, mechanical, morphological, and thermal decomposition properties as a function of NPG content [35]. In their study, authors basically synthesized the polyols for the fabrication of PURNCFs and at the same time, they varied the NPG content from 0.1 to 0.5 wt% in PURNCFs. According to experimental data, it was pointed out that inclusion of only 0.25 wt% NPG enhanced compressive strength and modulus values of PURNCFs [35]. Based on these findings, it was concluded that NPG is a much more effective nanoparticle in terms of enhancing mechanical properties compared to other nanoparticles due to its special two-dimensional geometry and higher specific surface area [35].

Figure 3.24 shows compressive (a) strength and (b) modulus of pure PURF and PURNCFs consisting of nanoporous graphene with different contents. According to results, it was shown that incorporation of nanoporous graphene improves reasonably compressive strength and modulus values of PURNCFs [35]. As shown in Fig. 3.24a, b, it was found out that compressive strength and modulus values of PURNCFs were enhanced quite effectively with inclusion of only 0.25 wt% NPG. Thus, this result showed that NPG is a very effective nanofiller in terms of enhancing the mechanical properties compared to other nanofillers such as clay, nanosilica, CNT, etc. [2, 3, 35, 41, 51].

The reason behind the superb performance of NPG in relative to other nanofillers was discussed such that NPG has a higher specific surface area in comparison with nano-sized graphene and graphene oxide. Furthermore, NPG having larger specific surface area results in quite successful improvement of mechanical properties [35]. In addition, the combination of lower density with higher mechanical strength values also causes much higher values in terms of reduced compressive strength and modulus in NPGs in comparison with other classical nanofillers [2, 3, 35, 41, 51]. Also, PURNCFs consisting of NPGs with much higher values of reduced compressive

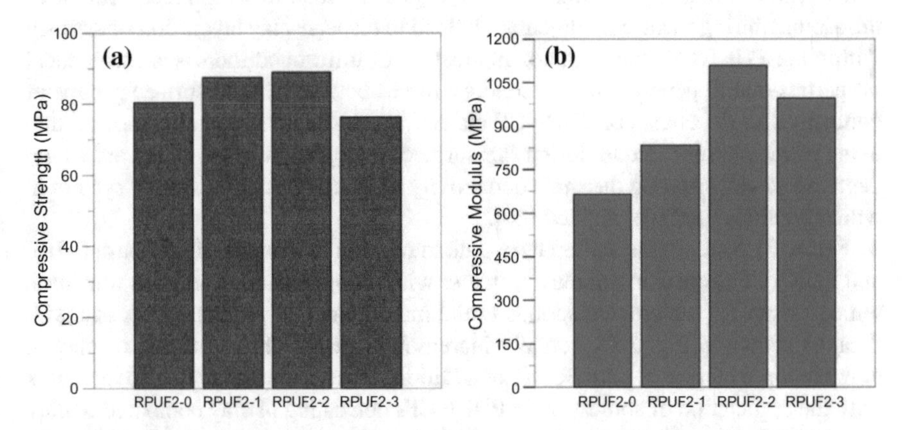

Fig. 3.24 Compressive **a** strength and **b** modulus of pure PURF and PURNCFs consisting of NPG with different contents. Reproduced with permission from Ref. [35]. Copyright 2017 Wiley Periodicals, Inc.

modulus and strength values do perform much more better in terms of mechanical properties relative to PURNCFs consisting of classical nanoparticles such as nanosilica, clay, CNT, graphene oxide, etc. with much weaker reduced compressive modulus and strength values. Based on these results, it is quite worthwhile to incorporate NPGs which are modified with various surface functional groups (small molecules, oligomers or even polymers) into PURNCFs so that the properties of final PURNCFs should be extensively improved compared to nanocomposite foams that are loaded with bare NPGs [35].

Based on the data in Fig. 3.24, it was also revealed that the compressive strength of PURNCFs slightly decreased with the addition of 0.5 wt% NPG [35]. The reason behind this reduction in compressive strength was explained such that in this particular sample, a plasticizing effect occurred due to the use of excess amount of surfactant in PURNCF consisting of 0.5 wt% NPG. Thus, based on this result, it was shown that inclusion of 0.5 wt% NPG diminished both compressive strength and modulus in PURNCFs [35].

3.4 Thermal and Thermomechanical Properties

PURNCFs consisting of clay which was modified by polymeric diphenylmethane diisocyanate (PMDI) were synthesized by using the method of ultrasound [20]. Based on TEM results, authors showed that the interlayer spacing between PMDI surface modified clays in PURNCFs was increased with the application of ultrasound. Based on mechanical test results, it was shown that the maximum values in terms of tensile and flexural strengths of PURNCFs were obtained with the incorporation of 3 wt% nanoclay based on PMDI amount [20]. Thus, the reason behind the increase of tensile and flexural strengths of PURNCFs was explained such that the incorporation of ultrasound during surface modification helped to disperse clay layers homogenously within the PURNCF matrix. Also, application of ultrasound increased mechanical properties due to generation of secondary forces between PMDI surface groups of nanoclay and PU chains of PURNCF matrix. In addition, authors also showed that using ultrasound technique during the surface modification of nanoclay led to the decrease of cell size and thermal conductivity of PURNCFs compared to system in which no ultrasound was applied [20].

Figure 3.25 shows the shifted tan δ data in relation to temperature for pure PURF and PURNCFs consisting of clay with and without ultrasound. The glass transition temperature (T_g) which corresponds to the maximum value of tan δ peak in DMA data, as shown in Fig. 3.25, actually increases in PURNCFs consisting of clay in comparison with that of pure PURF. In addition, it was shown that T_g also increases with the application of sonication in PURNCFs consisting of clay compared to clay based PURNCF system in which no sonication was applied during the sample preparation. Basically, it was revealed that T_g values of PURNCFs containing nanoclay that were prepared with and without sonication were 7 and 5 °C higher compared to that of pure PURF, respectively [20].

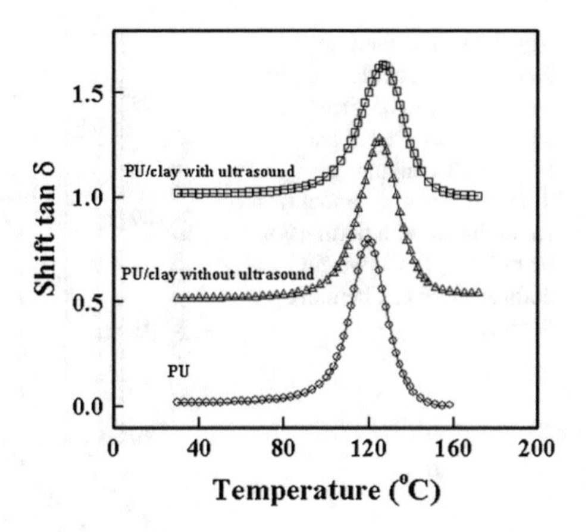

Fig. 3.25 Shifted tan δ versus temperature for pure PURF and PURNCFs consisting of clay with and without ultrasound treatment: (○) pure PURF; (Δ) PURNCF consisting of clay without ultrasound; (□) PURNCF consisting of clay with ultrasound. Reproduced with permission from Ref. [20]. Copyright 2006 Wiley Periodicals, Inc.

In addition to tan δ data with respect to temperature, the results for storage modulus with changing temperature were pointed out to be in very good consensus with the behavior that PURNCF system consisting of clay was highly mechanically reinforced with the application of sonication [20]. Furthermore, it was basically revealed that storage modulus values in PURNCFs were much higher compared to that of pure PURF. In addition, it was also pointed out that storage modulus values in PURNCFs that were prepared via sonication method was observed to be much higher than those that were prepared without sonication [20].

The reason behind the increase of storage modulus in PURNCFs consisting of nanoclay with the application of sonication can be explained such that the aggregation behavior of nanoclay basically decreased with sonication since sonication is a very effective and easier method in order to break apart the agglomerates of nanoparticles such as clay, CNT, nanosilica, graphene oxide, etc. Thus, the reduction of clay aggregation in the PURF system, homogenous dispersion of separated nanoclay layers with much higher surface areas in either polyols or isocyanates and uniform distribution of all components during reactive foaming process lead to elevation of suitable forces between nanoclay fillers and PURNCF chains. The higher number of intermolecular interactions between PURNCF chains and nanoclay fillers basically lowers down polymer movements due to exfoliated nature of clay particles with higher surface areas. Thus, this physical mechanism provides suitable conditions for the increase of T_g and storage modulus of PURNCFs that were prepared with sonication method since molecular motions of PU chains become more restricted and the polymer matrix is converted into a much more rigid matrix due to the presence of exfoliated nature of clay nanoparticles with higher surface areas.

Polyurethane rigid foams containing nanoclay dispersions were prepared, previously [8]. During sample preparation, authors used both pristine nanoclay and nanoclays containing various aspect ratios that were surface modified with organic functional groups. In addition, nanofillers were dispersed in polyol and polyisocyanate components, separately [8]. Authors characterized these nanodispersions by using X-ray scattering, SEM, cryo-electron microscopy, FT-IR, thermal

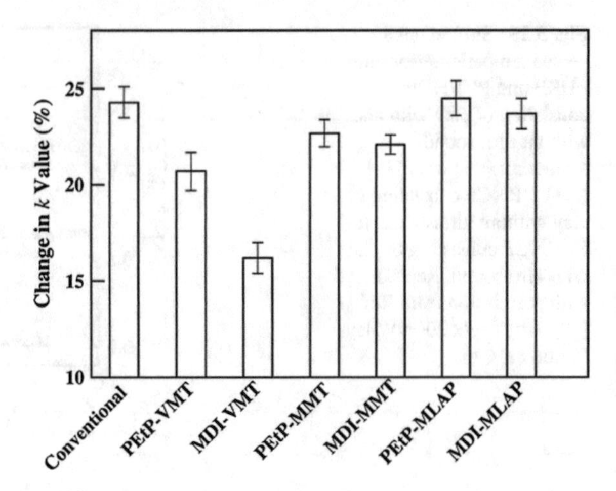

Fig. 3.26 The thermal conductivity coefficient data with respect to accelerated aging of pure PURF and PURNCFs containing different types of nanoclay. Reproduced with permission from Ref. [8]. Copyright 2009 American Chemical Society

conductivity tests and rheology. The effect of reactive foaming process on exfoliation was investigated by comparing the conditions of filler dispersions in components and in PURFs. Based on these experimental findings, it was pointed out that properties of PURNCFs are directly correlated with physicochemical and rheological conditions of nanodispersions [8].

Figure 3.26 shows thermal conductivity coefficient data with respect to accelerated aging of pure PURF and PURNCFs consisting of nanoclay [8]. According to results, it is clearly noticed that PURNCFs which were prepared from MDI-VMT blends display the least percentage change in terms of thermal conductivity coefficient. In addition, it was also revealed that barrier properties of PURNCFs that were prepared from dispersions of MMT in MDI component and VMT in polyol component perform more superior compared to other PURNCFs that were prepared from dispersions of MLAP in MDI component and MLAP in Polyol component. The reason behind the minimum percentage change of thermal conductivity coefficient of PURNCFs that were prepared from MDI-VMT dispersions was explained such that much better dispersion of VMT clay in isocyanates provides suitable conditions for the increase of surface area which basically slows down the diffusion of gases throughout the PURNCF matrix [8, 52]. In addition, the better dispersion of VMT in MDI leads to higher number of heterogenous nucleation sites during the reactive foaming process, and basically this mechanism causes the formation of smaller cell sizes in PURNCFs with lower thermal conductivity values [8, 10, 53].

PURNCFs filled with nanoclay were fabricated using a specific method in which clay nanolayers were intercalated and exfoliated with the help of neutralized dimethylol butanoic acid (DMBA) [9]. In accordance with experimental data, it was shown that values of foam density and compression strength decreased while gel, tack-free and cream time increased with increasing amounts of nanoclay into PURFs. On the other hand, it was also reported that the thermal conductivity, cell size, the amount of closed cells and the volume change as a function of temperature decreased with the incorporation of nanoclay into PURF system. Based on thermomechanical and

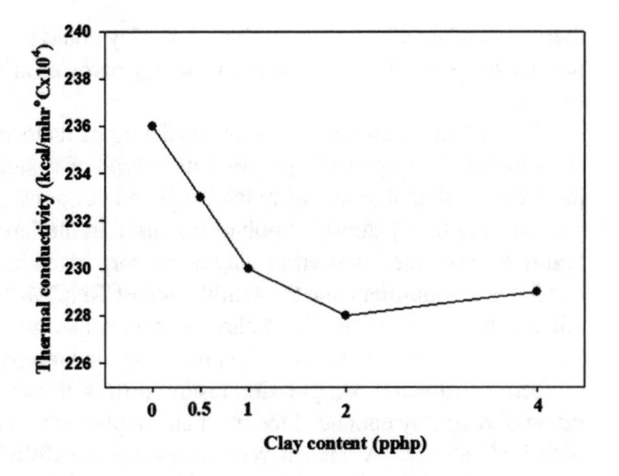

Fig. 3.27 Thermal conductivities of PURNCFs containing different amounts of nanoclay. Reproduced with permission from Ref. [9]. Copyright 2010 Wiley Periodicals, Inc.

thermal stability results, it was shown that nanoclay incorporation enhanced temperatures such as glass transition and thermal decomposition due to reduced mobility of PU chains by barrier effects of nanoclay platelets [9].

Figure 3.27 shows thermal conductivities of PURNCFs containing different amounts of nanoclay. According to results, it was revealed that thermal conductivity values of PURNCFs decrease systematically with increasing clay concentration up to incorporation of 1% clay since cell size was also found out to be decreasing up to the addition of 2 pphp clay [9]. However, on the other hand, as shown in Fig. 3.27, it was also reported that the thermal conductivity of PURNCFs increases with the addition of more than 2 pphp clay. The thermal conductivity dependence of PURNCFs on clay content was explained such that the thermal conductivity decreased with decreasing cell size up to 2 pphp clay, but above this clay concentration, thus thermal conductivity increased due to the increase of cell size in PURNCFs [9]. Thus, based on these results, it is concluded that cell size is the major driving mechanism that determines the value of thermal conductivity in PURFs or PURNCFs consisting of nanoparticles. In agreement with other previously published works [3, 8, 10, 52, 53], there are two fundamental ways of reducing the thermal conductivity coefficient in PURFs. Basically, the first method is the reduction of cell size since there is directly proportional relationship between thermal conductivity and cell size. The second well-known method is to decrease the diffusion of gas molecules throughout PURF matrix via incorporation of fillers such that these fillers behave as blockades against the transport of gas molecules within the closed-cellular structure of PURFs [3]. In addition, it is well-known that using nanoclays in PURFs can decrease thermal conductivity by acting as barriers against diffusion of gas molecules and also by directly reducing the average cell size of foams [2, 3, 8, 10, 52]. Here in Fig. 3.27, thermal conductivity decreases up to the addition of 2 pphp nanoclay since nanoclay uses its superior properties such as physical barrier property against diffusion of gas molecules and cell size reducing agent property due to its nano-sized particles with high surface areas. However, after 2 pphp nanoclay incorporation into PURFs, thermal conductivity increases since nanoclay could not use its outstanding properties due to aggregation behavior of nanoclay particles at higher clay contents in rigid foams.

PU rigid nanocomposite foams consisting of nanoclay were fabricated by substituting some of synthetic polyol with a type of castor oil [4]. Basically, authors produced diethanol amide from hydroxylated castor oil by using the transamidation process, and the synthesized polyol was used in the fabrication of water-blown PU foams. Surface-modified MMT clay was incorporated in the amounts of 0.5, 1.0, 2.0, and 5.0% as nanofillers into the synthesized PURNCFs. Authors investigated systematically the effects of modified clay on mechanical properties such as compression strength, compression modulus, foam density, and morphology of PURNCFs. Based on foam density and compression testing results, it was confirmed that the incorporation of nanoclay enhanced the mechanical properties of PURNCFs. In accordance with DSC and DMA data, it was revealed that exfoliation and plasticizing effect occurred with the addition of 1 wt% of nanoclay [4].

Figure 3.28 shows the change in the storage modulus values of clay-based PURNCFs as a function of temperature. According to results, it was pointed out that storage modulus values of PURNCFs containing 1.0, 2.0 and 5.0% clay were observed to be much higher compared to storage modulus data of pure PURF within all the temperature range of experiments. In addition, it was also pointed out that storage modulus values in PURNCFs containing 0.5% clay were observed to be much lower compared to storage modulus data of other PURNCFs at all temperatures [4]. In agreement with the storage modulus data of PURNCF consisting of 0.5% clay, it

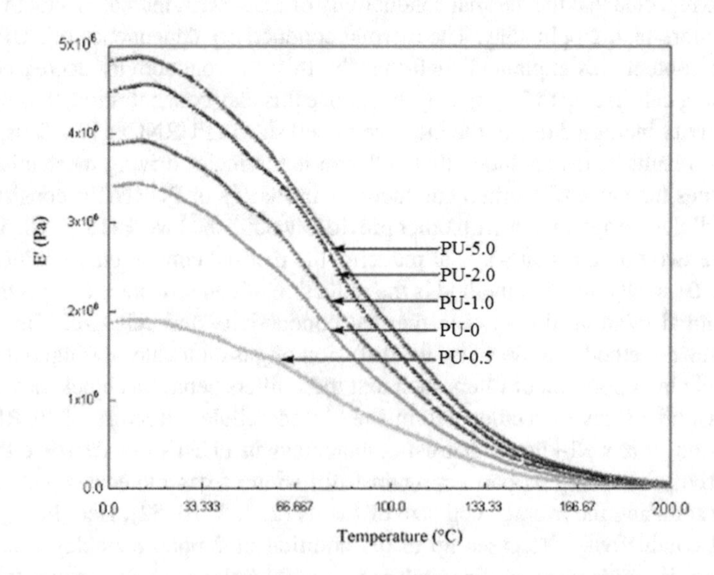

Fig. 3.28 The change of storage modulus values of PURNCFs filled with diverse clay concentrations in relation to temperature. Reproduced with permission from Ref. [4]. Copyright 2013 Society of Plastics Engineers

was revealed that the viscosity measurements of intercalated polyol-clay dispersion consisting of 0.5% clay also resulted in lower storage modulus values. However, on the other hand, it was also shown that PURNCFs containing 1% clay content or more have reinforcing effects due to the exfoliated nature of nanoclay within the polymer matrix and this behavior leads to the increase of storage modulus values in these PURNCFs [4]. Thus, based on these results, it is noted that PURNCFs in which polymer chains are intercalated within nanoclay galleries exhibit lower storage modulus due to the less amount of interactions between the polymer chains and less amount of available surface area of clay. However, on the other hand, exfoliated PURNCFs display higher storage modulus values in comparison with intercalated ones due to higher amount of interactions between the polymer chains and higher amount of available surface area of clay. In conclusion, the morphological specifications in PURNCFs consisting of nanoclay clearly dictate and control thermomechanical properties. The intercalated PURF/nanoclay system in which one polymer chain is inserted between two nanolayers of clay does not have the suitable conditions (higher number of favorable interactions between PU chains and nanoclay surfaces) for the enhancement of storage modulus compared to pure PURF system. On the other hand, the exfoliated PURF/nanoclay system in which many polymer chains are located between exfoliated nanolayers of clay provide the necessary conditions for the enhancement of storage modulus since favorable interactions between PU chains and exfoliated nanolayers of clay are maximized in these types of PURNCFs.

PURNCFs containing nanoclay were fabricated with the aim of investigating the effects of three different nanofillers such as MMT surface-modified with quaternary ammonium salt, a synthetic layered silicate and Bentonite clay on the properties of PURNCFs [50]. Nanoclay based PURNCFs were prepared by performing a single shot laboratory based process based on a foaming reaction involving 2:1 ratio of NCO to OH groups. Also, it was reported that required contents of catalysts, water, nanoclays, oligoether polyol and PMDI were present in the reaction mixture during the reactive foaming process. Authors explored thermal and thermomechanical properties of PURNCFs by using DMA, TGA, LOI and thermal conductivity measurements. Based on the experimental results, it was shown that PURNCFs that were prepared with selected nanofillers had higher mechanical strength and improved fire resistance properties [50].

Figure 3.29 shows storage modulus data of pure PURF and PURNCFs consisting of 3 and 9 wt% of nanofillers with respect to temperature. Based on the results, it was shown that PURNCFs consisting of nanofillers have much higher storage modulus values compared to pure PURF. In addition, it was also reported that among PURNCFs filled with nanofillers, PURNCFs consisting of MMT surface-modified with quaternary ammonium salt has the highest storage modulus values [50]. The improvement of storage modulus in PURNCFs filled with MMT surface-modified with quaternary ammonium salt was explained such that favorable intermolecular interactions between isocyanates of PMDI and hydroxyls of MMT surface-modified with quaternary ammonium salt resulted in the increase of modulus in these rigid nanocomposites [50].

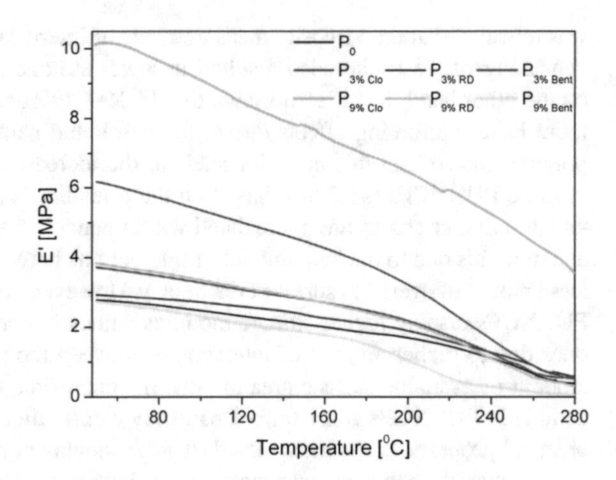

Fig. 3.29 The storage modulus data of pure PURF and PURNCFs consisting of 3 and 9 wt% of nanofillers. Reproduced with permission from Ref. [50]. Copyright 2013 Wiley Periodicals, Inc.

Moreover, the storage modulus values of PURNCFs consisting of synthetic layered silicate or Bentonite clay were observed to be much lower than the storage modulus data of PURNCF modified by MMT which was surface-modified with quaternary ammonium salt. The reason behind the observation of lower storage modulus values in PURNCFs filled with synthetic layered silicate or Bentonite was explained such that larger particle sizes and different surface physicochemical structures of these clays basically lead to lower storage modulus values in PURNCFs consisting of these clays [50]. According to results, it is apparent that surface functional groups of clays and their particle sizes are playing vital roles in the adjustment of storage modulus values in PURNCFs. Basically, smaller nanoparticle sizes provide higher surface areas for establishment of compatible intermolecular contact between nanoparticles and polymer chains. Thus, this property results in improvement of both mechanical and thermomechanical properties in PURNCFs consisting of these smaller sized nanoparticles. However, on the other hand, nanoparticles consisting of much larger particle sizes have much lower surface areas compared to smaller sized ones, thus this property leads to the decrease in number of intermolecular interactions between polymer chains and surface functional groups of larger particles. Eventually, this outcome will decrease both mechanical and thermomechanical properties of PURNCFs filled with larger particles.

PURNCFs consisting of different types of nanoclay with various concentrations were synthesized, previously [1]. The densities of PURNCFs were adjusted to be 140–160 kg/m^3 so that they can be used in structural applications such as construction and underwater buoyancy. Authors used wide-angle XRD, SEM and TEM in order to confirm the nanocomposite structures of PURNCFs. In addition, DMA and mechanical compressive tests were performed to understand the detailed mechanisms behind mechanical and thermomechanical properties of PURNCFs containing nanoclay. In accordance with mechanical and thermomechanical properties data, it was confirmed that inclusion of nanoclay fillers to PURF system enhanced properties such as storage and compressive modulus but reduced mean cell size in PURNCFs.

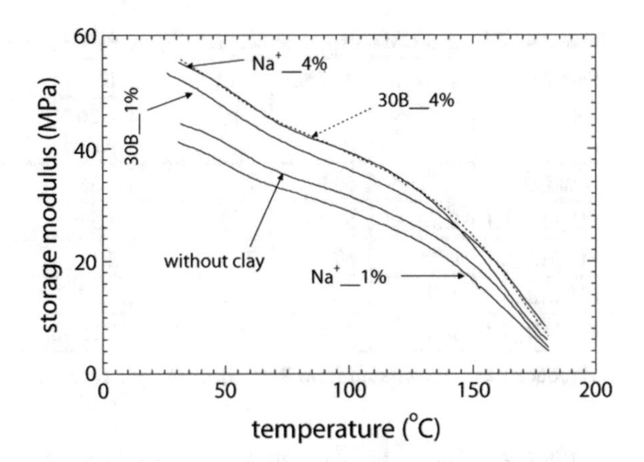

Fig. 3.30 Storage modulus data of pure PURF and PURNCFs consisting of different nanoclays with different concentrations. Reproduced with permission from Ref. [1]. Copyright 2006 Wiley Periodicals, Inc.

In contrast, it was pointed out that the hydraulic resistance of PURNCFs was measured to be much lower compared to that of pure PURF [1].

Figure 3.30 shows the storage modulus data of pure PURF and PURNCFs consisting of different nanoclays with different concentrations. According to results, it was pointed out that incorporating different types of nanoclay with various amounts to pure PURF basically increases storage modulus values within the studied temperature region. In addition, it was also reported that at around room temperature, the storage modulus of PURNCF consisting of 4% clay is about 25% higher than that of pure PURF [1]. Thus, this result shows that the incorporation of 4% clay into PURFs clearly induces reinforcing effect in comparison with pure PURF without any clay. Moreover, storage modulus value of PURNCF consisting of 1% MMT surface-modified with quaternary ammonium salt was reported as much larger compared to PURNCF filled with 1% Na^+ MMT [1]. The reason behind this difference can be explained such that hydroxyls of MMT which was surface-modified with quaternary ammonium salt interact much more better with isocyanates of PMDI component, thus improving thermomechanical properties much more better compared to PURNCF system consisting of Na^+ clay.

Previously, a new polymerization procedure in PURFs was found out for vermiculite (VMT) dispersion in PURFs [6]. In their study, authors showed that VMT dispersion in MDI was mainly enhanced when VMT was surface-modified with cetyltrimethylammonium bromide (CTAB) via the cation exchange reaction. In addition, it was also reported that the application of a high intensity dispersive mixing method onto polyol-clay blend and combining this polyol blend with the VMT dispersion in MDI increased the dispersion degree of VMT in nanocomposite foams [6]. Authors performed DSC, DMA and tensile tests to understand the details of thermal, thermomechanical and mechanical properties of PURFs with incorporation of VMT. Based on mechanical and permeability test results, PURNCFs which were prepared with polyol-assisted VMT dispersion or master batch (MB) method exhibited about 85% improvement in terms of elastic modulus and 40% reduction in terms of CO_2 permeability at 3.3 wt% nanoclay loadings [6].

Table 3.4 DSC and DMA experimental data of pure PURF and PURNCFs

Sample	ϕ_{clay} (%)	T_g (°C)	tan δ peak (°C)	E'		
				−20 °C (GPa)	25 °C (GPa)	120 °C (MPa)
Neat PU	0	66.6	73.3	1.36	0.98	10.5
PU_MB_1.1	0.39	67.7	72.5	1.57	1.19	15.0
PU_MB_2.2	0.79	69.2	76.5	2.12	1.49	18.1
PU_MB_3.3	1.19	70.7	77.6	2.57	1.81	22.3
PU_MB_4.4	1.60	74.4	76.7	2.75	2.01	29.2

Reproduced with permission from Ref. [6]. Copyright 2013 American Chemical Society

Table 3.4 shows the clay volume fractions (ϕ_{clay}), glass transition temperatures (T_g) based on DSC experiments, temperatures correspond to tan δ peaks and storage modulus values of pure PURF and PURNCFs. Based on these results, it was shown that glass transition temperatures that were obtained both from DMA and DSC results increases systematically with increasing amounts of nanoclay in PURNCFs. Moreover, the addition of CTAB-VMT to PURFs increased the storage modulus in a large extent [6]. The reason behind this increase in storage modulus was explained such that the high aspect ratio, uniform dispersion and high exfoliation of CTAB-VMT nanoplatelets lead to better interactions between CTAB functional groups of VMT and polymer chains, thus leading to the increase of storage modulus values in PURNCFs [6].

In agreement with previously published other works [1, 4, 50], it was shown that the thermal mobility of PU chains in nanoclay filled PURNCFs are obstructed since nanoclay particles act as rigid surfaces against polymer mobility. As a consequence of this specific characteristics of nanoclay, incorporation of nanoclay increased T_g in PURCFs. Also, favorable forces between CTAB molecules of VMT nanolayers and PU chains contribute to homogenous dispersion of CTAB-VMT nanoparticles in PURNCF matrix. Thus, this mechanism in nanoclay filled PURNCFs leads to the elevation of glass transition temperature and storage modulus compared to the situation in pure PURF and PURNCFs consisting of nanoclay which is not surface modified with compatible surface functional groups.

The influences of using graphene on the improvement of thermal insulation properties of PURFs were analyzed by preparing and characterizing pure PURF and PURNCFs consisting of 0.3 and 0.5 wt% graphene [11]. According to experimental data, radiative component related to thermal conductivity at the beginning diminished with incorporation of 0.3 wt% graphene via mechanisms of cell size reduction and elevation of extinction coefficient. Based on thermal conductivity measurements as a function of time, it was also revealed that slower aging rates were observed in graphene-based PURNCFs compared to that of pure PURF [11].

Figure 3.31 shows the thermal aging of pure PURF and graphene-filled PURNCFs with different graphene contents. According to results, it was revealed that PURNCFs containing both 0.3 and 0.5 wt% graphene exhibited slower thermal aging rates

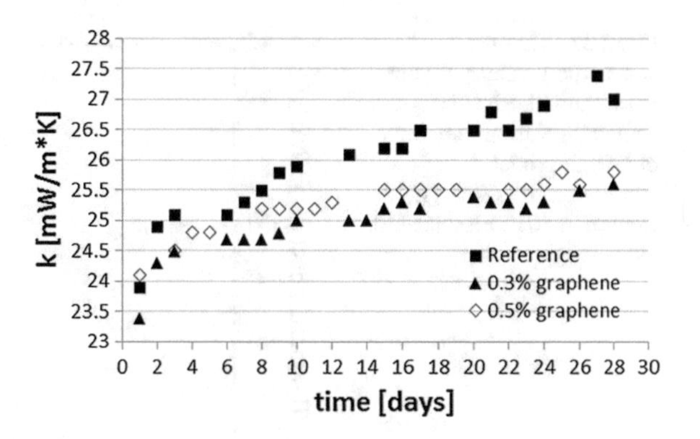

Fig. 3.31 Thermal aging of pure PURF and graphene-filled PURNCFs consisting of different graphene contents. Reproduced with permission from Ref. [11]. Copyright 2015 John Wiley & Sons, Ltd.

compared to pure PURF in a period of 4 weeks [11]. The reason behind this difference between graphene filled PURNCFs and pure PURF was explained such that the addition of graphene reduces cell sizes of foams, and the reduction in cell sizes leads to increase in the amount of material within cell windows. Thus, for this reason, the rate of aging was substantially decreased since cell windows are mainly responsible for acting as barriers against diffusion [11, 54]. On the other hand, the diffusion of gases could be hindered by 2-dimensional plate-like nanoparticles such as graphene, clays, etc. by delaying or slowing down the diffusion process [1–4, 8, 11, 52, 54].

Water blown PURFs were produced by using in situ polymerization method via functionalization of polyols with graphene oxide (GO) [12]. Authors investigated the effects of using GO functionalized with polyol on the closed-cellular morphology, foaming kinetics, thermal conductivity and compressive mechanical properties of PURFs. Based on experimental results, it was shown that about 33% cell reduction occurred with the additions of small amounts of GO into the PURF system [12]. Moreover, incorporation of GO reduced thermal conductivity values of PURNCFs. In contrast, based on mechanical testing data, it was pointed out that inclusion of GO particles basically reduced mechanical properties of PURNCFs [12].

Figure 3.32 shows thermal conductivity values of pure PURF and PURNCFs consisting of GO with different contents. Based on the results, it was shown that the thermal conductivity of PURNCFs consisting of very small amounts of GO (0.017–0.083 wt%) improves substantially, and its minimum value was obtained with the addition of 0.033 wt% GO [12]. In addition, it was also revealed that compared to pure PURF, thermal conductivity of PURNCFs was reduced about 5%. Furthermore, it was also shown that the PURNCF consisting of 0.083 wt% GO has the smallest cell size, but its thermal conductivity is much higher than expected. The reason behind this observation was explained such that the higher thermal conductivity in this nanocomposite foam might be the result of its higher density, heterogenous cellular structure with higher open cell content [12].

In the literature, it is well-known that using plate-like nanoparticles in PURFs can decrease thermal conductivity by acting as barriers against diffusion of gas molecules and also by directly reducing the average cell size of foams [2, 3, 8–10, 52]. Here in Fig. 3.32, thermal conductivity of PURNCFs decreases up to the addition of 0.04 wt%

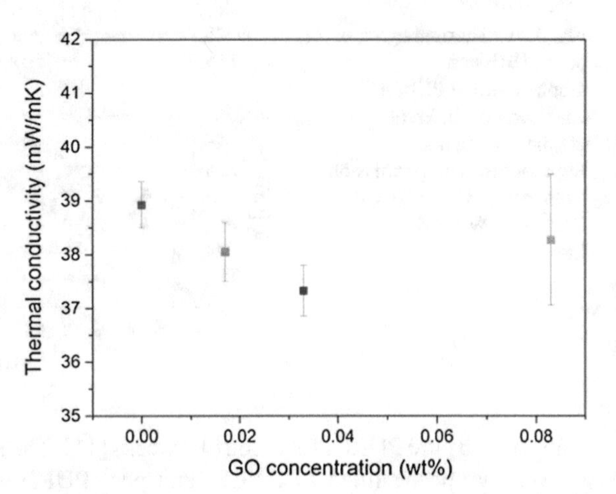

Fig. 3.32 Thermal conductivity values of pure PURF and PURNCFs containing various amounts of GO. Reproduced with permission from Ref. [12]. Copyright 2017 Elsevier Ltd.

GO since GO nanoparticles could act as physical blockades against diffusion of gas molecules and also function as cell size reducing agents due to their nano-sized particles with high surface areas. However, after 0.04 wt% GO nanoclay incorporation into PURFs, thermal conductivity increases since GO could not use its outstanding properties due to aggregation behavior of GO particles at higher amounts of GO in PURNCFs.

Recently, nanoparticle enhanced PURFs consisting of different amounts of graphite and nano-sized graphene oxide particles were produced for solving the industrial problems that were faced by cargo systems in which liquefied natural gas is transported [13]. Authors used SEM and XRD analysis to comprehend the effects of using nanoparticles on the closed-cellular morphology of developed PURNCFs. In addition, the mechanical properties of PURNCFs were investigated at both room and at 110 K which correspond to boiling temperatures of LNG. Furthermore, based on morphology, thermal conductivity and macroscopic failure tests, it was found out that the morphology and mechanical strength of PURNCFs were improved and highly dependent on weight percent of nano-graphene oxide particles compared to pure PURFs [13].

Figure 3.33 shows the thermal conductivity of pure PURF, PURCFs consisting of graphite and PURNCFs consisting of GO with different contents that were measured at ambient (298 K) temperature. Based on the results, it was shown that thermal conductivity does not change much with the addition of graphite to PURFs. However, on the other hand, it was reported that thermal conductivity diminished approximately by 3 and 8% with incorporation of 0.05 and 0.1 wt% of GO, respectively compared to that of pure PURF [13]. Thus, the reduction of thermal conductivity in GO filled PURNCF consisting of 0.1 wt% of GO addition was explained such that very small amounts of GO act as nucleation agents in reducing cell sizes to very small dimensions, thus decreasing thermal conductivity values, accordingly [13, 28, 55]. Based on these results, it is noted that thermal conductivity of PURNCFs consisting of 2-dimensional nanoparticles was mainly controlled by cell size reduction and

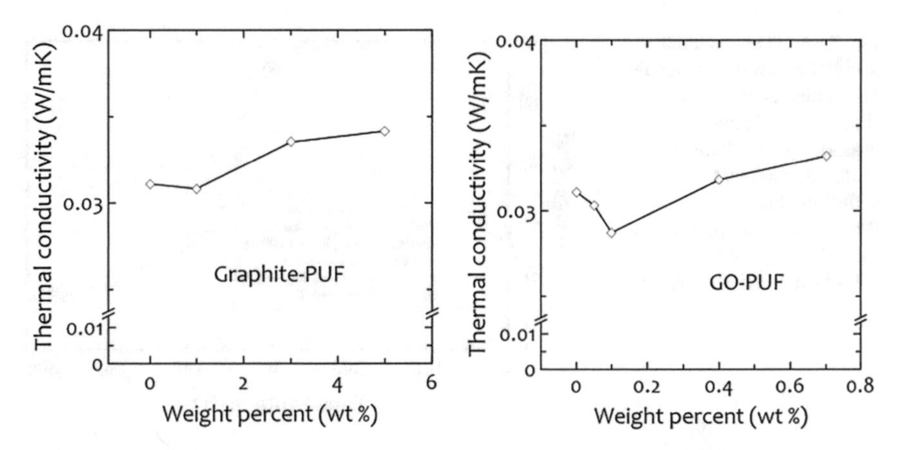

Fig. 3.33 Thermal conductivity of pure PURF, PURCFs consisting of graphite and PURNCFs consisting of GO with different contents measured at 298 K. Reproduced with permission from Ref. [13]. Copyright 2017 Elsevier Ltd.

degree of barrier effect coming from nanoparticles. For achieving higher degrees of barrier effect in plate-like nanoparticles, their geometrical parameters such as sizes, surface areas and surface functional groups should be adjusted such that minimum levels in terms of thermal conductivity may be reached with the addition of very small quantities from these plate-like nanoparticles. In agreement with other plate-like nanoparticle filled PURNCF works [2, 3, 8–10, 12, 52], thermal conductivity of PURNCFs decreases up to the addition of 0.05 wt% GO since GO nanoparticles could act as physical barriers against transport of gas molecules and also function as cell size reducing agents due to their nano-sized particles. However, after 0.05 wt% GO incorporation into PURFs, thermal conductivity increases since GO could not function as an effective plate-like nanoparticle due to their aggregation properties at higher amounts of GO in PURNCFs [12, 13].

3.5 Thermal Degradation and Flammability

PURNCFs containing organoclay particles with improved thermal insulation properties with high thermal insulation properties were synthesized, previously [7]. In their PURNCFs, authors used organoclay as the plate-like nanofiller which was surface modified with polymeric diphenylmethane diisocyanate (PMDI) by using a silane coupling agent [7]. Based on mechanical testing results of PURNCFs, it was shown that there is no big difference in terms of compressive and flexural strengths between PURNCFs consisting of organoclay layers that were prepared with and without silane coupling agent. On the other hand, in accordance with SEM and thermal conductivity data, it was found out that in terms of thermal conductivity and cell size, PURNCFs consisting of organoclay-silane coupling agent system had much lower

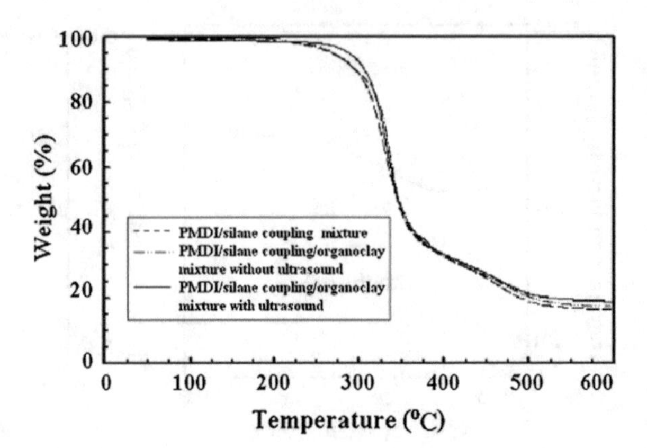

Fig. 3.34 TGA results of **a** PMDI and silane molecule blend, blends of PMDI, silane molecule and organoclay with **b** no ultrasonication and **c** ultrasonication. Reproduced with permission from Ref. [7]. Copyright 2008 Wiley Periodicals, Inc.

values compared to PURNCFs consisting of organoclay without the silane coupling agent. In conclusion, authors interpreted these findings such that using silane coupling agents during surface modification of organoclay led to the improved exfoliation of organoclay nanolayers within the PURNCF matrix, and thus decreased the values of thermal conductivity and cell size in comparison with the system that did not contain any silane coupling agent [7].

Figure 3.34 shows TGA results of (a) PMDI and silane molecule blend, blends of PMDI, silane molecule and organoclay with (b) no ultrasonication and (c) ultrasonication. Based on results, it was shown that the blend of PMDI with silane molecule and organoclay that was prepared with application of ultrasound degrades at a higher temperature compared to that of PMDI and silane molecule blend [7]. The difference in terms of thermal degradation between these two blends was explained such that the presence of organoclay with high dispersion degree or exfoliation character in PURNCFs leads to reduction in the discharge of oxygen and volatile products through PURNCF matrix. Furthermore, it was also reported that the organoclay consisting of silane coupling agent effectively breaks apart the aggregates of organoclay and leads to the exfoliation of nanoclay within the PURNCF matrix. Furthermore, it was also suggested that intercalation of organoclay due to chemical reactions between organoclay surface functional groups and coupling agents attached to PMDI might increase degradation temperature in the blend of PMDI/silane coupling agent/organoclay [7]. Based on these results, the thermal degradation temperatures of PURNCFs consisting of nanoclay greatly depend on the interactions between the surface functional groups of clay and polymer chains, surface modification agents, and the dispersion levels (exfoliation or intercalation) of clays within the PURNCF matrix.

PURNCFs containing diphosphonium montmorillonite which was synthesized via insertion of quaternary diphosphonium salt within nanoclay galleries were prepared and characterized, previously [5]. This surface modified nanoclay as a plate-like nanofiller was used in the preparation and thermal behavior analysis of PURNCFs [5]. Based on TGA data, it was revealed that weight loss in diphosphonium-MMT was observed to be around 18% at about 400 °C. In general, during experiments, 5 wt% of diphosphonium-MMT was used as the nano-plate filler in the fabrication of PURNCFs. In addition to incorporation of diphosphonium based MMT into PURFs, commercial MMTs such as ammonium-modified MMT and unmodified pristine MMT were applied as plate-like nanofillers with phosphinate (IPA) in PURNCFs [5].

TGA and DTG results that were obtained in air atmosphere for pure PURF, PURF filled with phosphinate, PURNCF filled with phosphinate and diphosphonium-MMT, PURNCF filled with phosphinate and unmodified MMT and PURNCF filled with phosphinate and ammonium-modified MMT are shown in Figs. 3.35 and 3.36, respectively. According to results, it was pointed out that mass loss as a function of temperature for each PURNCF was totally different from each other. In addition, it was reported that thermal decomposition temperatures at 5 wt% mass loss were quite similar for unfilled and phosphinate filled PURFs. In contrast, this thermal decomposition temperature was observed to be about 10 °C higher for PURNCFs consisting of nanoclay due to barrier effect of nanoclay against diffusion gases [5].

In agreement with other published works in the literature [5, 7, 9, 10, 50], the presence of plate-like nanoparticle such as nanoclay besides a flame retardant agent in PURNCFs improves thermal decomposition temperatures and thermal stability much more better compared to PURFs consisting of only flame retardant agent. The reason behind this observation can be explained such that clay nanolayers act as extra thermal barriers or obstacles against the diffusion of volatile gases during combustion process [5, 7], thus leading to higher thermal stability values in PURNCFs filled with

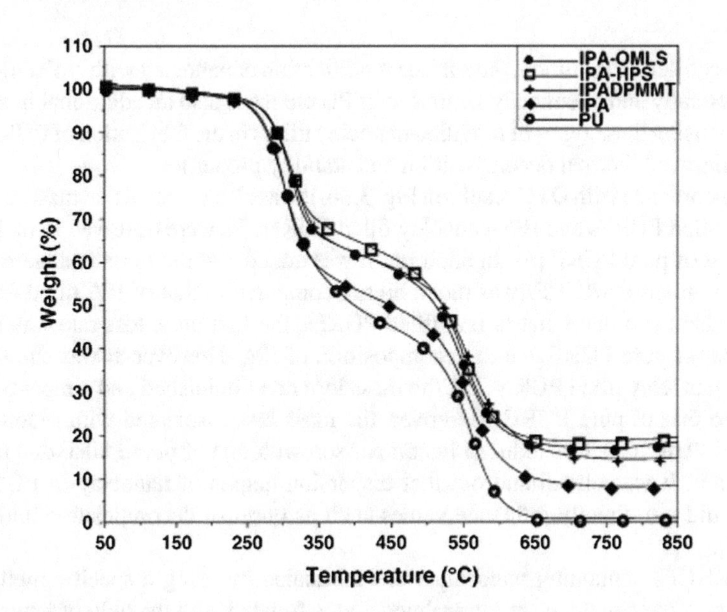

Fig. 3.35 TGA results under air atmosphere for pure PURF, PURF filled with phosphinate, PURNCF filled with phosphinate and diphosphonium-MMT, PURNCF filled with phosphinate and unmodified MMT and PURNCF filled with phosphinate and ammonium-modified MMT. Reproduced with permission from Ref. [5]. Copyright 2009 Elsevier Ltd.

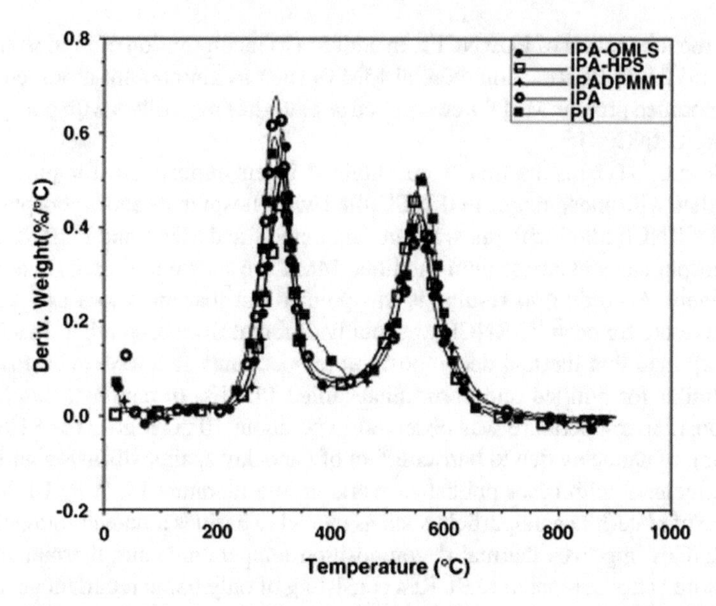

Fig. 3.36 DTG results under air atmosphere for pure PURF, PURF filled with phosphinate, PURNCF filled with phosphinate and diphosphonium-MMT, PURNCF filled with phosphinate and un modified MMT and PURNCF filled with phosphinate and ammonium-modified MMT. Reproduced with permission from Ref. [5]. Copyright 2009 Elsevier Ltd.

nanoclay fillers. In addition, the surface modification of nanoclay with molecules that are physically and chemically similar with PU chains is also an additional benefit in terms of using these types of modified nanoclay fillers in the fabrication of PURNCFs with improved thermal decomposition and stability properties.

In accordance with DTG results in Fig. 3.36, it was also revealed that mass residues in IPA filled PURFs and IPA-nanoclay filled PURNCFs were observed to be higher than that of pure PURF [5]. In addition, it was added that the mass residue of IPA-nanoclay filled PURNCFs was much higher compared to that of IPA filled PURFs. It was also mentioned that in IPA filled PURFs, the first mass loss rate was higher than that of pure PURF due to decomposition of IPA. However, it was shown that in IPA-nanoclay filled PURNCFs, the mass loss rate diminished and almost became equal to that of pure PURF. Moreover, the mass loss associated with second step in filled PURNCFs was reduced in comparison with that of pure PURF due to char formation. It was also found out that dispersion degree of nanoclay in PURNCF matrix did not directly influence values such as onset of decomposition and mass loss [5].

PURNCFs containing nanoclay were fabricated by using a specific method in which clay nanolayers were intercalated and exfoliated with the help of neutralized dimethylol butanoic acid (DMBA) [9]. In accordance with experimental data, it was shown that values of foam density and compression strength decreased while gel, tack-free and cream time increased with increasing amounts of nanoclay into PURFs.

On the other hand, it was also reported that the thermal conductivity, cell size, the amount of closed cells and the volume change as a function of temperature decreased with the incorporation of nanoclay into PURF system. Based on thermomechanical and thermal stability results, it was shown that nanoclay incorporation enhanced temperatures such as glass transition and thermal decomposition due to reduced mobility of PU chains by barrier effects of nanoclay platelets [9].

Figure 3.37 shows the TGA results of pure PURF and PURNCFs consisting of nanoclay with different contents. Based on the results, it was shown that the 10% weight loss temperature in PURNCFs consisting of nanoclay increased about 10 °C in comparison with pure PURF due to the barrier effect of nanoclay [9]. In addition, it was also reported that the final weight loss in PURNCFs consisting of nanoclay becomes smaller with the addition of clay. The reason behind this finding was explained such that nanoclay acts as a physical barrier against transport of oxygen and volatile products throughout PURNCF matrix, thus extending the pathway of diffusion gases remarkably during the degradation process [9].

Previously, phenol-urea-formaldehyde based glass fiber/nanoclay composites were fabricated to improve flame retardancy, toughness and compression strength of phenolic foams [49]. Authors investigated closed-cellular morphology, flame retardant, mechanical and thermal decomposition properties of phenolic foam nanocomposites in detail. Based on these results, it was shown that pulverization rate of phenolic foams reduced extensively with incorporation of nanoclay and glass fibers to pristine PURF [49]. Based on thermal stability results, inclusion of nanoclay enhanced thermal stability but diminished flame retardant properties such as heat release rate and smoke release values of PURNCFs. So overall, based on these results, it was confirmed that using both nanoclay and glass fiber has synergetic

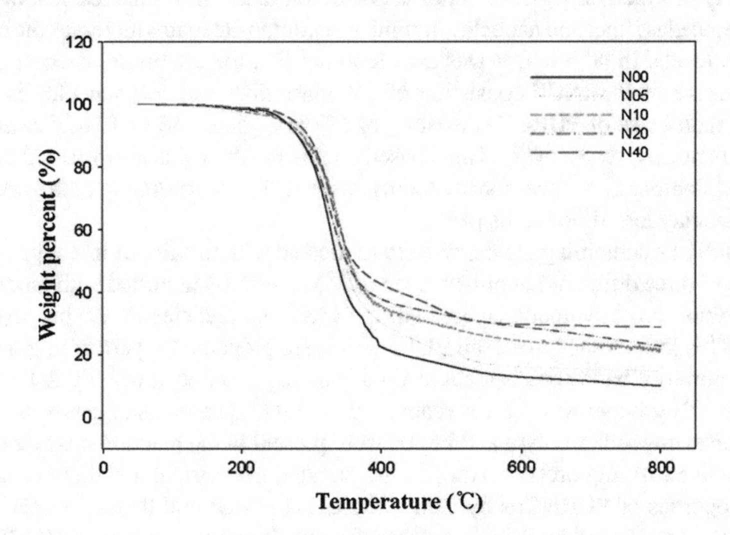

Fig. 3.37 TGA results of pure PURF and PURNCFs consisting of nanoclay with different amounts. Reproduced with permission from Ref. [9]. Copyright 2010 Wiley Periodicals, Inc.

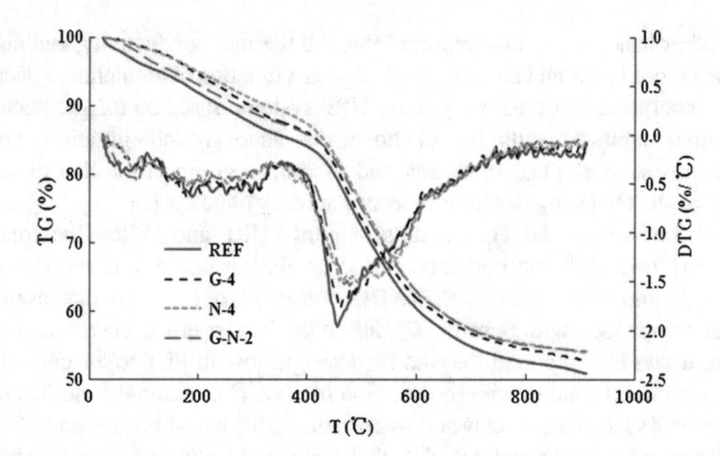

Fig. 3.38 TGA results of pure PURF, PURCF consisting of 5% glass fiber, PURNCF consisting of 5% nanoclay and PURNCF consisting of 5% glass fiber and 3% nanoclay. Reproduced with permission from Ref. [49]. Copyright 2015 Society of Plastics Engineers

effects in terms of improving mechanical and flame retardant properties and thermal stability of rigid nanocomposite foams [49].

Figure 3.38 shows TGA results of pure PURF and PURCF consisting of 5% glass fiber, PURNCF consisting of 5% nanoclay and PURNCF consisting of 5% glass fiber and 3% nanoclay. According to results, it was shown that the remaining carbon amount of PURNCF consisting of 5% glass fiber and 3% nanoclay was slightly higher than that of PURCF consisting of 5% glass fiber or PURNCF consisting of 5% nanoclay in which only glass fiber or nanoclay was used [49]. This result shows that using both glass fiber and nanoclay in rigid nanocomposite foams increases the carbon residue levels. In addition, it was also reported that the maximum decomposition temperature of PURNCF consisting of 5% glass fiber and 3% nanoclay is much higher than those of PURCF consisting of 5% glass fiber and PURNCF consisting of 5% nanoclay, respectively. Thus, based on this result, it was also verified that the thermal stability of nanocomposite foams increases by incorporating both glass fiber and nanoclay into rigid foams [49].

PURNCFs containing nanoclay were fabricated with the aim of investigating the effects of three different nanofillers such as MMT surface-modified with quaternary ammonium salt, a synthetic layered silicate and Bentonite clay on the properties of PURNCFs [50]. Nanoclay based PURNCFs were prepared by performing a single shot laboratory based process based on a foaming reaction involving 2:1 ratio of NCO to OH groups. Also, it was reported that required contents of catalysts, water, nanoclays, oligoether polyol and PMDI were present in the reaction mixture during the reactive foaming process. Authors investigated the thermal and thermomechanical properties of PURNCFs by using DMA, TGA, LOI and thermal conductivity measurements. Based on the experimental results, it was shown that PURNCFs that were prepared with selected nanofillers had higher mechanical strength and improved fire resistance properties [50].

Figure 3.39 shows the TGA and DTG results of pure PURF and PURNCFs consisting of (a) MMT surface-modified with quaternary ammonium salt, (b) synthetic layered silicate and (c) Bentonite with clay contents of 3, 6, and 9 wt%. Based on the results, it was shown that the temperature of 5% mass loss in PURNCFs increased with incorporation of 3 and 9 wt% from synthetic layered silicate and MMT surface-modified with quaternary ammonium salt [50]. Furthermore, it was also reported that the temperature associated with 5% mass loss in PURFs increased by about 7 °C with inclusion of only 9 wt% Bentonite. In contrast, it was also revealed that PURNCF consisting of 9 wt% MMT surface-modified with quaternary ammonium salt had the highest thermal stability such that its temperature corresponding to 50% mass loss was reported as 574 °C and its thermal resistance was improved by 41 °C relative to pure PURF. Based on these results, it was concluded that using nanofillers in PURFs reduces heat flow into PURNCF matrix and improves thermal stability due to homogenous distribution of clay nanolayers within PURNCF matrix [50].

PURNCFs consisting of unmodified vermiculite (VMT) clay were synthesized by distribution of nanoclay fillers either in polyol or isocyanate component before the blending process [10]. In accordance with experimental data, it was shown that incorporation of 5 pphp nanoclay increases the viscosity of polyol and decreases the gel and rise times during the reactive foaming process. In addition, it was found

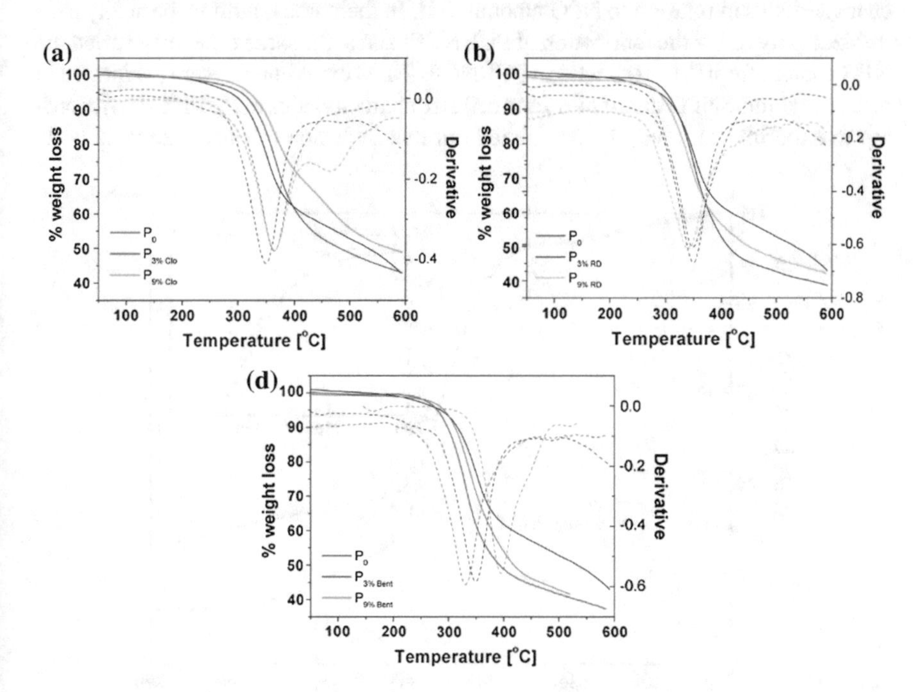

Fig. 3.39 TGA and DTG results of pure PURF and PURNCFs consisting of **a** MMT surface-modified with quaternary ammonium salt, **b** synthetic layered silicate and **c** bentonite with clay contents of 3, 6, and 9 wt%. Reproduced with permission from Ref. [50]. Copyright 2013 Wiley Periodicals, Inc.

out that the presence of clay in PURNCFs leads to the generation of gas bubbles which also drive the formation of smaller cells with almost identical sizes in final PURNCFs. Moreover, based on SEM results, it was shown that the closed cell contents in PURNCFs increase slightly with increasing amount of nanoclay. Based on the mechanical testing results, it was reported that the best mechanical properties were obtained when 2.3 wt% of nanoclay was dispersed in only isocyanate component before the reactive foaming process. Based on thermal conductivity results, it was revealed that the thermal conductivity of PURNCFs consisted of nanoclay was observed to be 10% lower than that of pure PURF [10].

Figure 3.40 shows the TGA results of pure PURF and PURNCFs consisting of 5 pphp VMT. Based on the results, it was shown that the initial starting temperature for degradation increases about 10 °C with the addition of 5 pphp VMT to pure PURF [10]. In addition, it was also reported that the final weight loss of PURNCFs consisting of 5 pphp VMT is lower by about 20% at 700 °C compared to pure PURF. Based on these results, it is noted that thermal decomposition behaviour of PURNCFs enhanced with inclusion of VMT to PURF because of uniform distribution of VMT nanolayers throughout PURNCF matrix [10].

PURNCFs containing nanoporous graphene (NPG) were fabricated, and authors investigated the foam density, mechanical, morphological, and thermal-resistant characteristics in relation to NPG amount [35]. In their study, authors basically synthesized polyols for the fabrication of PURNCFs and at the same time, they varied the NPG content from 0.1 to 0.5 wt% in PURNCFs. SEM experiments were performed in order to examine NPG distribution and cell size characteristics in PURNCFs. According to experimental data, it was pointed out that inclusion of only 0.25 wt% NPG

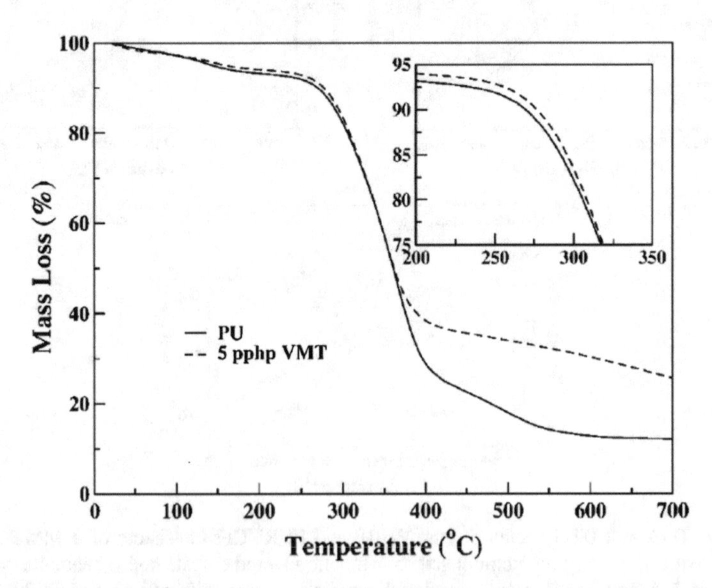

Fig. 3.40 TGA results of pure PURF and PURNCFs consisting of 5 pphp VMT. Reproduced with permission from Ref. [10]. Copyright 2008 Society of Plastics Engineers

enhanced compressive strength and modulus values of PURNCFs [35]. In accordance with TGA data, it was shown that degradation temperatures of PURNCFs were moderately improved with the increase of NPG content. Based on these findings, it was concluded that NPG is a much more effective nanoparticle in terms of enhancing mechanical properties compared to other nanoparticles due to its special two-dimensional geometry and higher specific surface area [35].

Figure 3.41 shows the TGA results of pure PURF and PURNCFs consisting of NPG. Based on the results, it was shown that thermal decomposition temperatures associated with 10, 50, and 80% weight loss do not increase significantly with increasing amounts of NPG. However, there are some minor differences in terms of thermal decomposition temperatures [35]. In addition, it was pointed out that all thermal decomposition temperatures of rigid foam samples, excluding temperature associated with 80% weight loss in PURNCF consisting of 0.1 wt% NPG and 35 phr of polyol, were improved based on the TGA data. The reason behind this improvement in decomposition temperatures was explained such that the cell size decreased with increasing percentage of NPG thus enlarging the surface area of NPG that was revealed to thermal decomposition [35].

In consensus with other published works in the literature [5, 7, 9, 10, 49, 50], using plate-like nanofillers such as nanoporous graphene in PURNCFs improves thermal decomposition temperatures and thermal stability much more better compared to that of pure PURF. The reason behind this observation can be explained such that NPG nanolayers function as extra thermal blockades against the diffusion of volatile gases during combustion process [5, 7], thus paving the way for improved thermal decomposition behavior in PURNCFs filled with NPG fillers. In

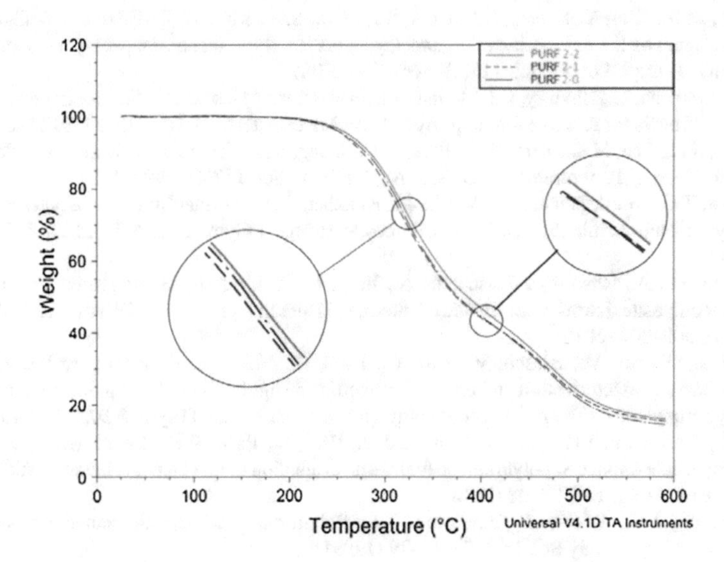

Fig. 3.41 TGA results of pure PURF and PURNCFs containing different amounts of NPG. Reproduced with permission from Ref. [35]. Copyright 2017 Wiley Periodicals, Inc.

addition, more enhancement in thermal stability of NPG filled PURNCFs can be provided if NPG nanolayers are surface modified with functional molecules that are both physically and chemically well-matched with PU chains. Furthermore, using both surface modified NPG nanofillers and chemically compatible flame retardant agents in PURNCFs is also another alternative method in terms of using these types of modified NPG nanofillers in the fabrication of PURNCFs with extra enhanced thermal decomposition and stability properties.

References

1. Mondal, P., Khakhar, D.V.: Rigid polyurethane-clay nanocomposite foams: preparation and properties. J. Appl. Polym. Sci. **103**(5), 2802–2809 (2007)
2. Cao, X., Lee, L.J., Widya, T., Macosko, C.: Polyurethane/clay nanocomposites foams: processing, structure and properties. Polymer **46**(3), 775–783 (2005)
3. Widya, T., Macosko, C.W.: Nanoclay-modified rigid polyurethane foam. J. Macromol. Sci. Part B Phys. **44**(6), 897–908 (2005)
4. Palanisamy, A.: Water-blown polyurethane-clay nanocomposite foams from biopolyol effect of nanoclay on the properties. Polym. Compos. **34**(8), 1306–1312 (2013)
5. Semenzato, S., Lorenzetti, A., Modesti, M., Ugel, E., Hrelja, D., Besco, S., Michelin, R.A., Sassi, A., Facchin, G., Zorzi, F., Bertani, R.: A novel phosphorus polyurethane foam/montmorillonite nanocomposite: preparation, characterization and thermal behaviour. Appl. Clay Sci. **44**(1–2), 35–42 (2009)
6. Park, Y.T., Qian, Y.Q., Lindsay, C.I., Nijs, C., Camargo, R.E., Stein, A., Macosko, C.W.: Polyol-assisted vermiculite dispersion in polyurethane nanocomposites. ACS Appl. Mater. Interfaces **5**(8), 3054–3062 (2013)
7. Han, M.S., Kim, Y.H., Han, S.J., Choi, S.J., Kim, S.B., Kim, W.N.: Effects of a silane coupling agent on the exfoliation of organoclay layers in polyurethane/organoclay nanocomposite foams? J. Appl. Polym. Sci. **110**(1), 376–386 (2008)
8. Harikrishnan, G., Lindsay, C.I., Arunagirinathan, M.A., Macosko, C.W.: Probing nanodispersions of clays for reactive foaming. ACS Appl. Mater. Interfaces **1**(9), 1913–1918 (2009)
9. Kim, S.H., Lee, M.C., Kim, H.D., Park, H.C., Jeong, H.M., Yoon, K.S., Kim, B.K.: Nanoclay reinforced rigid polyurethane foams. J. Appl. Polym. Sci. **117**(4), 1992–1997 (2010)
10. Patro, T.U., Harikrishnan, G., Misra, A., Khakhar, D.V.: Formation and characterization of polyurethane-vermiculite clay nanocomposite foams. Polym. Eng. Sci. **48**(9), 1778–1784 (2008)
11. Lorenzetti, A., Roso, M., Bruschetta, A., Boaretti, C., Modesti, M.: Polyurethane-graphene nanocomposite foams with enhanced thermal insulating properties. Polym. Adv. Technol. **27**(3), 303–307 (2016)
12. Santiago-Calvo, M., Blasco, V., Ruiz, C., Paris, R., Villafane, F., Rodriguez-Perez, M.A.: Synthesis, characterization and physical properties of rigid polyurethane foams prepared with poly (propylene oxide) polyols containing graphene oxide. Eur. Polym. J. **97**, 230–240 (2017)
13. Kim, J.M., Kim, J.H., Ahn, J.H., Kim, J.D., Park, S., Park, K.H., Lee, J.M.: Synthesis of nanoparticle-enhanced polyurethane foams and evaluation of mechanical characteristics. Compos. Part B Eng. **136**, 28–38 (2018)
14. LeBaron, P.C., Wang, Z., Pinnavaia, T.J.: Polymer-layered silicate nanocomposites: an overview. Appl. Clay Sci. **15**(1–2), 11–29 (1999)
15. Giannelis, E.P., Krishnamoorti, R., Manias, E.: Polymer-silicate nanocomposites: model systems for confined polymers and polymer brushes. Adv. Polym. Sci. **138**, 107–147 (1999)
16. Giannelis, E.P.: Polymer layered silicate nanocomposites. Adv. Mater. **8**(1), 29–35 (1996)

17. Hanemann, T., Szabo, D.V.: Polymer-nanoparticle composites: from synthesis to modern applications. Materials **3**(6), 3468–3517 (2010)
18. Santiago-Calvo, M., Tirado-Mediavilla, J., Ruiz-Herrero, J.L., Rodriguez-Perez, M.A., Villafane, F.: The effects of functional nanofillers on the reaction kinetics, microstructure, thermal and mechanical properties of water blown rigid polyurethane foams. Polymer **150**, 138–149 (2018)
19. Seo, W.J., Sung, Y.T., Han, S.J., Kim, Y.H., Ryu, O.H., Lee, H.S., Kim, W.N.: Synthesis and properties of polyurethane/clay nanocomposite by clay modified with polymeric methane diisocyanate. J. Appl. Polym. Sci. **101**(5), 2879–2883 (2006)
20. Seo, W.J., Sung, Y.T., Kim, S.B., Lee, Y.B., Choe, K.H., Choe, S.H., Sung, J.Y., Kim, W.N.: Effects of ultrasound on the synthesis and properties of polyurethane foam/clay nanocomposites. J. Appl. Polym. Sci. **102**(4), 3764–3773 (2006)
21. Seo, W.J., Jung, H.C., Hyun, J.C., Kim, W.N., Lee, Y.B., Choe, K.H., Kim, S.B.: Mechanical, morphological, and thermal properties of rigid polyurethane foams blown by distilled water. J. Appl. Polym. Sci. **90**(1), 12–21 (2003)
22. Yao, K.J., Song, M., Hourston, D.J., Luo, D.Z.: Polymer/layered clay nanocomposites: 2 polyurethane nanocomposites. Polymer **43**(3), 1017–1020 (2002)
23. Osman, M.A., Mittal, V., Morbidelli, M., Suter, U.W.: Polyurethane adhesive nanocomposites as gas permeation barrier. Macromolecules **36**(26), 9851–9858 (2003)
24. Choi, W.J., Kim, S.H., Kim, Y.J., Kim, S.C.: Synthesis of chain-extended organifier and properties of polyurethane/clay nanocomposites. Polymer **45**(17), 6045–6057 (2004)
25. Chang, J.H., An, Y.U.: Nanocomposites of polyurethane with various organoclays: thermomechanical properties, morphology, and gas permeability. J. Polym. Sci. Part B Polym. Phys. **40**(7), 670–677 (2002)
26. Tan, S.Q., Abraham, T., Ference, D., Macosko, C.W.: Rigid polyurethane foams from a soybean oil-based polyol. Polymer **52**(13), 2840–2846 (2011)
27. Tien, Y.I., Wei, K.H.: High-tensile-property layered silicates/polyurethane nanocomposites by using reactive silicates as pseudo chain extenders. Macromolecules **34**(26), 9045–9052 (2001)
28. Kim, Y.H., Choi, S.J., Kim, A.M., Han, M.S., Kim, W.N., Bang, K.T.: Effects of organoclay on the thermal insulating properties of rigid polyurethane foams blown by environmentally friendly blowing agents. Macromol. Res. **15**(7), 676–681 (2007)
29. van Maris, R., Tamano, Y., Yoshimura, H., Gay, K.M.: Polyurethane catalysis by tertiary amines. J. Cell. Plast. **41**(4), 305–322 (2005)
30. Grimminger, J., Muha, K.: Silicone surfactants for pentane blown rigid foam. J. Cell. Plast. **31**(1), 48–72 (1995)
31. Zhang, X.D., Macosko, C.W., Davis, H.T., Nikolov, A.D., Wasan, D.T.: Role of silicone surfactant in flexible polyurethane foam. J. Colloid Interface Sci. **215**(2), 270–279 (1999)
32. Harikrishnan, G., Patro, T.U., Khakhar, D.V.: Polyurethane foam-clay nanocomposites: nanoclays as cell openers. Ind. Eng. Chem. Res. **45**(21), 7126–7134 (2006)
33. Xu, Z.B., Tang, X.L., Gu, A.J., Fang, Z.P.: Novel preparation and mechanical properties of rigid polyurethane foam/organoclay nanocomposites. J. Appl. Polym. Sci. **106**(1), 439–447 (2007)
34. Wang, Z.Z., Li, X.Y.: Mechanical properties and flame retardancy of rigid polyurethane foams containing SiO_2 nanospheres/graphene oxide hybrid and dimethyl methylphosphonate. Polym. Plast. Technol. Eng. **57**(9), 884–892 (2018)
35. Hoseinabadi, M., Naderi, M., Najafi, M., Motahari, S., Shokri, M.: A study of rigid polyurethane foams: the effect of synthesized polyols and nanoporous graphene. J. Appl. Polym. Sci. **134**(26), 45001 (2017)
36. Novoselov, K.S., Geim, A.K., Morozov, S.V., Jiang, D., Zhang, Y., Dubonos, S.V., Grigorieva, I.V., Firsov, A.A.: Electric field effect in atomically thin carbon films. Science **306**(5696), 666–669 (2004)
37. Chiappone, A., Roppolo, I., Naretto, E., Fantino, E., Calignano, F., Sangermano, M., Pirri, F.: Study of graphene oxide-based 3D printable composites: effect of the in situ reduction. Compos. Part B Eng. **124**, 9–15 (2017)

38. Li, Y.L., Wang, S.J., Wang, Q.: Enhancement of tribological properties of polymer composites reinforced by functionalized graphene. Compos. Part B Eng. **120**, 83–91 (2017)
39. Lin, F., Xiang, Y., Shen, H.S.: Temperature dependent mechanical properties of graphene reinforced polymer nanocomposites—a molecular dynamics simulation. Compos. Part B Eng. **111**, 261–269 (2017)
40. Bernal, M.M., Martin-Gallego, M., Molenberg, I., Huynen, I., Manchado, M.A.L., Verdejo, R.: Influence of carbon nanoparticles on the polymerization and EMI shielding properties of PU nanocomposite foams. RSC Adv. **4**(16), 7911–7918 (2014)
41. Yan, D.X., Xu, L., Chen, C., Tang, J.H., Ji, X., Li, Z.M.: Enhanced mechanical and thermal properties of rigid polyurethane foam composites containing graphene nanosheets and carbon nanotubes. Polym. Int. **61**(7), 1107–1114 (2012)
42. Jing, Q.F., Liu, W.S., Pan, Y.Z., Silberschmidt, V.V., Li, L., Dong, Z.L.: Chemical functionalization of graphene oxide for improving mechanical and thermal properties of polyurethane composites. Mater. Design **85**, 808–814 (2015)
43. Mitsunaga, M., Ito, Y., Ray, S.S., Okamoto, M., Hironaka, K.: Intercalated polycarbonate/clay nanocomposites: nanostructure control and foam processing. Macromol. Mater. Eng. **288**(7), 543–548 (2003)
44. Colton, J.S., Suh, N.P.: The nucleation of microcellular thermoplastic foam with additives. 1. Theoretical considerations. Polym. Eng. Sci. **27**(7), 485–492 (1987)
45. Wood, G.: The ICI Polyurethane Handbook, 2nd edn. Wiley, New York (1990)
46. Oertel, G.: Polyurethane Handbook. Hanser, New York (1993)
47. Klempner, D., Frisch, K.C.: Handbook of Polymeric Foams and Foam Technology. Oxford University Press, New York (1991)
48. Xia, H.S., Song, M.: Intercalation and exfoliation behaviour of clay layers in branched polyol and polyurethane/clay nanocomposites. Polym. Int. **55**(2), 229–235 (2006)
49. Hu, X.M., Cheng, W.M., Nie, W., Wang, D.M.: Flame retardant, thermal, and mechanical properties of glass fiber/nanoclay reinforced phenol-urea-formaldehyde foam. Polym. Compos. **37**(8), 2323–2332 (2016)
50. Danowska, M., Piszczyk, L., Strankowski, M., Gazda, M., Haponiuk, J.T.: Rigid polyurethane foams modified with selected layered silicate nanofillers. J. Appl. Polym. Sci. **130**(4), 2272–2281 (2013)
51. Dolomanova, V., Rauhe, J.C.M., Jensen, L.R., Pyrz, R., Timmons, A.B.: Mechanical properties and morphology of nano-reinforced rigid PU foam. J. Cell. Plast. **47**(1), 81–93 (2011)
52. Bharadwaj, R.K.: Modeling the barrier properties of polymer-layered silicate nanocomposites. Macromolecules **34**(26), 9189–9192 (2001)
53. Theon, J.A.: In: Wilson, F.G., Ulrich, H., Riese, W. (eds.) Reaction Polymers, 2nd edn. Hanser Publishers, Munich, Germany (1992)
54. Smits, G.F.: Effect of cellsize reduction on polyurethane foam physical properties. J. Build. Phys. **17**(4), 309–329 (1994)
55. Harikrishnan, G., Singh, S.N., Kiesel, E., Macosko, C.W.: Nanodispersions of carbon nanofiber for polyurethane foaming. Polymer **51**(15), 3349–3353 (2010)

Chapter 4
PU Rigid Nanocomposite Foams Containing Cylindrical Nanofillers

4.1 Introduction

For cylindrical nano-sized fillers, carbon nanotubes, carbon nanofibers and bio-based cellulosic nanofibers can be given as the nanocylinder examples that have been used in PURNCFs [1–14]. However, among PURF/cylindrical nano-sized filler studies that were published in the literature, the most extensively researched additive material is carbon nanotubes (CNTs) due to their outstanding properties in PURFs such as thermal and electrical conductivity, thermal stability, mechanical, thermal and thermomechanical properties [1–11].

Carbon nanotubes are generally classified under two main groups such as multi walled carbon nanotubes (MWCNTs) and single walled carbon nanotubes (SWCNTs) according to their specific wall structures. Typically, within their geometrical structures, MWCNTs consist of multiple nanotubes in which smaller carbon nanotubes are surrounded by other bigger nanotubes, and each carbon nanotube contains an external planar sheet consisting of typical rotated graphene composition [15]. Normally, CNTs are composed of typical graphite layers which are basically rotated into cylinders which have diameters of several nanometers. In contrast, SWCNTs have a single sheet of rotated planar graphene structure in which an exact cylindrical geometry is constructed consisting of 1 nm and several centimetres in terms of its base diameter, and length, respectively [16, 17]. On the other hand, MWCNTs generally have diameters of up to 100 nm and lengths of several micrometers [18]. MWCNTs are composed of multiple nanocylinders that are surrounded by each other and set apart by a gap of 0.35 nm which is almost equal the distance between planar sheets in graphite [18, 19]. In general, CNTs have outstanding physical and mechanical properties in comparison with other nanofillers. The density of CNTs is very low which is about 1.3 g/cm^3 and their Young's modulus value is higher than 1 TPa which is considered to be a very high value in comparison with that of carbon fibers [6, 20]. Furthermore, the maximum tensile strength of CNTs was measured as 63 GPa [6, 21]. Thus, because of these facts, CNTs have been extensively used as nanoadditives for

E. Burgaz, *Polyurethane Insulation Foams for Energy and Sustainability*,
Advanced Structured Materials 111,
https://doi.org/10.1007/978-3-030-19558-8_4

the fabrication of PURFs with enhanced flame retardant, morphological, mechanical and thermomechanical properties [22].

The effects of CNT addition to PURFs have been investigated by several researchers in terms of morphology, density, thermal and electrical conductivity, thermal stability, mechanical, thermal and thermomechanical properties [1–11]. However, the most important problems that should be solved in the preparation of PURNCFs consisting of CNTs are aggregation and dispersion problems of carbon nanotubes throughout PURF matrix. Geometrical large aspect ratios and strong π-π type of interactions are mainly responsible for the non-uniform distribution of CNTs within PURF matrix [1, 23–25]. Consequently, too much aggregation in PURF/CNT systems leads to the decrease of interfacial area and thus thermal and mechanical properties. For this reason, many works have been carried out for solving the non-homogenous distribution and aggregation problems of CNTs [1, 23–26]. For this purpose, the compatibility between CNTs and polymer matrix was improved by performing surface modification of CNTs such as silane modification [2, 27–30], amination [1], hydroxylation [2, 31], carboxylation [1], fluorination [32] and isocyanate treatment [33].

Although the surface modified CNTs has been used in PURFs, the physicochemical compatibility between the matrix and CNTs is still not satisfactory due to the presence of weak interactions between organic functional groups of CNTs and PURF matrix [1]. However, new methods for surface modification of CNTs have been developed and these methods are more superior compared to previous ones in terms of increasing interfacial adhesion between PURF matrix and CNTs and consequently the dispersion degree of CNTs within the matrix. Only very recently, PEO-*graft*-MWCNTs were synthesized from ethylene oxide by using the ring-opening polymerization method, and these grafted MWCNTs were used in the fabrication of PURNCFs to examine the effects of using grafted MWCNTs on thermal stability, morphological and mechanical properties of PURNCFs [1]. It was shown that the cell size decreases and mechanical properties as well as glass transition temperatures of PURNCFs increase with incorporation of PEO-*graft*-MWCNTs [1]. Thus, the improvement in mechanical properties and reduction of cell size are clear indications for enhanced physicochemical compatibility and strong interfacial adhesion between surface-modified MWCNTs and PURF matrix. Previously, hydroxylated MWCNTs were surface modified with 3-Triethoxysilylpropylamine (APTS) and dipodal silane (DSi), and these modified MWCNTs were used in the fabrication of PURNCFs [2]. Based on morphological and mechanical results, the values of cell density, tensile strength and elastic modulus increased in PURNCFs consisting of 1.5 wt% of DSi-MWCNT more than those containing the same amount of APTS-MWCNT [2]. Therefore, this result also proves the formation of strong and favorable interactions between DSi-MWCNTs and PU chains in PURNCFs.

Because of high thermal conductivity values of MWCNTs, it is mainly accepted that thermal conductivity values of PURFs typically increase with the addition of MWCNTs [2, 29, 30]. Many studies have been performed to comprehend influences of MWCNT incorporation on the thermal conductivity behavior of PURFs. It was previously reported that thermal conductivity of PURFs typically increases with incorporation of low amounts (0.3 wt%) MWCNTs [34]. Moreover, thermal conductivity values of PURNCFs do not increase that much if the concentration of MWCNTs is fixed in the range of 0.05–0.5 wt%. [35]. Although thermal, mechanical and morphological properties of PURNCFs consisting of CNTs have been systematically investigated, no detailed analysis has been reported about the flame retardance or fire resistance behavior of these materials in the literature.

Very recently, PURNCFs consisting of different amounts of MWCNTs were fabricated and their thermal conductivity, thermal and flammability behaviors were investigated [4]. It was reported that the flame behavior of PURNCFs consisting of MWCNTs is strongly dependent on the cellular morphology due the effects of MWCNT presence in the foam structure. In addition, it was also found out that the flame propagation speed of PURNCFs increases with the addition 0.1 wt% MWCNT, but it further decreases for MWCNT contents above 1 wt% [4]. The flammability and the peak heat release rate of PURNCFs consisting of CNTs were reduced in a great extent due to favourable properties of CNTs. Specifically, the decrease of flammability properties and increase of fire resistance properties in PURNCFs were observed since CNT network structure formed a protective layer on the surface of foams [36]. Moreover, the maximum heat release rate was reduced by 20% by CNTs which were used as nanocylindrical fillers in PURF coating for natural fabrics [37]. It was previously shown that very small additions of CNTs enhance the mechanical strength of PURNCFs by 20% due to favourable properties of CNTs, homogenous distribution and smaller size of pores in PURNCFs. In addition, thermal stability and flammability of PURNCFs consisting of CNTs increase due to the insulation property of carbon layer that is formed on PURNCF surface [5]. Moreover, limiting oxygen index (LOI) values of PURNCFs improved moderately with incorporation of very small amounts (0.05–0.1 wt%) of CNTs [6].

Mechanical properties of PURNCFs containing minor contents (less than 2 wt%) of MWCNTs were examined particularly by focusing on the correlation between morphology and anisotropy of foam structure through the use of well-established micromechanical models. Based on experimental and modelling results, the mechanical properties of nanocomposite rigid foams were found out to be very strongly dependent on the anisotropic nature and morphology associated with the unit cell of foams [7].

Besides thermal, mechanical and flammability properties of PURNCFs, electrical properties of PURNCFs consisting of CNTs have been studied at various CNT loadings, foam densities and temperatures by several researchers [11, 38, 39]. The electrical properties of PURCNFs displayed a percolation threshold around 1.2 wt%, and the conductive PURCNFs exhibited outstanding electrical stability over a wide temperature region that spans from room temperature up to 180 °C [11]. Besides dependence of electrical properties on various CNT loadings, changing the foam density has a substantial effect on the electrical behavior of PURNCFs consisting of CNTs. For this purpose, the conductivity behavior of PURNCFs consisting of CNTs with various densities were investigated for a fixed amount of CNT addition. Based on the experimental data, a density-dependent conductor-insulator transition which is correlated to the morphology of PURNCFs was observed [38]. Besides the relationship between foam density and electrical properties, changing the foam temperature has a considerable influence on the electrical behavior of PURNCFs consisting of CNTs. In PURNCFs consisting of 1 wt% CNT and density range of 200–550 kg/m^3, the electrical resistivity of PURNCFs was studied with respect to temperature, and it was shown that upon heating, the electrical resistivity of PURNCFs decreases between 25 and 100 °C since the diffusion of CO_2 gas into closed cells increases the pressure. Also, the decrease in volume resistivity was observed in PURNCFs due to the increase of conductivity by the presence of straighter and closely packed CNTs [39]. In another study, the relationship between the electrical conductivity and foaming conditions of PURNCFs consisting of different contents of MWCNTs and densities was established. Based on experimental data and modelling results based on statistical Percolation laws, it was shown that electrical properties of PURNCFs depends on morphological variables [9].

Using one type of nanofiller such as CNTs, nanoclay or nanosilica, etc in PURNCFs improves thermal stability, flame retardancy, mechanical, thermomechanical and morphological properties. However, besides using one type of nanofiller in PURNCFs, hybrid nanofiller systems consisting of at least two kinds of nanofillers with distinct properties can be used in order to fabricate PURNCFs with more superior properties compared to those consisting of only one type of nanofiller. Previously, hybrid nanofiller systems consisting of MMT and CNTs were fabricated and distributed in PURFs from 0.25 up to 1 wt% by in situ polymerization [8]. The cell sizes of foams reduced and their compressive modulus values increased with the incorporation of hybrid nanofillers into PURFs. The enhancement in morphological and mechanical properties was observed because of the fact that the dispersion of MMTs in PURNCFs is facilitated by the presence of CNTs and positive synergy is provided by combining favourable mechanical and physical properties of CNTs and MMTs [8].

Besides using CNTs as nanocylinders in PURNCFs, there are also a few studies in the literature about the incorporation carbon nanofibers into PURFs to enhance thermal stability, flame retardancy and mechanical properties. Previously, PURNCFs consisting of CNFs were fabricated, and it was shown that incorporation of only 1 wt% CNFs into PURFs increases tensile, compression and flexural modulus values by 86, 40 and 45% whereas tensile, compression and flexural strengths increase by 35, 57 and 40% compared to pure PURF, respectively. In addition, the thermal decomposition temperature of PURNCF consisting of only 1 wt% CNF increases about 18 °C in comparison with pure PURF [13]. In another study of PURNCFs consisting of vapor grown CNFs, it was reported that the reaction kinetics and conditions for the reactive foaming process were not affected by inclusion of CNFs, and the closed-cellular morphology becomes more uniform without any defects with the addition of CNFs to PURFs. Moreover, fire resistance properties and thermal conductivity values of PURNCFs consisting of less than 0.5 wt% were observed to be much more better compared to pure PURF [14]. In addition to synthetic CNFs, a few studies have used bio-based nanofibers in PURFs in order to increase the inspiration for using cheaper renewable materials from biomass for the development of PURNCFs with improved properties. Recently, the influences of cellulose nanofiber modification specifically concentrating on hydrophobic modification with latex on morphology, foam density and height, mechanical and water vapor properties of lignin and soy-based green PURFs were studied [12]. Impact and compressive mechanical properties of soy-based green PURFs enhanced via incorporation of 1 wt% hydrophobic cellulosic nanofiber. Also, open cell content of green PURNCFs was significantly reduced from 90 to 12% with cellulosic nanofiber and lignin reinforcement [12].

4.2 Morphology

The incorporation of nanocylindrical particles in the form of nanotubes and nanofibers into PURFs in a large extent improves the physical and mechanical properties. However, the major problem to overcome during the mixing process of nanofibers and nanotubes with PURFs is the aggregation behavior of these nanoparticles. Nanofibers and carbon nanotubes have a great tendency towards aggregation due to their high surface areas. Most of the time, the particle sizes of nanocylindrical particles are practically larger than those of their individual primary particles due to their highly aggregation behavior. Thus, for this reason, over the years, researchers both in academia and industry have developed so many techniques such as various surface modification techniques and ultrasonication in order to reduce the aggregation behavior of cylindrical nanoparticles in PURNCFs. Among these techniques, ultrasound sonication has been one of the most promising and easiest methods that has been used in the preparation of PURNCFs consisting of homogenous distribution of nanoparticles within the PU matrix. However, the experimental conditions of using ultrasonication during the fabrication of PURNCFs should be carefully optimized in order to develop highly efficient and advanced PURNCFs with improved physical and mechanical properties.

Kabir et al. examined influences of using sonication method consisting of various technical parameters on the preparation conditions and characteristics of PURNCFs that were strengthened with carbon nanofibers (CNFs) [25]. Authors used pseudo-static compression tests to explore mechanical properties of PURNCFs consisting of CNFs, and they particularly measured the compressive yield strength of PURNCFs which were prepared with different experimental conditions. Specifically, authors studied influences of various experimental parameters such as sonication temperature, selection of either polyol or polyisocyanate as the sonication medium, sonication time and power, concentration of nanoparticles and total amount of foam on the dispersion properties of CNFs in PURFs. Based on the results, it was found out that the application of ultrasound to CNF dispersions in the polyisocyanate part leads to improved properties. In addition, it was also reported that there is an optimum time limit for the sonication, and this time limit is closely dependent on the sonication power, concentration of CNFs in the dispersions and the foam amount that is used during the experiments [25].

Figure 4.1 shows SEM images of pure CNF and PURNCFs consisting of 0.5 wt% CNF with various degrees of sonication time such as (a) pure CNF, (b) PURNCF

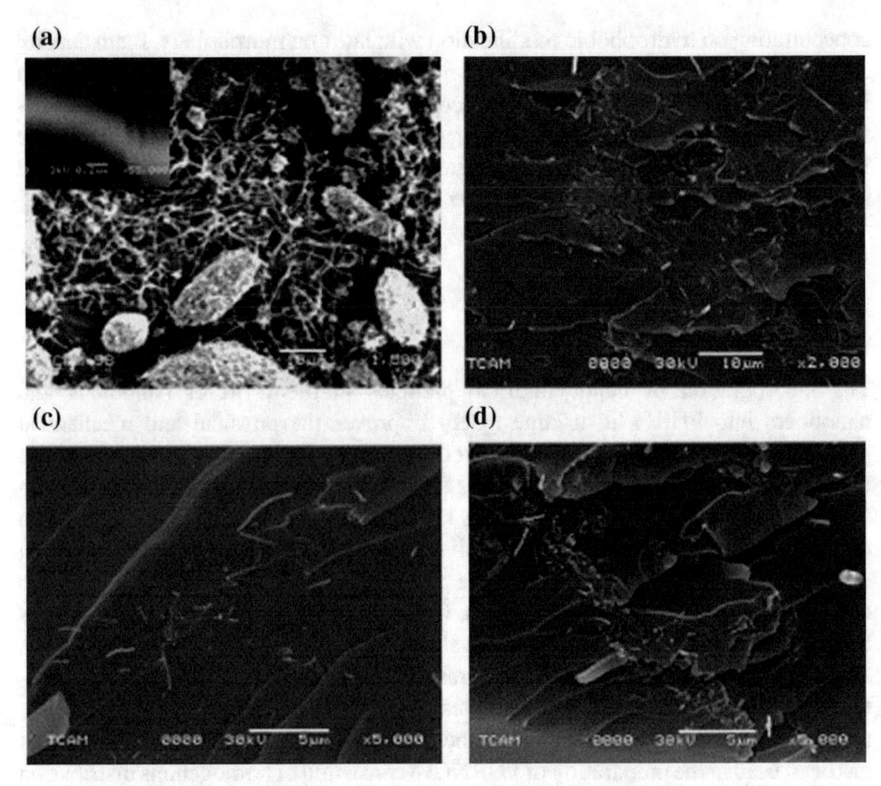

Fig. 4.1 SEM images of pure CNF and PURNCFs consisting of 0.5 wt% CNF with various degrees of sonication time such as **a** pure CNF, **b** PURNCF consisting of 0.5 wt% CNF without sonication, **c** PURNCF consisting of 0.5 wt% CNF with sonication time of 20 min, **d** PURNCF consisting of 0.5 wt% CNF with sonication time of 40 min. Reproduced with permission from Ref. [25]. Copyright 2007 Elsevier Ltd.

consisting of 0.5 wt% CNF without sonication, (c) PURNCF consisting of 0.5 wt% CNF with sonication time of 20 min, (d) PURNCF consisting of 0.5 wt% CNF with sonication time of 40 min. Based on the results, as depicted in Fig. 4.1a, the mean diameter of CNFs was noticed as 200 nm [25]. Based on the SEM image of PURNCF consisting of 0.5 wt% with no sonication (Fig. 4.1b), it was observed that aggregates of CNFs are visible within the PURNCF matrix since CNFs consisting of this small diameter have high adhesive forces between themselves due to their high surface areas. As shown in Fig. 4.1c, it was pointed out that the dispersion of CNFs turns out to be homogenous after the application of ultrasound for 20 min. Furthermore, it was also shown that the application of 40 min sonication resulted in the formation of aggregation of several CNFs [25]. The reason behind the increase in the formation of aggregates containing several CNFs as a function of sonication time was explained such that CNF agglomerates cannot be separated if the sonication time or energy is not that high enough, however, on the other hand, aggregates consisting of several CNFs can be formed if the sonication time or energy is too high. Therefore, based on these results, it was concluded that the most appropriate sonication time has to be applied to maximize the distribution capacity and performance of CNFs in either polyol or polyisocyanate component of PURNCFs [25].

The homogenous distribution of nanofillers within PURF components such as polyol and polyisocyanates before the reactive foaming process is one of the most important criteria in the fabrication of PURNCFs with advanced physical and mechanical properties. During sample preparation of PURNCFs with nano-sized fillers, the aggregation problem usually occurs due to very high surface areas of nanofillers. Thus, before forming the solid PURNCF product, the mixing of nano-sized fillers within polyol or polyisocyanate component should be successfully achieved by benefiting from the less viscous medium property of these liquid components. During the mixing process of nanofillers in liquid medium of polyisocyanate or polyol component, homogenous distribution of nanofillers with decreased amount of aggregation problem is achieved by applying ultrasonication with specified amount of time and temperature. However, for each PURNCF/nanofiller system, the ultrasound conditions might vary due to different physical and chemical properties of nano-sized fillers that are used as nanoadditives in the fabrication of PURNCFs with improved properties. Thus, for this reason, the optimum conditions for ultrasonication method should be found out for each PURCNF/nanofiller combination to eliminate the chance of aggregation problem in nanofiller/polyol or nanofiller/polyisocyanate liquid mixtures before the reactive foaming process of PURNCFs.

Zhou et al. used carrot nanofibers (CtNFs) to strengthen physical and mechanical properties of biobased castor oil polyol based PURNCFs [40]. In their study, authors utilized the procedure of well-distribution of CtNFs within polyol component, and they also examined the characteristics of CtNF reinforced PURNCFs. In addition, during the preparation of PURNCFs, the homogenous distribution of CtNFs in the polyol component was accomplished, and PURNCFs consisting of various amounts of CtNFs were fabricated by using the co-solvent-assisted mixing method. Based on the results, it was found out that CtNF reinforced PURNCFs displayed homogenous distribution of cell sizes. Moreover, compressive strength and modulus values of PURNCFs were extensively improved compared to those of pure PURF. In addition, the best compression properties were obtained in PURNCFs consisting of 0.5 phr CtNF. Based on the theoretical calculations, the modulus of solid PURNCFs was improved compared to pure PURF. In practical terms, based on compression property results with respect to foam density, fabricated biobased PURNCFs consisting of CtNFs performed equal to the performance level of commercial PURFs. Based on these results, it was concluded that these biobased PURNCFs consisting of CtNF could be used in applications such as core materials for lightweight sandwich composites [40].

Figure 4.2 shows (a) the real picture of pure biobased PURF, the microstructures of (b) pure biobased PURF and (c) biobased PURNCF with 0.25 phr CtNF. Based on the results, it was found out that closed-cellular structure and cellular shapes of biobased PURNCFs did not change that much with increasing amount of CtNFs. In addition, it was shown that most of the cells have the closed-cellular morphology with spherical shapes of cell geometry [40]. Furthermore, it was also reported that the

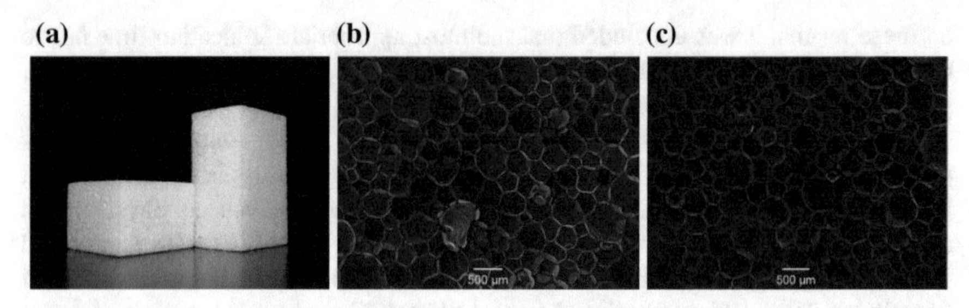

Fig. 4.2 **a** Real photograph of pure biobased PURF, the microstructures of **b** pure biobased PURF and **c** biobased PURNCF with 0.25 phr CtNF. Reproduced with permission from Ref. [40]. Copyright 2016 Elsevier Ltd.

morphological parameters of cell walls and struts in biobased PURNCFs consisting of 0.25 and 0.5 phr of CtNFs have almost the same morphology due to the uniform dispersion of CtNF and their good interactions with the PURNCF matrix at these low amounts of CtNF compared to PURCNF consisting of 1 phr of CtNF. Moreover, aggregates of CtNFs were observed in the cell struts and cell walls. It was shown that CtNF has a good dispersion in the polyol component and it has also uniform distribution within the PURNCF matrix. Based on the experimental observations, it was revealed that increasing the CtNF content in PURNCFs led to a slightly narrower cell size distribution, and formed a much more uniform PURNCF with smaller cell size compared to pure PURF. Thus, according to SEM data, it was confirmed that the existence of CtNFs in the reaction medium led to emergence of homogenous cell size dispersions and smaller cell sizes since CtNFs function as multiple numbers of cell nucleating agents during the reactive foaming process [40]. It is very well known that using nanofillers in PURFs decreases cell sizes and enables the formation of closed-cellular morphology with uniform distribution of cell sizes and higher cell densities since nanofillers form favorable interactions with PU chains due to their higher surface areas and roles as bubble nucleation centers during the reactive foaming process [7–9, 11, 25, 38–40]. Furthermore, the fabrication of PURNCFs with improved properties is mainly dependent on the establishment of favorable interactions between the surface of nanofillers and PU chains. Thus, for this reason, various surface modification methods should be employed to chemically attach surface functional groups onto nanofillers that are physically and chemically compatible with PU matrix [1, 2, 30, 32, 33]. By this way, the number of favorable interactions between surface modified nanofillers and PURNCF matrix could be maximized to produce PURNCFs with improved properties such as reduced cell sizes and higher cell densities.

Faruk et al. fabricated green PURFs consisting of completely soy polyol and lignin and optimized experimental conditions such as amounts of components during sample preparation to improve physical and mechanical properties of green rigid foams [12]. Authors also incorporated the cellulosic nanofibers into green PURFs to increase the rigidity of system. They systematically investigated influences of important experimental parameters such as various types of surface modified cellulose fibers on morphology, foam density, water vapor and mechanical properties of PURFs. SEM was used to analyze the change in closed-cellular morphology of PURFs with respect to incorporation of cellulose nanofibers. In addition, they performed odor tests on PURFs to analyze the odor concentration of 100% soyol based PURFs consisting of lignin and nanofiber. In accordance with experimental data, it was found out that mechanical properties of PURNCFs were enhanced due to beneficial effects created by incorporation of surface modified cellulose fibers. Furthermore, odor concentrations of PURNCFs consisting of nanofibers were diminished greatly compared to absolutely synthetic based polyol PURFs due to the replacement of isocyanate content. In accordance with results, it was concluded that PURFs with advanced properties could be fabricated by replacing some portion of isocyanate with lignin and addition of nanofiber [12].

Figure 4.3 shows the closed-cellular morphology of pure PURF and soy-based PURNCFs consisting of different amounts of lignin and surface modified cellulose fibers. In accordance with SEM results, it was shown that incorporation of cellulose micro and nanofibers that were not gone through any kind of treatment into PURFs diminished the cell sizes of rigid foams [12]. In addition, it was shown that the reactive foaming process and foam rising of PURNCFs was retarded and hindered due to inclusion of cellulose fibers to PURF system. Authors tried to look at the nanofiber structures by using SEM and TEM techniques but they could not obtain any good results since the SEM resolution was not good enough to see nanofibers. Authors performed measurements of characteristic features of closed-cellular morphology based on counted 400 cells from actual SEM micrographs. It was shown that incorporation of both lignin and surface modified cellulose nanofiber diminished cell sizes down to 410–190 μm. In addition, it was also reported that the spherical geometry and shape of cells in PURFs were not affected that much with introduction of lignin and cellulosic nanofibers into PURFs [12].

In agreement with previously published other PURNCF/NF works [25, 40], favorable interactions between the surface of NFs and PU chains led to homogenous distribution of NFs within PURNCF matrix. Also, homogenous distribution of NFs throughout PURNCF matrix paved the way to maximize the number of nucleation centers which enable the formation of cells with smaller sizes, higher cell density and undisturbed spherical cell geometries.

Harikrishnan et al. prepared PURNCFs consisting of very small amount of carbon nanofibers (CNFs) by creating a network structure of CNFs in dispersions of polyisocyanate and polyol components of PURFs [14]. Authors benefited from the reactive foaming process of these CNF dispersions in PURF components to produce PURFs with improved properties. According to results, it was pointed out that the conditions of polymerization kinetics and reactive foaming process in PURFs were not negatively influenced with incorporation of fillers. In addition, it was shown that CNF inclusion led to the emergence of more homogeneous closed-cellular morphology in PURNCFs. Furthermore, thermal conductivity and fire resistance properties of PURNCFs were improved with incorporation of very low amount of filler which was less than 0.5 wt%. Also, it was reported that open cells or structural defects

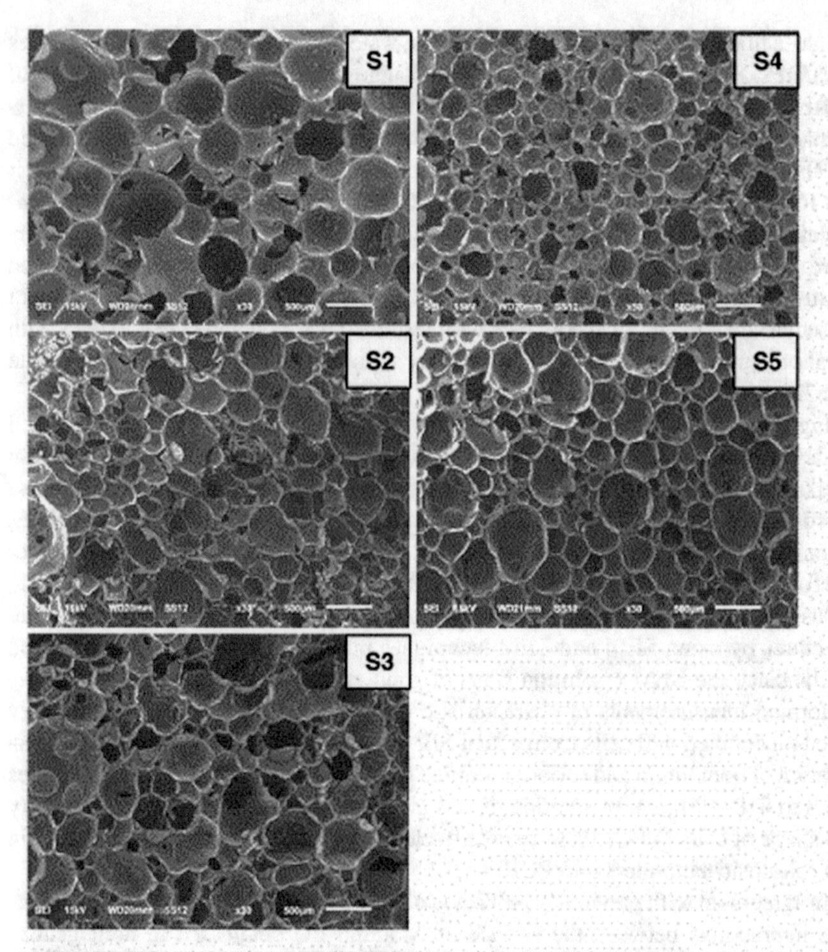

Foam Sample	Notation
Control PU foam	S1
PU foam, 5% lignin, 1% microfiber, isocynate reduced 7%	S2
PU foam, 5% lignin, 1% nanofiber, isocynate reduced 7%	S3
PU foam, 5% lignin, 1% Enzyme treated nanofiber, isocynate reduced 7%,	S4
PU foam, 5% lignin, 1% Hydrophobic (Latex) treated nanofiber, isocynate reduced 7%	S5

Fig. 4.3 Closed cellular morphology of pure PURF and soy-based PURNCFs consisting of lignin and cellulose fibers. Reproduced with permission from Ref. [12]. Copyright 2013 Springer Science + Business Media

within the closed-cellular morphology were not formed with the addition of CNFs to PURFs [14].

Figure 4.4 displays SEM micrographs of (a) pure CNF (b) pure PURF (c) PURNCF which was prepared from 1% CNF dispersion in polyol (d) PURNCF which was prepared from 1% CNF in MDI (e) magnified pure PURF (f) magnified PURNCF which was prepared from 1% CNF in MDI. According to results, it was shown that incorporation of CNF into PURFs moderately diminished the cell size of foams [14]. As shown in Fig. 4.4e, f, it was observed that nanofibers are primarily located on the cell windows of rigid foams. In addition, the closed cellular morphology (the amount of closed cell contents and other features of closed cells) in both pure PURF and PURNCFs is basically equal to each other and shares the same characteristics. Based on this experimental finding, it was confirmed that incorporation of CNFs into PURFs did not lead to the formation of open cells and did not change the closed-cellular morphology in PURNCFs. Moreover, the inclusion of CNFs into PURFs did not clear the way for escape of blowing agent since CNFs did not damage the closed-cellular morphology during the reactive foaming process [14]. In accordance with these results, it is well-noted that introducing very small amounts of CNFs into PURFs did not change the characteristics of closed cellular morphology and thus, these optimum contents (1%) of CNFs are basically satisfactory for the fabrication of PURNCFs with enhanced mechanical and thermal insulation properties.

During the period of last twenty years, researchers both in academia and industry have focused on the development of sustainable PURFs using biologically renewable materials to decrease the application of synthetic polyols and polyisocyanates in the fabrication of PURFs. Recently, Septevani et al. studied on the fabrication of cellulose nanocrystal (CNC) filled PURNCFs which were prepared from a 80:20 mixture of polyether and bio-based polyester polyols to improve mechanical properties and thermal insulation characteristics of PURFs [41]. Based on the results, it was shown that thermal conductivity values of PURNCFs diminished by 2.4% with inclusion of 0.4 wt% CNC to PURFs. In addition, the same authors also previously observed that 0.4 wt% introduction of acid-hydrolysed CNC into PURFs which had the same type of polyether polyol reduced thermal conductivity values by 5% due to improved nucleation property of modified CNC. In accordance with SEM results, it was shown that incorporation of bio-based polyester polyols which do not have ionic nature prevented CNC from functioning as a nucleation agent during the reactive foaming process. In addition to this finding, specific compressive strength values diminished compared to the system having synthetic polyols due to unfavorable intermolecular interactions between CNC and bio-based polyester polyols. However, in contrast, it was revealed that the specific modulus perpendicular to foam rise was extensively improved compared to pure PURF due to effective orientation of CNC parallel to foam rise [41].

Figure 4.5 shows the morphological changes of PURNCFs with respect to addition of CNC and ultrasonication treatment into the polyol system which is a mixture of bio- and synthetic-based polyols. According to results, it was shown that CNC incorporation into PURFs prepared from bio-based polyol did not change

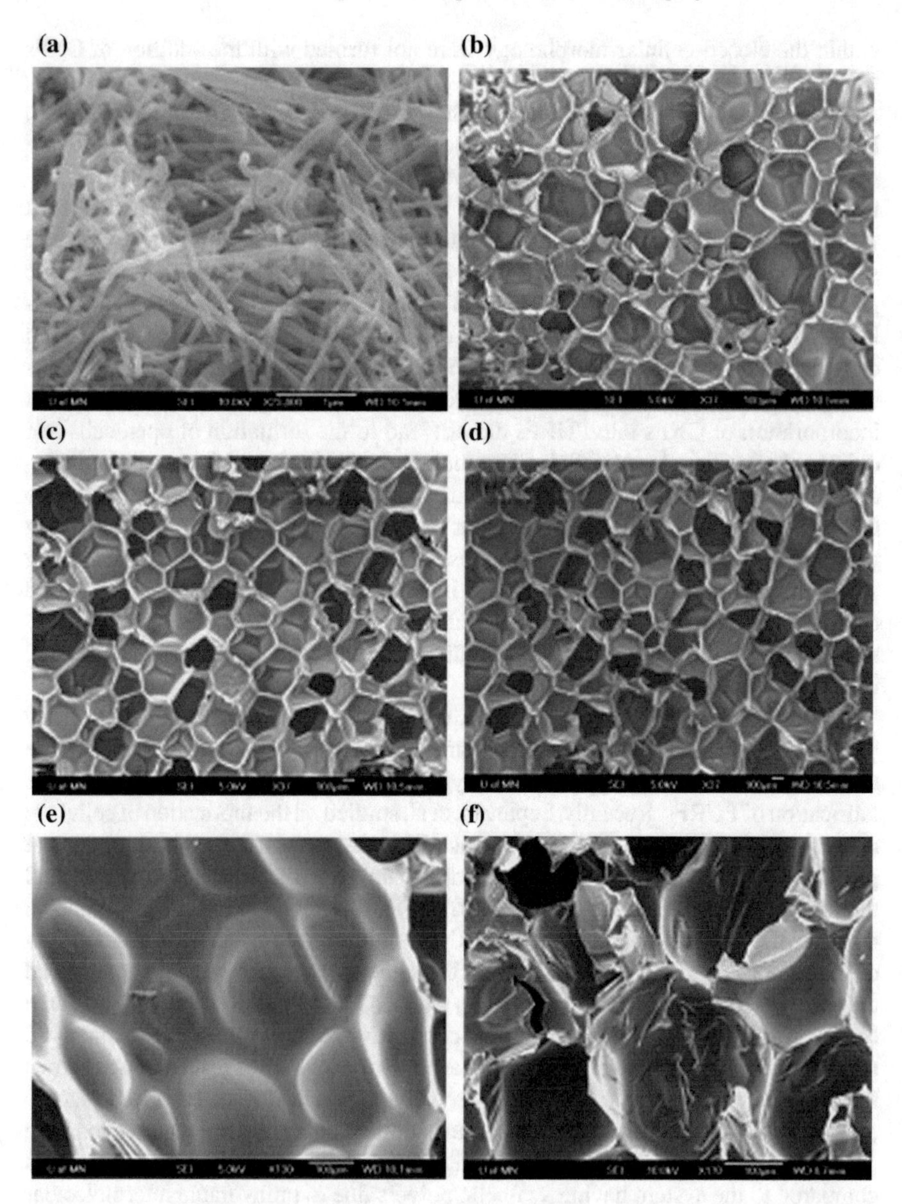

Fig. 4.4 SEM images of **a** pure CNF, **b** pure PURF, **c** PURNCF which was prepared from 1% CNF dispersion in polyol, **d** PURNCF which was prepared from 1% CNF in MDI, **e** magnified pure PURF, **f** magnified PURNCF which was prepared from 1% CNF in MDI. Reproduced with permission from Ref. [14]. Copyright 2010 Elsevier Ltd.

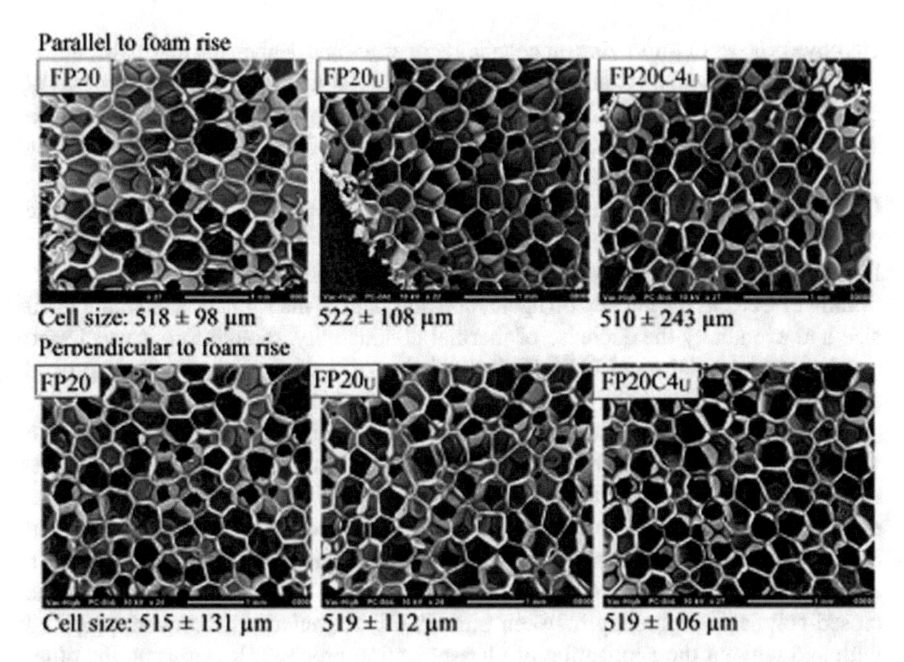

Fig. 4.5 Morphological changes of PURNCFs with respect to addition of CNC and ultrasonication treatment into the polyol system which is a mixture of bio- and synthetic-based polyols. Reproduced with permission from Ref. [41]. Copyright 2017 Elsevier Ltd.

closed-cellular morphology of PURNCFs compared to pure PURF [41]. More specifically, as shown in Fig. 4.5, it was shown that no big difference was observed in the closed-cellular morphology (cell size and cell content) of PURNCFs in both two directions to foam rise. However, another study that was published by Septevani et al. [42] showed that average cell size of foams which are parallel to foam rise direction was reduced due to the presence of CNC which acted as a nucleation agent during the reactive foaming process. However, in Fig. 4.5, the reason behind similarities between the closed-cellular morphology was explained such that the presence of bio-based polyol in sample FP20C4U decreased the job of CNC as the nucleation agent, and thus led to the formation of cell sizes that are similar to pure PURF [41]. It is well-known that the favorable interactions between nanofillers and isocyanate component during the reactive foaming process enhances the dispersion of nanofillers throughout the dispersion medium, thus leading to formation of smaller cell sizes with uniform morphology due to the heterogenous nucleation ability of nanofillers in the reaction medium. Thus, in the system as shown in Fig. 4.5, the nucleation ability of CNCs was suppressed due to the presence of bio-based polyol component in PURF formulation, thus eliminating the suitable environment for the reduction of cell size in PURNCFs consisting of CNCs [41].

Previously, Septevani et al. performed a cheaper and continuous method to improve thermal insulation properties of PURFs [42]. Their study was focused on an effective process of improving mechanical properties and decreasing thermal conductivity of PURFs with no change in foam density via inclusion of very low amounts of cellulose nanocrystals (CNCs) into PURFs by using an optimized solvent-free ultrasonication method. Based on the results, it was shown that the thermal conductivity of PURFs was reduced by 5% compared to pure PURF with incorporation of 0.4 wt% CNC which was not gone through any specific method of surface modification. The rationale behind this reduction in thermal conductivity was explained such that the favorable interaction between CNC and the polyol component increased the functioning of CNC as the nucleation agent, and thus caused the formation of smaller cell size and eventually the decrease of thermal conductivity. In addition, it was found out that modulus values of PURNCFs consisting of optimized conditions improved perpendicular to foam rise compared to that of pure PURF [42].

Figure 4.6 presents SEM micrographs of cross sectional sections of pure PURF with and without ultrasonication and PURNCFs consisting of CNC and ultrasonication parallel to foam rise. Based on the results, it was shown that the ultrasonication process affected CNC's dispersion quality in the polyol component and also changed thermal conductivity values and closed-cellular morphology of PURNCFs [42]. In addition, it was also reported that no big difference was observed in terms of the closed-cellular morphology between pure PURF formulations that were prepared with and without the application of ultrasonication process. However, on the other hand, it was revealed that the cell size was changed in ultrasonicated PURNCFs consisting of 0.4 and 0.8 wt% CNC parallel to foam-rise direction compared to pure PURF that was prepared with the application of ultrasonication process. The reason behind the change in cell size with the addition of CNC was explained such that CNC orientation parallel to foam rise during reactive foaming event decreased cell size slightly which is in agreement with other studies in which carbon nanofibers are oriented in polystyrene foams parallel to foam rise direction [42, 43]. Thus, based on these findings, the application of ultrasonication and incorporation of CNC in PURFs resulted in the decrease of cell size because CNCs acted as heterogenous nucleation centers. In addition, CNC orientation parallel to foam rise direction acted as blockades against the growth of cells during the reactive foaming process, thus leading to the observation of slightly smaller sizes in PURNCFs consisting of CNC in comparison with pure PURF that was prepared with ultrasonication.

Chen et al. synthesized MWCNTs which were grafted with poly(ethylene oxide) (grafted MWCNTs) via ring-opening polymerization of ethylene oxide [1]. In addition, they fabricated PURNCFs containing MWCNTs and grafted MWCNTs to explore influences of grafted PEO molecules on morphological and mechanical properties and reactive foaming process of PURNCFs [1]. Based on the results, it was shown that using grafted MWCNTs had a little amount of impact on viscosity of polyol. However, on the other hand, it was reported that the pore size of PURNCFs was decreased with inclusion of grafted MWCNTs. Furthermore, it was revealed that incorporation of nanofiller increased the glass transition temperature and mechanical properties of PURNCFs due to favorable interactions between grafted MWCNTs and PU chains [1].

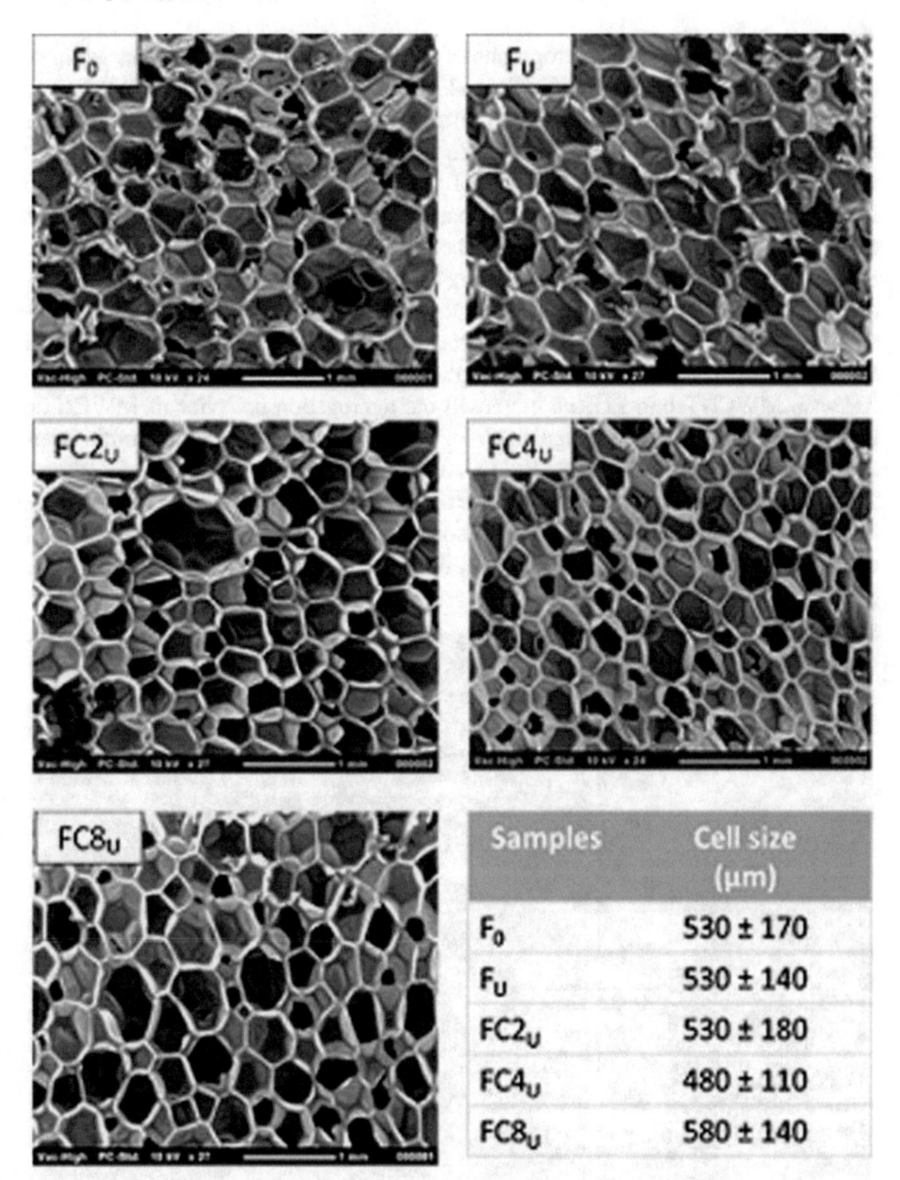

Samples	Cell size (μm)
F_0	530 ± 170
F_U	530 ± 140
$FC2_U$	530 ± 180
$FC4_U$	480 ± 110
$FC8_U$	580 ± 140

Fig. 4.6 SEM images of specimen areas of pure PURF with and without ultrasonication and PURNCFs consisting of CNC and ultrasonication along the foam rise direction. Reproduced with permission from Ref. [42]. Copyright 2017 Elsevier Ltd.

Figure 4.7 shows SEM micrographs of (a) pure PURF, PURNCFs consisting of (b to e) 0.5 up to 3.0 wt% MWCNTs, and PURNCFs consisting of (f to i) 0.5 up to 3.0 wt% grafted MWCNTs. Based on results, it was shown that the closed-cellular morphology of pure PURF was found to be regular and spherical. In addition, it was also pointed out that the cell size of PURFs was observed to be smaller with inclusion of 1.0 wt% MWCNTs in comparison with that of pure PURF. The reason behind this finding was discussed such that 1 wt% MWCNTs acted as heterogeneous nucleation centers during reactive foaming event [1, 44]. However, on the other hand, as shown in Fig. 4.7d, e, the closed cellular morphology was observed to be damaged in PURNCFs consisting of more than 2.0 wt% MWCNT. The reason behind the collapse of cellular structure was explained such that the addition of more than 2.0 wt% MWCNT into PURFs increased the aggregation behavior of MWCNTs, thus reducing the nucleation agent role of MWCNTs due to the unfavorable interactions between MCWNTs and the PURF matrix. On the contrary, it was shown that the closed cellular morphology of PURNCFs consisting of grafted MWCNTs is substantially much better in comparison with those consisting of MWCNTs at the same MWCNT concentrations [1]. Specifically, it was reported that the struts

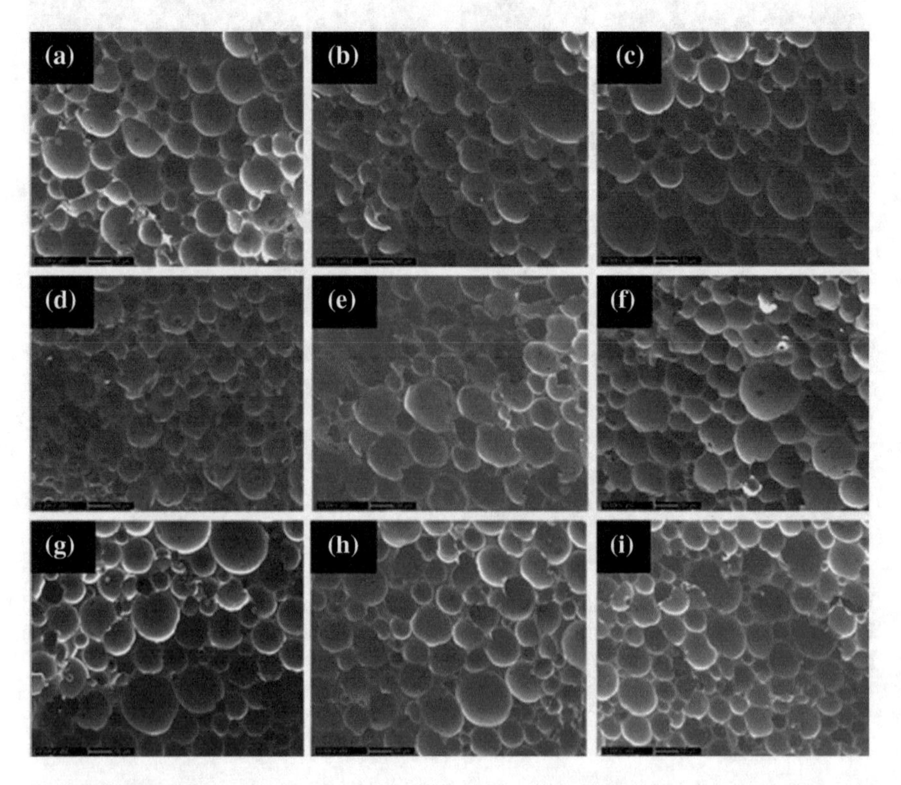

Fig. 4.7 SEM micrographs of **a** pure PURF, PURNCFs consisting of **b–e** 0.5 up to 3.0 wt% MWCNTs, and PURNCFs consisting of **f–i** 0.5 up to 3.0 wt% grafted MWCNTs. Reproduced with permission from Ref. [1]. Copyright 2018 Society of Chemical Industry

preserved their smooth appearance and the cells maintained their spherical shapes with incorporation of up to 3.0 wt% grafted MWCNTs to PURFs. The reason behind this finding was explained such that favorable interactions between grafted MWCNTs and PU chains led to uniform distribution of grafted MWCNTs throughout PURF matrix, and thus did not cause any damage to the closed-cellular morphology in PURNCFs [1].

In accordance with results from other previously published works [2, 12, 30, 32, 33], it is very well known that using surface modified nanofillers that were treated with various surface modification methods in PURFs generally leads to fabrication of PURNCFs with improved properties. Thus, the grafting technique is also one of the mostly used surface modification techniques for nanoparticles, and this technique has been employed to chemically attach surface functional groups onto nanofillers which are physically and chemically compatible with PU matrix. By this way, the number of favorable interactions between surface modified nanofillers and PURNCF matrix could be maximized to produce PURNCFs with improved properties such as very low cell sizes and very high cell densities. Thus, based on differentiation of SEM results in Fig. 4.7, it is clearly seen that using grafted MWCNTs has many advantages over using bare MWCNTs in PURFs since the detailed features of closed-cellular morphology (cell size, density and shape) are positively affected with incorporation of grafted MWCNTs to PURFs [1].

Yaghoubi et al. surface modified multi-walled carbon nanotubes consisting of hydroxyl functional groups by using 3-aminopropyltriethoxysilane (Si) and dipodal silane [2]. Then, authors prepared PURNCFs consisting of silanized MWCNTs to explore influences of these silane functionalized MWCNTs on thermal, morphological and mechanical properties of PURNCFs [2]. Based on morphological results, it was shown that incorporation of silanized MWCNTs into PURFs enhances the cell density in PURNCFs. More specifically, according to SEM and mechanical testing data, it was reported that mechanical properties of PURNCFs containing 1.5 wt% MWCNT modified with dipodal silanes were improved much more in comparison with those of PURNCFs containing 1.5 wt% Si-MWCNT Based on mechanical testing results, it was also shown that tensile strength and modulus values were positively improved because of favorable intermolecular interactions between PU chains and DSi-MWCNTs in PURNCFs [2].

Figure 4.8 displays SEM micrographs of pure PURF and DSi-MWCNT-filled PURNCFs: (a) pure PURF and PURNCFs consisting of (b) 1.5 and (c) 3 wt% MWCNT modified with dipodal silane. Based on the results, it was shown that the cell density increased and cell size reduced with incorporation of higher amounts of Si-MWCNT in both types of silanized PURNCFs [2]. The reason behind the increase of cell density in PURNCFs was explained such that silanized MWCNTs acted as nucleation centers and increased the nucleation and growth of cells during the reactive foaming process. Previously, it was shown that the bubble nucleation process was enhanced with the inclusion of small amount of well-dispersed nanoparticles which acted as extra nucleation centers within the PURF dispersions [2, 45]. Based on the cell measurements of SEM micrographs, it was revealed that the distribution of MWCNTs in PURNCFs consisting of 1.5 wt% MWCNTs modified with dipodal

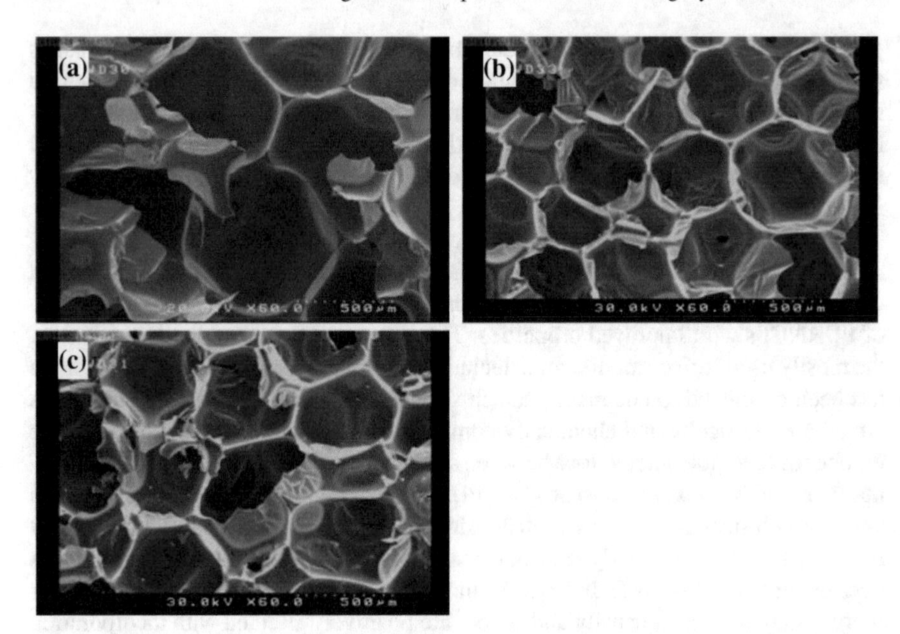

Fig. 4.8 SEM micrographs of pure PURF and PURNCFs consisting of various amounts of MWCNTs modified with dipodal silane: **a** pure PURF and PURNCFs consisting of **b** 1.5 and **c** 3 wt% MWCNTs modified with dipodal silane. Reproduced with permission from Ref. [2]. Copyright 2018 Elsevier Ltd.

silane was found out to be much more better in comparison with those in PURNCFs consisting of 1.5 wt% silanized MWCNTs [2]. On the contrary, it was also reported that extensive aggregations of MWCNTs were observed in both types of silanized PURNCFs with inclusion of 3 wt% silanized MWCNT since highly interfacial adhesion forces were present between nanoparticles [2].

Espadas-Escalante et al. investigated thermal conductivity and fire resistance properties of PURNCFs containing MWCNTs which have densities of about 45 kg/m^3 and MWCNT concentrations of from 0.1 up to 2 wt% [4]. Based on thermal conductivity measurement results, it was shown that increasing amount of MWCNTs elevated thermal conductivity values of PURNCFs moderately due to occurrence of interfacial thermal resistance in these systems. Moreover, it was observed that the content of MWCNTs in PURNCFs strongly affected fire resistant properties and closed-cellular morphology. More specifically, it was revealed that incorporation of 0.1 wt% MWCNTs to PURFs increases the flame propagation speed. However, in contrast, the flame propagation speed was observed to decrease while the amount of MWCNTs was further increased in PURNCFs. Furthermore, it was also pointed out that inclusion of MWCNTs to PURFs reduced the mass loss of PURNCFs after flame extinction. Thus, this observation suggest that the existence of MWCNTs increases resistance against fire in PURNCFs [4].

Figure 4.9 shows optical and SEM micrographs of pure PURF and PURNCFs consisting of different MWCNT contents. Based on the results, it was shown that in both

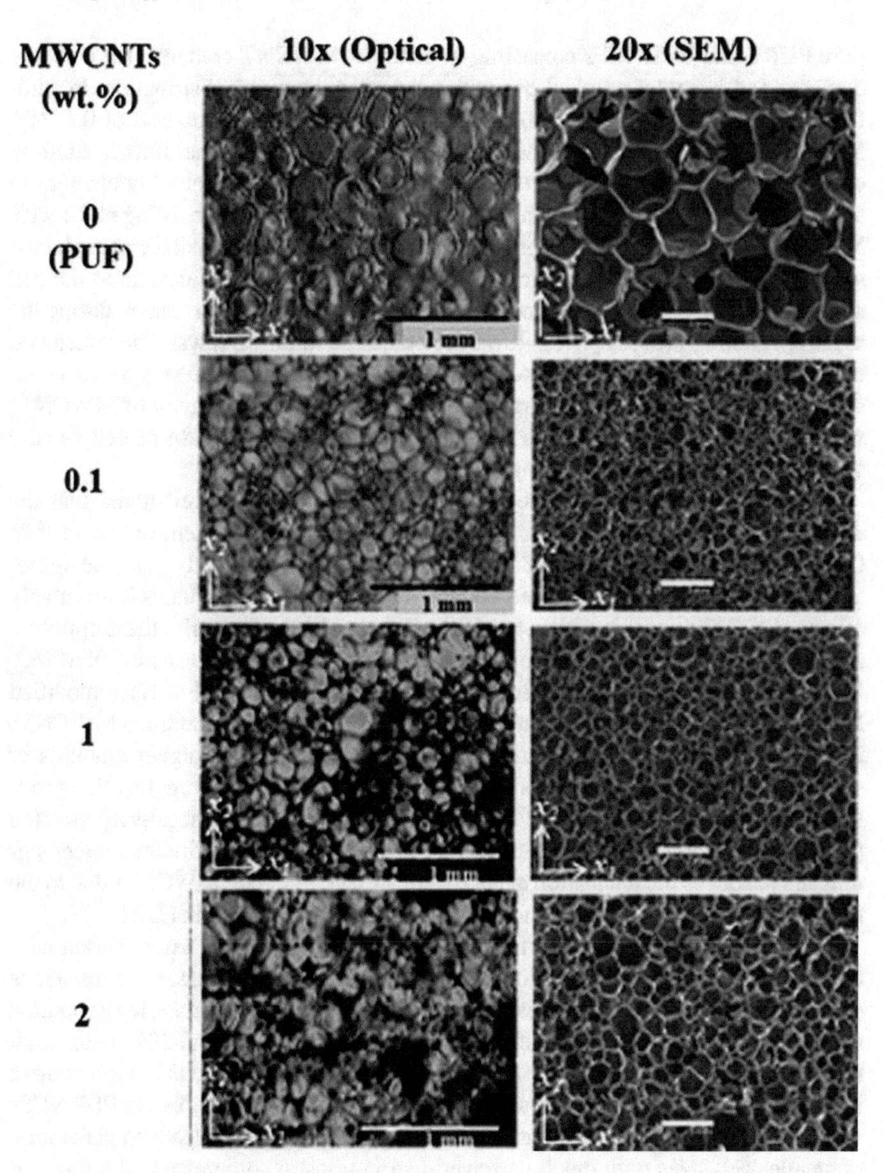

Fig. 4.9 Optical and SEM micrographs of pure PURF and PURNCFs containing various amounts of MWCNT. Reproduced with permission from Ref. [4]. Copyright The Author(s) 2016

pure PURF and PURNCFs consisting of different MWCNT contents, the cells are composed of irregular polyhedrons with a heterogeneous size distribution. In addition, it was reported that the cell density increases with incorporation of 0.1 wt% MWCNTs in PURNCFs [4]. But, in contrast, it was revealed that further addition of MWCNTs such as 1 and 2 wt% to PURNCFs did not cause any big changes in the closed-cellular morphology in comparison with PURNCFs consisting of 0.1 wt% MWCNT. The elevation of cell density because of MWCNTs inclusion was explained such that MWCNTs acted as nucleation centers for bubbles and decreased the cell size accordingly via the generation of more bubbles or nucleation centers during the reactive foaming process [4, 46–48]. Based on these findings, it was also concluded that in PURNCFs consisting of 1 and 2 wt% MWCNTs, highly and very less concentrated regions of MWCNTs emerged due to the aggregation behavior of MWCNTs which led to heterogeneous distribution of cell sizes and decrease of cell density compared to PURNCFs consisting of 0.1 wt% MWCNT [4].

Based on the experimental results given in Fig. 4.9, it is well noted that the optimum concentration of MWCNTs in PURNCFs should be kept not more than 0.1 wt% since higher contents of MWCNTs (1 and 2 wt%) usually generate aggregation problem in which the closed-cellular morphology of rigid foams is negatively affected [4]. Thus, for this reason, in PURNCF/bare MWCNTs works, these optimum amounts should be kept in mind for the fabrication of PURNCFs consisting of MWCNTs with improved properties. However, in contrast, in PURNCF/surface modified MWCNTs works [1, 2], the optimum concentrations of surface-modified MWCNTs in PURNCFs should be kept not more than 1–1.5 wt% since higher amounts of surface-modified MWCNTs (more than 2 wt%) usually generate aggregation problem in which the closed-cellular morphology of rigid foams is negatively affected [1, 2]. Also, using surface modified MWCNTs more than the optimum content significantly reduces the nucleation agent role of surface modified MWCNTs due to the unfavorable interactions between MCWNTs and the PURF matrix [1, 2].

Ciecierska et al. investigated the reinforcement of PURFs by using carbon nanotubes (CNT) or graphite as nano/micron-sized additives to enhance fire resistance thermal and mechanical properties [5]. According to experimental data, it was pointed out that mechanical properties of PURNCFs improved by almost 20% with small additions of CNTs with maximum of 0.05 wt%. The reason behind this improvement in mechanical strength was attributed such that the strengthening effect in PURNCFs was mainly coming from CNTs and also incorporation of CNTs leads to generation of smaller cell sizes with much narrower distributions in comparison with those of pure PURF. Furthermore, the size of porous structure did not change but the characteristic porosity of foams substantially increases with higher CNT contents which leads to a decrease of mechanical strength in PURNCFs. Based on thermal stability results, it was reported that the flammability reduces and thermal stability increases with incorporation of CNTs because of protective carbon coating generation on the exterior of PURNCFs [5].

Figure 4.10 displays SEM images of PURNCFs consisting of CNTs with different contents: (a)–(d) PURNCFs consisting of 0.1% CNT; (e) and (f) PURNCFs consisting of 0.01% CNT. Based on the results, two types of pores such as large pores with a size

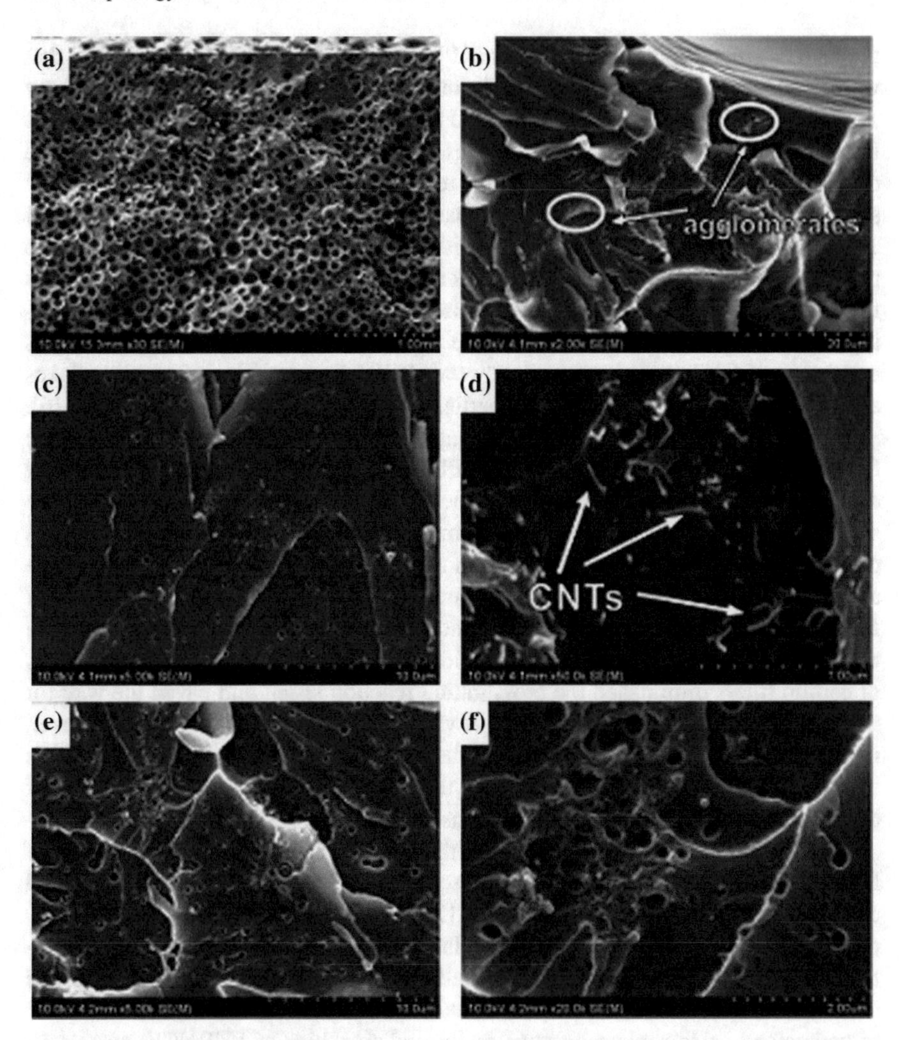

Fig. 4.10 SEM images of PURNCFs consisting of CNTs with different contents: **a–d** PURNCFs consisting of 0.1% CNT; **e** and **f** PURNCFs consisting of 0.01% CNT. Reproduced with permission from Ref. [5]. Copyright 2016 Elsevier Ltd.

span of 120–170 μm and small pores with a size span of 0.1–1 μm were observed [5]. In addition, it was also reported that the distribution of large pores was not altered by inclusion of CNTs, and in contrast, small pores were not homogeneously dispersed within the PURF matrix [5].

Ciecierska et al. explored the effects of using various types of carbon fillers on flammability and thermal decomposition properties of PURFs [6]. They incorporated MWCNTs and graphite sheets as fillers into PURFs. Authors used SEM in order to reveal the distribution state of fillers and closed-cellular structure of PURFs. In addition, they used TGA and TG including infrared spectroscopy (TG–IR) to comprehend thermal degradation behavior of PURFs and also analyzed the products that were produced as a result of thermal decomposition process. The activation energy of PURFs was also specifically measured based on the results from TGA analysis. Furthermore, authors performed specific tests in order to measure LOI and smoke density values of PURFs. Based on the results, it was found out that the thermal stability of PURNCFs did not change that much compared to that of pure PURF. The activation energy of PURNCFs consisting of CNTs was increased compared to pure PURF. Furthermore, based on TG–IR results, it was revealed that no differences in the nature of volatile products were observed in PURNCFs and pure PURF during the thermal degradation process. The flame retardancy behavior of PURNCFs consisting of CNTs was improved, however PURCFs consisting of graphite flakes exhibited better flame retardancy properties compared to CNT filled PURNCFs [6].

Figure 4.11 shows the SEM micrographs of rigid foams: (a) pure PURF; (b) PURNCF consisting of 0.1% CNT; (c) PURCF consisting of 5% graphite. Based on the SEM results, it was shown that the closed cellular morphology such as the size and shapes of cells in either nano- or micron-sized fillers containing PURFs was different compared to pure PURF [6]. In addition, it was also pointed out that the absence of any type of closed cellular morphology was observed on the bottom of foams, and the foam growth direction in mould was labelled with arrows. The reason behind observation of these pore-free regions was explained such that incorporation of carbon fillers diminished the crosslinking reaction speed and paved the way for forming these regions due to increase of viscosity in presence of CNTs. Moreover, it was previously shown that uniformly dispersed carbon nanotubes within PURFs usually have very high surface areas, and thus they have a high potential in terms of changing the viscosity of these dispersions [6, 49].

Espadas-Escalante et al. investigated the anisotropic compressive properties of PURNCFs consisting of 0.1, 1 and 2 wt% MWCNT with respect to principal directions of crosswise isotropic foams [7]. Specifically, their work was focused on the anisotropic property-structure relationships of PURNCFs which are controlled by specific features of closed cellular morphology such as cell size and density. Authors used microscopy and existent micromechanical models in order to understand anisotropic property-structure relationships of PURNCFs systemati-

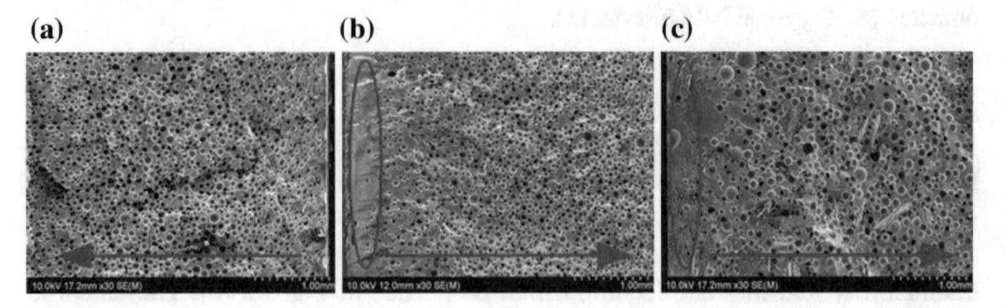

Fig. 4.11 SEM micrographs of rigid foams: **a** pure PURF; **b** PURNCF consisting of 0.1% CNT; **c** PURCF consisting of 5% graphite. Reproduced with permission from Ref. [6]. Copyright The Author(s) 2015

cally. Based on the results, it was shown that PURNCFs consisting of 0.1 and 1 wt% MWCNT exhibited much more substantial enhancement of compression properties along the main direction of foam free growth compared to pure PURF, while less property enhancement was observed along the diagonal direction of foam [7]. In addition, it was also emphasized that with inclusion of 0.1 wt% MWCNTs to PURF, structural anisotropy of PURNCFs increased and the cell structures of free growing foams were stretched out and observed to be more solid in the foam growth direction. Authors used two types of analytical models such as a simple parallelepiped cell model and a tetrakaidecahedron representation model in order to predict the rigidity and solidity ratios along two main anisotropic directions of PURNCFs. Based on experimental and modelling results, it was concluded that property-structure relationships between the anisotropic structure of unit cell and mechanical properties of final macroscopic PURNCFs are highly correlated with each other [7].

Figure 4.12 displays SEM micrographs of pure PURF and PURNCFs consisting of different MWCNT contents. Based on results, it was shown that cell shapes were irregular in the form of hexagons or pentagons, and they were more elongated along the foam free growth direction. Moreover, the shape of cells was shown to be more isotropic, spherical and they had also similar lengths in the plane which was perpendicular to foam free growth [7]. In addition, in both pure PURF and PURNCF consisting of MWCNTs, a good distribution of cell sizes was observed based on SEM micrographs. Furthermore, it was basically shown that the cell size was suddenly decreased with inclusion of 0.1 wt% MWCNTs to PURFs. However, in contrast, it was pointed out that the cell size or the closed-cellular morphology does not change any more with incorporation of 1 and 2 wt% MWCNTs. In addition, with inclusion of 1 and 2 wt% MWCNTs, the structural defects were formed in the closed-cellular morphology and the defects were observed to be oriented mostly along the plane which is perpendicular to foam free growth. Moreover, based on SEM results in Fig. 4.12, the cell density in PURNCFs was observed to be increasing with the addition of MWCNTs. This result shows that MWCNTs functioned as bubble forming agents during reactive foaming event, and thus led to elevation of cell density and reduction of cell size in PURNCFs consisting of MWCNTs [7]. The same phenomena was similarly observed in other works [1, 2, 4–6], in which MWCNTs were incorporated into PURFs, and with incorporation of small contents of MWCNTs, thermal and physical and mechanical properties of PURNCFs were substantially improved due to decrease of cell size and elevation of cell density in comparison with pure PURF.

Madaleno et al. investigated the effects of using nanohybrid particles of montmorillonite (MMT) and carbon nanotube (CNT) on the performance of PURNCFs [8]. They used chemical vapour deposition for the synthesis of hybrids, and then these MMT-CNT nanohybrid particles were incorporated into PURFs using an in situ polymerization process [8]. Authors used SEM, TEM and optical microscopy in order to evaluate the closed cellular morphology of PURNCFs. Based on the results, it was shown that smaller cell sizes and higher values of cell density were observed in PURNCFs in comparison with pure PURF. According to TGA data, it was revealed that thermal properties in PURNCFs were improved with the addition of nanopar-

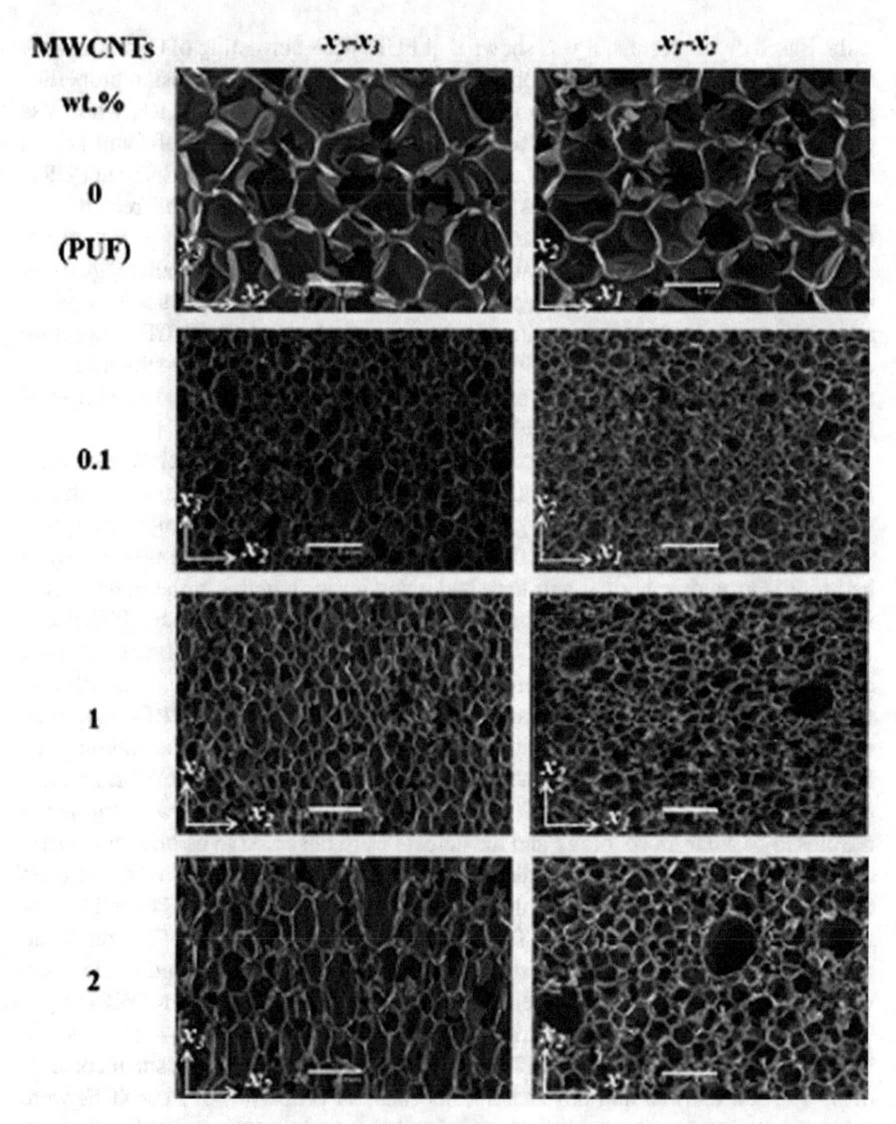

Fig. 4.12 SEM micrographs of pure PURF and PURNCFs consisting of different MWCNT contents. Scale bar is 1 mm Reproduced with permission from Ref. [7]. Copyright 2015 Elsevier Ltd.

ticle hybrids. In addition, the compressive properties of PURNCFs consisting of small amounts of montmorillonite carbon nanotube hybrids were also improved in comparison with that of pure PURF [8].

Figure 4.13 shows the SEM micrographs of (a) pure PURF. (b) PURNCF consisting of 0.25% hybrid nanoparticle (c) PURNCF consisting of 0.5% hybrid nanoparticle. (d) PURNCF consisting of 1% hybrid nanoparticle. Based on the results, it was shown that both pure PURF and PURNCFs share the typical cellular morphology consisting of a polygonal shape which has approximately the same dimensions in all directions, partially open cells and spherical holes on the faces of cells [8]. In addition, in agreement with other works in the literature [1, 2, 4–7], it was emphasized that cell size reduction and cell density elevation were observed when amount of hybrid nanoparticles was increased up to 1%. In addition, the viscosity of the reaction medium was observed to be significantly increased due to the incorporation of montmorillonite-carbon nanotube hybrids which could act as obstacles against the nucleation and growth of cells and resulted in smaller cell sizes in comparison with those of pure PURF [8].

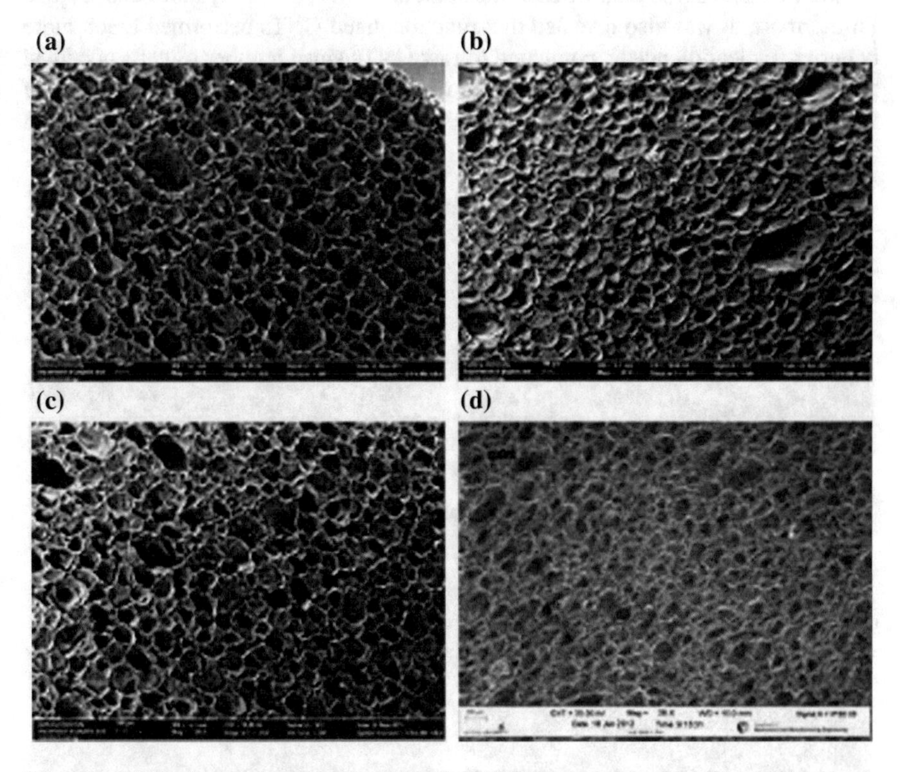

Fig. 4.13 SEM micrographs of **a** pure PURF. **b** PURNCF consisting of 0.25% hybrid nanoparticle. **c** PURNCF consisting of 0.5% hybrid nanoparticle. **d** PURNCF consisting of 1% hybrid nanoparticle. Reproduced with permission from Ref. [8]. Copyright 2012 Elsevier Ltd.

Caglayan et al. investigated the effects of homogeneous spreading of CNTs over foam matrix on the foamability, closed-cellular morphology, and mechanical properties of PURFs. Authors incorporated raw and functional CNTs into PURFs up to 0.2 wt%, and analyzed the mechanical properties of pure PURF and PURNCFs consisting of CNTs [9]. Based on the results, it was shown that the dispersion of CNTs in PMDI with low degree of thickness and density leads to greater distribution of CNTs, and also resulted in higher value of compressive strength in comparison with that of pure PURF. Finally, 30% elevation in terms of both shear and ultimate strength was obtained based on the enhancement in foam core properties of PURNCFs filled with CNTs [9].

Figure 4.14 shows the SEM micrographs of (a) pure PURF and PURNCFs consisting of (b) 0.1 wt% functional CNT in polyol, (c) 0.1 wt% raw CNT in PMDI, (d) 0.1 wt% functional CNT in PMDI. Based on the SEM results, it was shown that CNTs which were located on cell boundaries had very limited influence on the cell nucleation process in PURNCFs [9]. In addition, it was also highlighted that cell sizes are smaller and cell densities are larger in PURNCFs consisting of CNTs compared to pure PURF in agreement with results of other works [1, 2, 4–8] about this subject. Furthermore, it was also revealed that functionalized CNTs performed much more better as nucleation agents compared to raw CNTs since number of cells occupied per unit area was greater for PURNCFs which were prepared from dispersions of functionalized CNTs in both polyol and PMDI components [9]. The reason behind

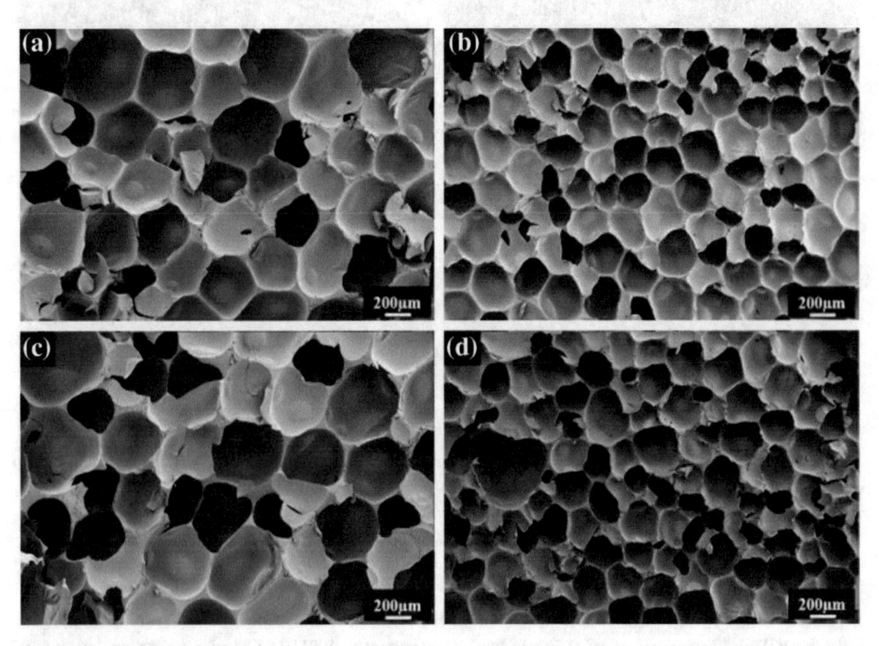

Fig. 4.14 SEM micrographs of **a** pure PURF and PURNCFs consisting of **b** 0.1 wt% functional CNT in polyol, **c** 0.1 wt% raw CNT in PMDI, **d** 0.1 wt% functional CNT in PMDI. Reproduced with permission from Ref. [9]. Copyright 2018 Elsevier Ltd.

this experimental finding was explained such that easier dispersions and distributions of functionalized CNTs in reaction medium because of their carboxylic acid functional groups and lower aspect ratios led to the formation of cell structures with smaller sizes and increased cell density during the reactive foaming process [9].

Yan et al. reported a comparative study of PURNCFs consisting of graphene nanosheets (GNSs) and carbon nanotubes (CNTs) [10]. Based on the rheological measurements, it was shown that the dispersion of 0.3 wt% GNS in polyol was the optimum condition for the reactive foaming of polyol with polyisocyanate. In accordance with SEM and TEM experimental data, it was revealed that homogeneous dispersion of GNSs and CNTs was observed in PURNCFs. Based on mechanical properties, it was highlighted that incorporation of 0.3 wt% CNTs enhanced compressive modulus values of PURNCFs. Moreover, it was reported that glass transition temperature elevated by 14 °C with incorporation of 0.3 wt% CNTs to PURFs. Based on these results, it was concluded that GNSs are more effective nanofillers compared to CNTs for PURFs in terms of thermal stability and mechanical property improvements. The reason behind superior performance of GNSs over CNTs was explained such that characteristic two-dimensional geometry, surface structure, and higher specific surface area of GNSs pave the way for the establishment of more solid intermolecular interactions and reduction in the mobility of polymer chains at the interface between GNSs and PURNCF matrix. Furthermore, based on thermal conductivity results, it was shown that thermal conductivity values of PURNCFs did not change that much with the incorporation of GNSs and CNTs [10].

Figure 4.15 shows the SEM micrographs of (a) pure PURF, (b) PURNCF consisting of 0.3 wt% GNSs and (c) PURNCF consisting of 0.3 wt% CNTs. Based on SEM results, it was shown that the addition of both 0.3 wt% GNSs and CNTs diminished cell sizes and elevated cell densities of PURNCFs moderately in comparison with pure PURF [10]. Previously, it was shown that GNSs are considered to be more effective nanoparticles than CNTs in terms of decreasing free energy for cell nucleation and generating more functional nucleation centers due to their higher surface areas and stronger intermolecular interactions with polymer chains in PURNCFs [10, 50, 51].

Yan et al. prepared PURNCFs consisting of low amounts of CNTs (about 1.2 wt%) with the establishment of uniform distribution CNTs throughout closed-cellular morphology of PURFs [11]. Based on the results, it was shown that PURNCFs with conductive properties exhibited superior electrical stability within temperature range of 20–180 °C which proves the use these PURNCFs in suitable applications for many years. According to DMA and mechanical property testing data, it was pointed out that incorporation of 2.0 wt% CNTs improved compression properties by 31% and increased storage modulus by 50% at room temperature. Furthermore, only 0.5 wt% CNT inclusion into PURFs improved thermal stability in such a way that thermal decomposition temperature at 50% weight loss elevated from 450 to 499 °C [11].

Figure 4.16 shows the SEM micrographs of (a) pure PURF, (b) PURNCF consisting of 2.0 wt% CNT, (c) high magnification image of PURNCF consisting of 2.0 wt% CNT. Based on SEM results, it was shown that the cell size slightly increased with inclusion of 2.0 wt% CNTs which proves that CNTs did not function prop-

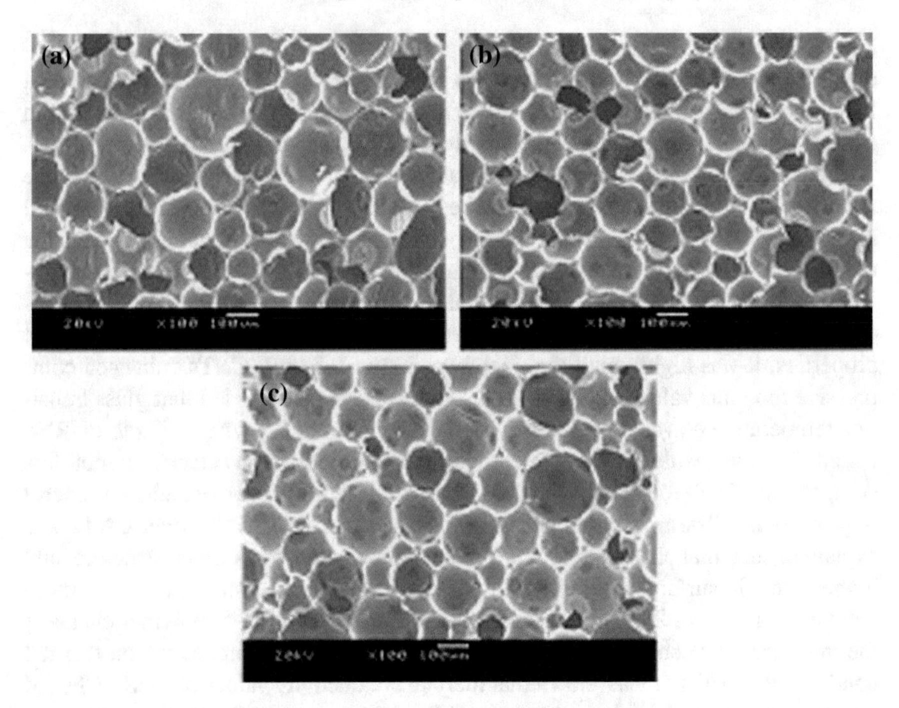

Fig. 4.15 SEM micrographs of **a** pure PURF, **b** PURNCF consisting of 0.3 wt% GNSs and **c** PURNCF consisting of 0.3 wt% CNTs. Reproduced with permission from Ref. [10]. Copyright 2012 Society of Chemical Industry

erly as nucleation centers during reactive PU foaming [11]. Moreover, as shown in Fig. 4.16b, it was revealed that PURNCF sample consisting of 2.0 wt% CNTs had closed cells with spherical shapes which shows that the addition of CNTs did not change the nature of cell formation in PURNCFs. In addition, as shown in 4.16c, it was also highlighted that many single and no aggregates of CNTs were observed in the cell struts which reveals the homogeneous distribution of CNTs within PURNCFs [11].

Based on the experimental results given in previously published works in the area of PURNCFs filled with CNTs, GNSs and MMTs [5–10], it is well noted that physical and chemical properties of nanofillers (aspect ratio, surface area, particle size, surface modification) dictate final properties of PURNCFs that are prepared with these types of nanofillers. Based on the variation of these specific properties of nanoparticles, the optimum amount of nanoparticles alters due to changes in intermolecular interactions between nanoparticles and PU chains [1, 2, 4]. In addition to optimum content of nanofillers, the second most important parameter in production of advanced PURNCFs with enhanced properties is surface modification of nanoparticles [9]. Usually, using surface modified nanoparticles such as functionalized CNTs rather than using unmodified CNTs improves closed-cellular morphology as well as mechanical properties of PURNCFs in comparison with those of pure

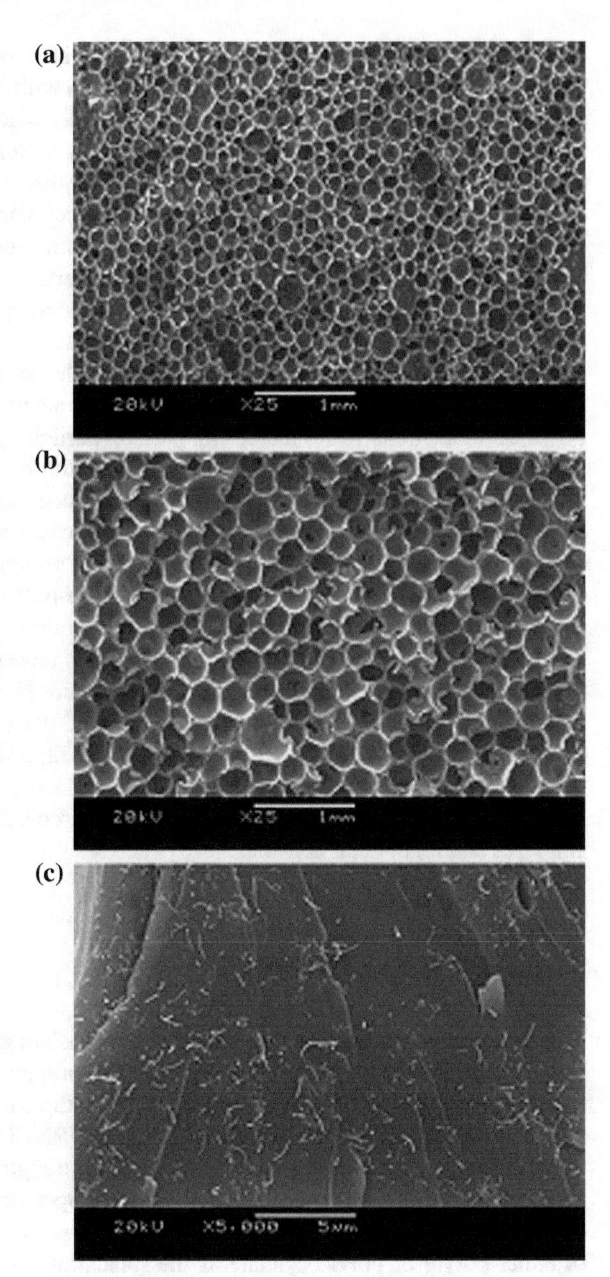

Fig. 4.16 SEM micrographs of **a** pure PURF, **b** PURNCF consisting of 2.0 wt% CNT, **c** high magnification image of PURNCF consisting of 2.0 wt% CNT. Reproduced with permission from Ref. [11]. Copyright 2011 Wiley Periodicals, Inc.

PURF [9]. The rationale behind structure-property improvements in PURNCFs filled with surface-modified nanoparticles in comparison with those of PURNCFs consisting of unmodified nanoparticles was explained very nicely in previously published works [7–9, 11, 25, 30, 32, 33, 38–40]. It is very well known that the fabrication of advanced PURNCFs with improved physical, thermal and mechanical properties depends strongly on the generation of closed-cellular morphology with uniform distribution of cell sizes and higher cell densities since these morphological parameters in rigid foams directly affect macroscopic performance of PURFs. Moreover, the creation of this kind of perfect closed cellular morphology in PU rigid foams is strongly correlated with the establishment of favorable interactions between the surface of nanofillers and PU chains. If nanoparticles are not surface modified, there is a highly potential risk for aggregation which ultimately reduces favorable particle-polymer interactions while increasing particle-particle type of interactions. On the other hand, if nanoparticles are surface modified with specific functional groups, the aggregation behavior of particles could be eliminated due to establishment of beneficial secondary type of intermolecular bondings between surface functional groups of nanoparticles and polymer chains. In that case, polymer-particle type of interactions could be dominant over unfavorable particle-particle type of interactions. By this way, the properties of PURNCFs consisting of these surface-modified nanoparticles can be maximized over PURNCFs containing unmodified nanoparticles. Thus, for this reason, various surface modification methods should be employed to chemically attach surface functional groups onto nanofillers which are physically and chemically compatible with PU matrix [1, 2, 30, 32, 33]. By this way, the number of favorable interactions between surface modified nanofillers and PURNCF matrix could be maximized to produce PURNCFs with improved properties such as reduced cell sizes and higher cell densities.

4.3 Mechanical Properties

The influences of using sonication technique consisting of various technical conditions were investigated on the preparation and properties of PURNCFs that were reinforced with carbon nanofibers (CNFs) [25]. Authors used pseudo-static compression tests to explore mechanical properties of PURNCFs consisting of CNFs, and they particularly measured the compressive yield strength of PURNCFs which were prepared with different experimental conditions. Specifically, authors studied influences of various experimental parameters such as sonication temperature, selection of either polyol or polyisocyanate as the sonication medium, sonication time and power, concentration of nanoparticles and total amount of foam on the dispersion properties of CNFs in PURFs. Based on the results, it was found out that the application of ultrasound to CNF dispersions in the polyisocyanate part leads to improved mechanical properties [25].

Figure 4.17 shows the compression test plots of pure PURF and PURNCFs consisting of 0.5 wt% CNFs with respect to various degrees of sonication. In Fig. 4.17, all

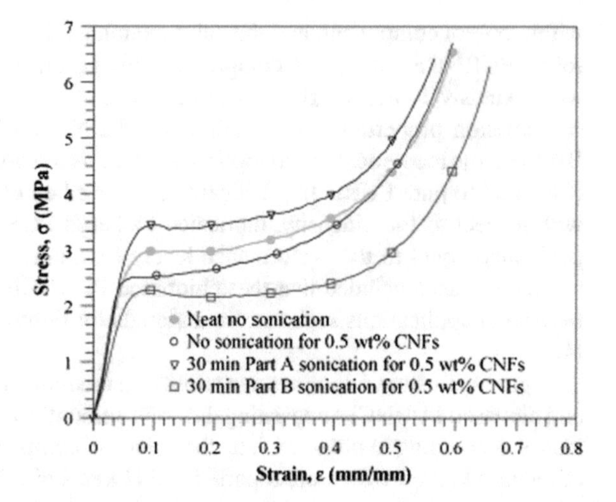

Fig. 4.17 Compression test plots of pure PURF and PURNCFs consisting of 0.5 wt% CNFs and various degrees of sonication. Reproduced with permission from Ref. [25]. Copyright 2007 Elsevier Ltd.

of the compression data exhibit three different types of distortion behaviour such as beginning linear deformation, linear plateau section, and finally the solid and thicker section. Based on compression test results, the average of three samples for each composition was obtained, and only compressive yield strength values of PURNCFs which were prepared by performing sonication of CNFs in polyisocyanate component were found out to be higher compared to that of pure PURF [25]. However, on the other hand, it was reported that sonication of CNFs in the polyol part of PURF leads to lower yield strength in comparison with PURNCFs consisting of the same amount of CNFs which were prepared without sonication. The reason behind this experimental finding was explained such that CNFs are mixed more homogenously via the sonication technique compared to the traditional mixing method, and also polyisocyanate part of PURF is less reactive and has lower viscosity compared to the polyol part which provides the medium for uniform distribution and mixing of CNFs [25].

Thus, based on these results, it is well-noted that obtaining better mechanical properties of PURNCFs consisting of nanocylindrical particles such as carbon nanofibers or carbon nanotubes greatly depends on choosing the right dispersion medium for the homogenously distribution of CNFs within the PURNCF matrix. Here, the polyisocyanate component is a much better choice than the polyol component due to its better dispersion properties such as lower viscosity and less reactivity. Consequently, the well dispersion of CNFs in the polyisocyanate component leads to favorable intermolecular interactions between CNFs and polymer chains during the reactive foaming process due to high surface areas of CNFs. Thus, this mechanism provides the formation of PURNCFs with lower cell size with uniform distribution and higher cell density and improved mechanical properties.

Previously, carrot nanofibers (CtNFs) were introduced into PURFs to improve physical and mechanical properties of biobased castor oil polyol based PURNCFs [40]. In their study, authors utilized the procedure of well-distribution of CtNFs within polyol component, and they also examined the characteristics of CtNF reinforced PURNCFs. Moreover, compressive strength and modulus values of PURNCFs were extensively improved compared to those of pure PURF. In addition, the best compression properties were obtained in PURNCFs consisting of 0.5 phr CtNF. Based on the theoretical calculations, the modulus of solid PURNCFs was improved compared to pure PURF. In practical terms, based on compression property results with respect to foam density, fabricated biobased PURNCFs consisting of CtNFs performed equal to the performance level of commercial PURFs. Based on these results, it was concluded that these biobased PURNCFs consisting of CtNF could be used in applications such as core materials for lightweight sandwich composites [40].

Figure 4.18 shows compressive stress and modulus data of pure biobased PURF and biobased PURNCFs consisting different amounts of CtNFs with respect to strain values. According to obtained data, the values of compressive strength and modulus of biobased pure PURF were reported as 241 kPa and 3.5 MPa, respectively [40]. In addition, it was also reported that the addition of 0.5 phr CtNF to PURFs led to the observation of highest compressive modulus and strength in biobased PURNCFs. Moreover, it was revealed that the PURNCFs consisting of higher amounts of CtNFs exhibited highly ductile behavior due to the fiber bridging effect caused by the high concentration of CtNFs. Furthermore, it was shown that when incorporated amount of CtNF was 1 phr in PURNCFs, the values compressive modulus and strength moderately diminished in comparison with PURNCF containing 0.5 phr CtNF due to aggregation behavior of CtNFs. Based on the modulus data, compression modulus values of PURNCFs consisting of 0.5 phr CtNF increased by 37% in comparison with that of pure PURF [40].

Green PURFs containing completely soy polyol and lignin were fabricated after optimization of experimental conditions such as amounts of components during sample preparation to enhance physical and mechanical properties [12]. Authors also incorporated the cellulosic nanofibers into green PURFs to increase the rigidity of system. They systematically investigated influences of important experimental

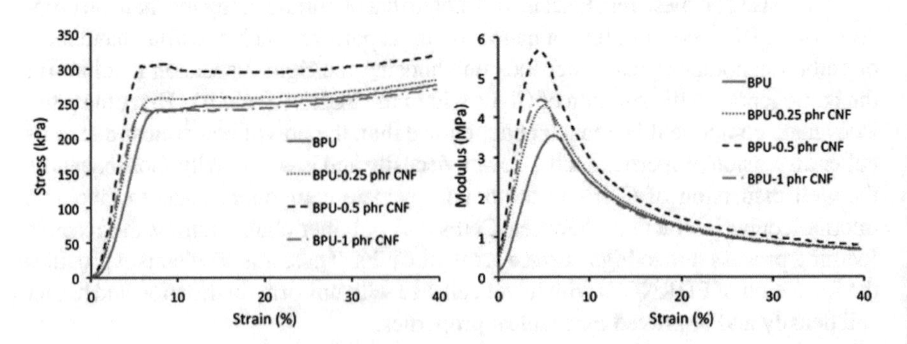

Fig. 4.18 Compressive stress and modulus data of pure biobased PURF and biobased PURNCFs consisting different amounts of CtNFs with respect to strain values. Reproduced with permission from Ref. [40]. Copyright 2016 Elsevier Ltd.

parameters such as various types of surface modified cellulose fibers on morphology, foam density, water vapor and mechanical properties of PURFs. In accordance with experimental data, it was found out that mechanical properties of PURNCFs were enhanced due to beneficial effects created by incorporation of surface modified cellulose fibers. Furthermore, odor concentrations of PURNCFs consisting of nanofibers were diminished greatly compared to absolutely synthetic based polyol PURFs due to the replacement of isocyanate content. In accordance with results, it was concluded that PURFs with advanced properties could be fabricated by replacing some portion of isocyanate with lignin and addition of nanofiber [12].

Figure 4.19 shows (a) compressive strength and (b) compressive modulus values of pure soy-based PURF and soy-based PURNCFs consisting of rigid lignin and cellulose nanofibers. In accordance with results, it was pointed out that compressive strength values of PURNCFs consisting of lignin and cellulose nanofibers exhibited almost identical values as that of pure soy-based PURFs even though isocyanate amount was reduced by 7% in PURNCFs [12]. Moreover, using cellulose nanofibers that were functionalized with enzyme and hydrophobic latex as nanofillers in PURFs increased the compressive strength compared to soy-based PURNCFs consisting of untreated nanofiber and pure soy-based PURFs. Furthermore, it was shown that chemical modification of nanofibers improved compressive strength values of PURNCFs consisting of these nanofibers due to the formation of favorable interactions between nanofibers and lignin with PURF components. Similarly, it was also revealed that the compressive modulus of PURNCFs consisting of lignin and hydrophobic latex treated cellulosic nanofiber increased by about 38% in comparison with pure soy-based PURF [12].

Previously, a cheaper and continuous method was used to improve thermal insulation properties of PURFs [42]. The study was focused on an effective process of improving mechanical properties and decreasing thermal conductivity of PURFs with no change in foam density by inclusion of very low amounts of cellulose nanocrystals (CNCs) into PURFs by using an optimized solvent-free ultrasonication method. In addition, it was found out that modulus values of PURNCFs consisting of optimized conditions improved perpendicular to foam rise compared to that of pure PURF [42].

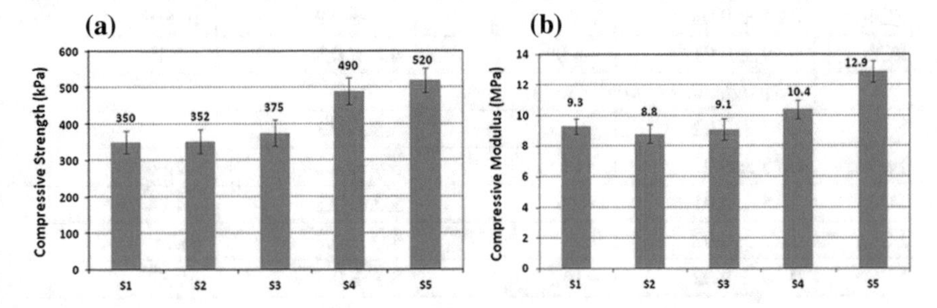

Fig. 4.19 **a** Compressive strength and **b** compressive modulus values of pure soy-based PURF and soy-based PURNCFs consisting of rigid lignin and cellulose nanofibers. Reproduced with permission from Ref. [12]. Copyright Springer Science + Business Media New York 2013

Mechanical properties data of pure PURF and ultrasonicated PURNCFs which were obtained parallel and perpendicular to foam rise direction are shown in Table 4.1. Based on results, it was shown that all foam samples had anisotropic properties such that they had larger compressive strength values along foam rise direction. In addition, it was also reported that no big change in foam density was observed between PURNCFs and pure PURF with incorporation of less than 0.8 wt% CNC [42]. As shown in Table 4.1, it was shown that no visible change in values of compressive strength or modulus between pure PURFs that were prepared with and without ultrasonication method. However, it was reported that compressive strength values of ultrasonicated PURNCF containing 0.4 wt% of CNC exhibited 1% improvement in foam growth direction and also 9% improvement in perpendicular to foam growth direction. In addition, it was also added that the same PURNCF consisting of 0.4 wt% of CNC also exhibited substantial enhancements of 11 and 17% in terms of modulus values in both two main directions of foams, respectively [42]. The considerable enhancement of modulus perpendicular to foam growth direction for PURNCF containing 0.4 wt% CNC was correlated to likely orientation of CNCs along the foam growth direction during reactive foaming process of PURNCFs [42, 52, 53].

Previously, MWCNTs which were grafted with poly (ethylene oxide) PEO-*graft*-MWCNTs (grafted MWCNTs) were synthesized via ring-opening polymerization of ethylene oxide [1]. In addition, authors fabricated PURNCFs containing MWCNTs and grafted MWCNTs to explore influences of PEO molecules on morphological and mechanical properties and reactive foaming process of PURNCFs [1]. Based on the

Table 4.1 Mechanical properties data of pure PURF and ultrasonicated PURNCFs which were obtained parallel and perpendicular to foam rise direction

Samples	Compressive strength (MPa)	% change/unit mass	Young's modulus (MPa)	% change/unit mass
	Parallel to foam rise			
F_0	0.21 ± 0.02	–	5.1 ± 0.3	–
F_U	0.21 ± 0.02	−1.1[a]	4.9 ± 0.7	−4.60[a]
$FC2_U$	0.18 ± 0.02	−14.5[b]	4.9 ± 0.8	0.69[b]
$FC4_U$	0.21 ± 0.03	1.2[b]	5.5 ± 0.9	10.80[b]
$FC8_U$	0.19 ± 0.01	−9.9[b]	4.8 ± 0.5	−1.18[b]
	Perpendicular to foam rise			
F_0	0.15 ± 0.02	–	3.5 ± 0.3	–
F_U	0.15 ± 0.01	−1.7[a]	3.4 ± 0.2	−2.96[a]
$FC2_U$	0.15 ± 0.01	−2.6[b]	3.2 ± 0.2	−4.84[b]
$FC4_U$	0.17 ± 0.02	9.2[b]	4.0 ± 0.2	16.74[b]
$FC8_U$	0.13 ± 0.02	−16.6[b]	3.1 ± 0.7	−9.37[b]

[a]In comparison with control F_0

[b]In comparison with control F_U

results, it was shown that using grafted MWCNTs had a little amount of impact on viscosity of polyol. However, on the other hand, it was reported that the pore size of PURNCFs was decreased with inclusion of grafted MWCNTs. Furthermore, it was revealed that incorporation of nanofiller increased the glass transition temperature and mechanical properties of PURNCFs due to favorable interactions between grafted MWCNTs and PU chains [1].

Figure 4.20 shows specific compressive (a) strength and (b) modulus of pure PURF, PURNCFs containing MWCNT and PURNCFs filled with grafted MWCNTs with respect to amount of nanofillers. In accordance with data, it was revealed that specific compressive strength and modulus values of PURNCFs consisting of MWCNTs diminish with increasing amounts of MWCNT [1]. Furthermore, lowest levels of specific compressive strength and modulus were obtained with incorporation of 3.0 wt% MWCNTs into PURFs. In accordance with this result, mechanical properties of PURNCFs consisting of MWCNTs were greatly influenced by the aggregation behavior of MWCNTs [1, 54]. In addition, as shown in Fig. 4.20, it was reported that specific compressive values of PURNCFs consisting of PEO-*graft*-MWCNTs are much more better compared to those of PURNCFs consisting of MWCNTs, especially at 2.0 or 3.0 wt% nanofiller content. The reason behind the marked difference between these two PURNCF systems was explained such that grafted PEO chains on MWCNTs provide the necessary conditions for superior affinity and enhanced interactions between PEO-*graft*-MWCNTs and PU chains, and thus enhancing the stress shift from PURF matrix to nanofillers, more effectively [1, 55, 56].

Previously, multi-walled carbon nanotubes consisting of hydroxyl functional groups were surface modified by using 3-aminopropyltriethoxysilane (Si) and dipodal silane [2]. Then, authors prepared PURNCFs consisting of silanized MWCNTs to explore influences of using these silane modified MWCNTs on thermal, morphological and mechanical properties of PURNCFs [2]. According to SEM and mechanical testing data, it was reported that mechanical properties of PURNCFs containing

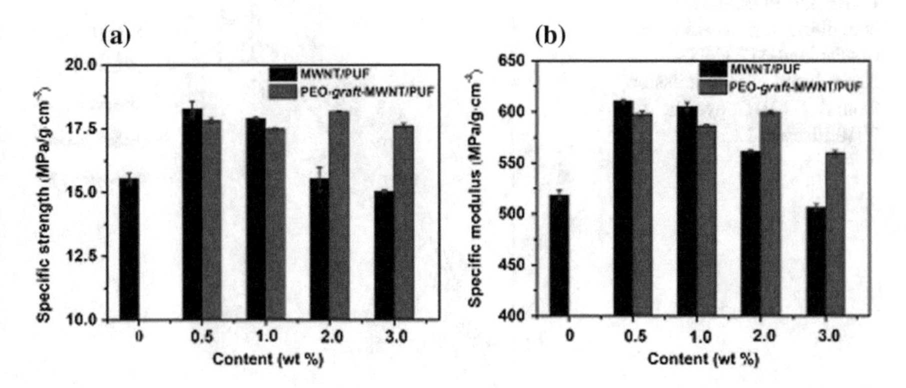

Fig. 4.20 Specific compressive **a** strength and **b** modulus of pure PURF, PURNCFs containing MWCNT and PURNCFs filled with grafted MWCNTs with respect to various amounts of nanofillers. Reproduced with permission from Ref. [1]. Copyright 2018 Society of Chemical Industry

1.5 wt% MWCNTs modified with dipodal silane improved much more in comparison with those of PURNCFs filled with 1.5 wt% Si-MWCNT. Based on mechanical testing results, it was also shown that tensile strength and modulus values increased because of favorable intermolecular interactions between PU chains and MWCNTs modified with dipodal silane in PURNCFs [2].

Figure 4.21 shows mechanical property results of pure PURF and PURNCFs containing different amounts of silanized-MWCNTs. Based on results, it was shown that pure foam and nanocomposite foams exhibited linear elastic deformation at smaller stress values and plastic deformation at larger stress values [2]. In addition, tensile strength and modulus values enhanced with inclusion of Si-MWCNT and dipodal silane-MWCNT into PURFs in comparison with those of pure PURF. Furthermore, tensile strength and modulus values of PURNCFs filled with 1.5 wt% MWCNTs modified with dipodal silane elevated by 53 and 20%, respectively in comparison with pure PURF. Moreover, it was added that PURNCFs consisting of dipodal silane-MWCNT have much better mechanical properties compared to those consisting of Si-MWCNT. The difference in terms of mechanical properties between the two types of PURNCFs was explained such that dipodal silane could be able to form more favorable interactions or bonds with CNTs than APTS due to its higher amount of silanol groups. Thus, this important physicochemical property of dipodal silane-MWCNTs provides the basis for stronger type of interactions between silanized-MWCNTs and PU chains [2, 57, 58].

Previously, reinforcement of PURFs was investigated by using carbon nanotubes (CNT) or graphite as nano/micron-sized additives to enhance fire resistance thermal and mechanical properties [5]. According to experimental data, it was pointed out that mechanical properties of PURNCFs improved by almost 20% with small addi-

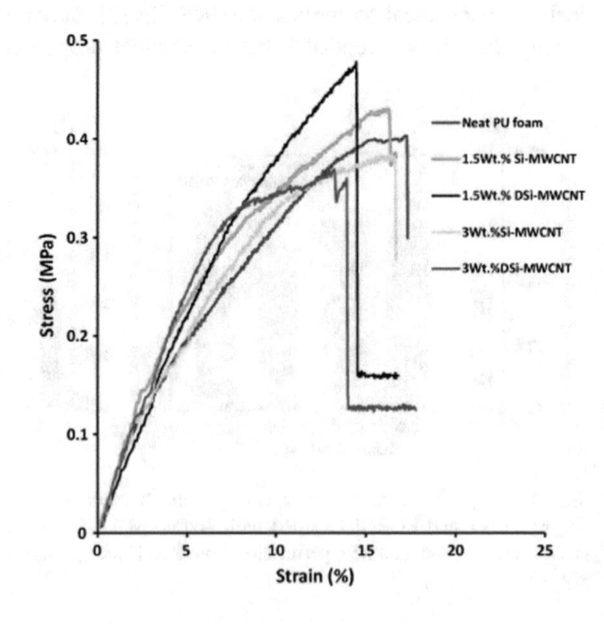

Fig. 4.21 Mechanical property test curves of pure PURF and PURNCFs containing various amounts of silanized-MWCNTs. Reproduced with permission from Ref. [2]. Copyright 2018 Elsevier Ltd.

tions of CNTs with maximum of 0.05 wt%. The reason behind this improvement in mechanical strength was attributed such that the strengthening effect in PURNCFs was mainly coming from CNTs and also incorporation of CNTs leads to generation of smaller cell sizes with much narrower distributions in comparison with those of pure PURF. Furthermore, the size of porous structure did not change but the characteristic porosity of foams substantially increases with higher CNT contents which leads to a decrease of mechanical strength in PURNCFs [5].

Figure 4.22 shows bending strength and modulus values of pure PURF and PURNCFs consisting of CNTs and schematical drawing of three-point bending test device. According to results, it was revealed that bending strength and modulus values of PURNCFs elevated with small additions of CNTs up to 0.05% to PURFs whereas the mechanical strength was reduced with higher additions of CNT [5]. Specifically, the highest values of mechanical properties were reported such that bending strength and modulus values enhanced by 21 and 6% with 0.01% inclusion of CNTs compared to pure PURF, respectively. The reason behind deterioration of mechanical properties at higher CNT contents was explained such that the increase of overall porosity in PURNCFs at higher CNT contents decreases the mechanical properties, and also for higher amounts of CNTs, the increase of viscosity provides poor mixing of polyol with the isocyanate component which decreases the mechanical properties of final PURNCFs [5].

Anisotropic compressive characteristics of PURNCFs consisting of 0.1, 1 and 2 wt% MWCNT were investigated with respect to principal directions of crosswise isotropic foams [7]. Specifically, the work was focused on anisotropic property-

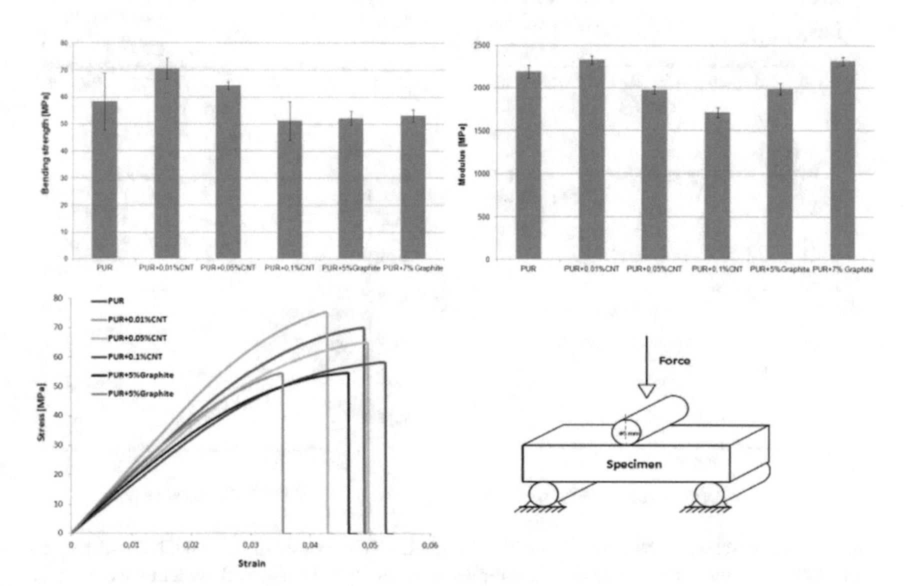

Fig. 4.22 Bending strength and modulus values of pure PURF and PURNCFs consisting of CNTs and schematical drawing of three-point bending test device. Reproduced with permission from Ref. [5]. Copyright 2016 Elsevier Ltd.

structure relationships of PURNCFs which were controlled by specific features of closed cellular morphology such as cell size and density. Based on micromechanical modelling results, it was shown that PURNCFs consisting of 0.1 and 1 wt% MWCNT exhibited much more substantial enhancement of compression properties along the main direction of foam free growth compared to pure PURF, while less property enhancement was observed along the diagonal direction of foam [7]. Authors used two types of analytical models such as a simple parallelepiped cell model and a tetrakaidecahedron representation model in order to predict the rigidity and solidity ratios along two main anisotropic directions of PURNCFs. Based on experimental and modelling results, it was concluded that property-structure relationships between anisotropic structure of unit cell and mechanical properties of final macroscopic PURNCFs are highly correlated with each other [7].

Figure 4.23 shows the compressive response of pure PURF and PURNCFs consisting of MWCNT with respect to MWCNT amount. Basically, Fig. 4.23a displays

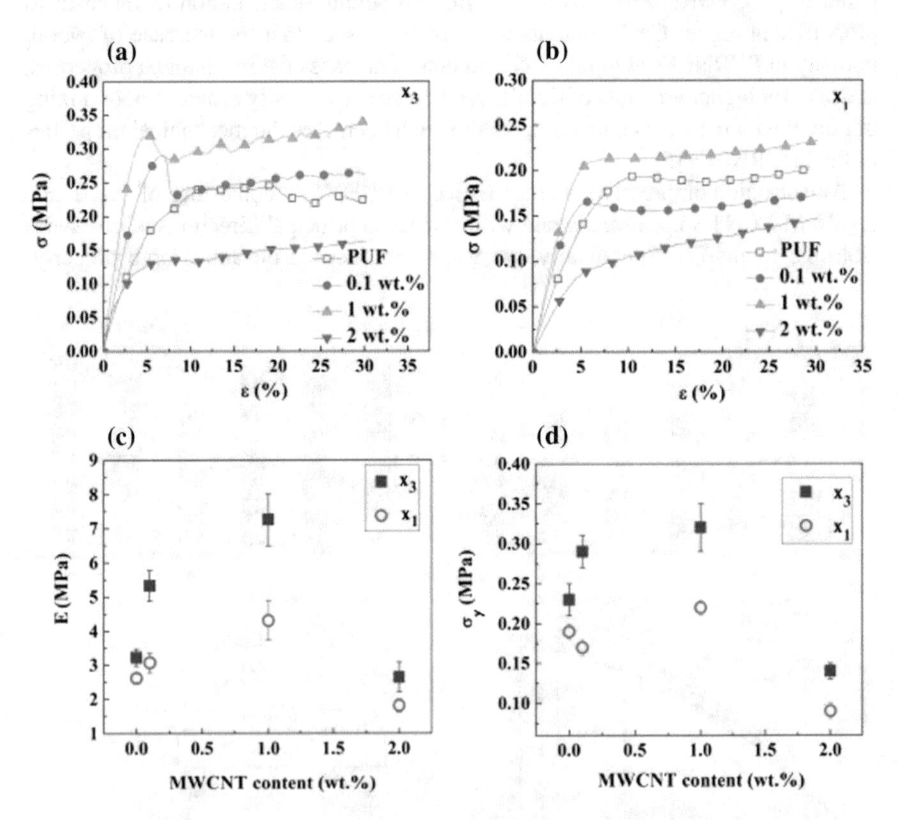

Fig. 4.23 Compressive behaviour of pure PURF and PURNCFs consisting of MWCNT with respect to MWCNT amount. Stress-strain data of samples tested along **a** foam growth direction and **b** transverse direction, **c** modulus values with respect to MWCNT content, **d** yield strength values with respect to MWCNT content. Reproduced with permission from Ref. [7]. Copyright 2015 Elsevier Ltd.

stress-strain data of samples tested along foam growth direction. Figure 4.23b shows stress-strain data of samples tested along transverse direction, (c) modulus values with respect to MWCNT content, (d) yield strength values with respect to MWCNT content. According to experimental data, it was noticed that elastic modulus and yield strength values elevated with incorporation of 0.1 and 1 wt% MWCNT along both foam growth and transverse directions [7]. In addition, the elevation of elastic modulus and yield strength values is closely connected to the modification of the closed-cellular morphology due to the presence of MWCNTs [59]. However, on the other hand, it was revealed that at 2 wt% MWCNT loading, the substantial reduction of elastic modulus and yield strength was detected because of yield strength reduction by the emergence of micron-sized defects at higher MWCNT amounts [7].

Thus, in agreement with other works [1, 2, 4–6, 8–11], it is well-noted that PURNCFs consisting of smaller amounts of CNTs (0.1 wt%) act as extra nucleation agents during reactive foaming process due to favorable interactions between larger surface areas of CNTs and PURF components, and increases the formation of more bubbles in the reaction medium. This mechanism causes the formation of smaller cells with higher density, and directly improves the physical, thermal and also mechanical properties of PURNCFs consisting of CNTs. But, on the other hand, at higher MWCNT loadings (about 2 wt%), the aggregation behavior of CNTs causes the formation of unfavorable interactions between CNTs and PURF components, thus increasing the cell size and damaging the closed cellular structure by inducing defects, and consequently decreases the mechanical properties.

Previously, effects of using nanohybrid particles of montmorillonite (MMT) and carbon nanotube (CNT) were investigated on the performance of PURNCFs [8]. Authors used chemical vapour deposition for the synthesis of hybrids, and then these MMT-CNT nanohybrid particles were incorporated into PURFs via in situ polymerization process [8]. Authors used SEM, TEM, mechanical testing, TGA and optical microscopy in order to evaluate the closed cellular morphology, mechanical and thermal decomposition properties of PURNCFs. Based on mechanical testing results, compressive properties of PURNCFs consisting of small amounts of montmorillonite carbon nanotube hybrids were also improved in comparison with that of pure PURF [8].

Figure 4.24 shows characteristic mechanical property data of pure PURF and PURNCFs containing various amounts of hybrid nanoparticles. According to results, compressive modulus values of PURFs were improved with addition of montmorillonite–carbon nanotube hybrids. In addition, it was pointed out that compressive modulus values increased by 31% with inclusion of 0.25% hybrid nanoparticles, but it did not exhibit any improvement with further addition of hybrid nanoparticles [8]. Furthermore, compressive strength values increased by 20% with incorporation of 0.25% hybrid nanoparticles, however they did not exhibit any improvement with further addition of hybrid nanoparticles due to the aggregation behavior of system. In agreement with previous studies [60], it was concluded that the exfoliation of silicate layers with the help of carbon nanotubes which were grown between nanoclay layers led to the homogenous dispersion of nanoclay throughout PURNCF matrix, and thus facilitated the fabrication of PURNCFs with superior mechanical properties [8].

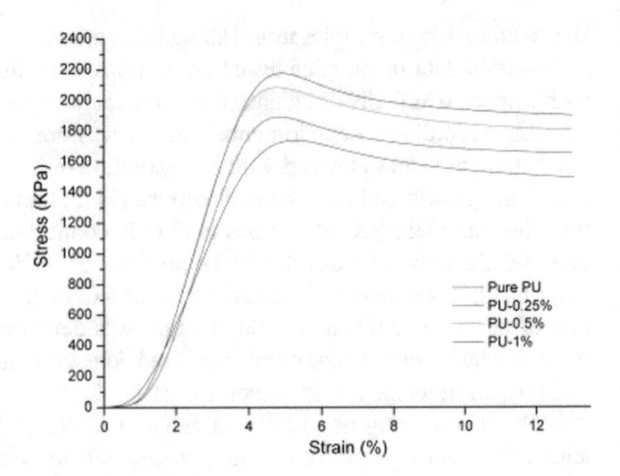

Fig. 4.24 Characteristic mechanical property data of pure PURF and PURNCFs containing various amounts of hybrid nanoparticles. Reproduced with permission from Ref. [8]. Copyright 2012 Elsevier Ltd.

Previously, the effects of homogeneous spreading of CNTs over foam matrix were explored in detail on the foamability, closed-cellular morphology, and mechanical properties of PURFs [9]. Authors incorporated raw and functional CNTs into PURFs up to 0.2 wt%, and analyzed the mechanical properties of pure PURF and PURNCFs consisting of CNTs [9]. Based on the results, it was shown that the dispersion of CNTs in PMDI with low degree of thickness and density leads to greater well distribution of CNTs, and also resulted in higher value of compressive strength in comparison with that of pure PURF. Finally, 30% elevation in terms of both shear and ultimate strength was obtained based on the enhancement in foam core properties of PURNCFs filled with CNTs [9].

Table 4.2 shows compressive strength results of pure PURF and PURNCFs consisting of CNTs with different contents. In accordance with compressive property data, it was emphasized that increasing CNT amount in PURNCFs does not always

Table 4.2 Compressive strength results of pure PURF and PURNCFs consisting of CNTs with various amounts

Sample	Dispersion media	Compressive stress (MPa)	Specific compressive stress (MPa m^3/kg)	Enhancement (%)
Neat PU		0.88 ± 0.03	0.0082 ± 0.0002	–
0.1 wt% R-CNT	Polyol	0.79 ± 0.01	0.0083 ± 0.0001	1.2
0.1 wt% F-CNT		0.97 ± 0.02	0.0089 ± 0.0002	8.5
0.2 wt% F-CNT		1.06 ± 0.03	0.0079 ± 0.0003	−3.6
0.05 wt% R-CNT	PMDI	1.13 ± 0.03	0.0089 ± 0.0002	8.5
0.1 wt% R-CNT		1.04 ± 0.01	0.0084 ± 0.0001	2.4
0.1 wt% F-CNT		1.19 ± 0.02	0.0093 ± 0.0002	13.4
0.2 wt% R-CNT		1.01 ± 0.02	0.0079 ± 0.0001	−3.6

Reproduced with permission from Ref. [9]. Copyright 2018 Elsevier Ltd.

lead to elevation of mechanical properties due to clustering behavior of CNTs [9]. In addition, the PURNCF system which was prepared from 0.1 wt% functionalized CNT dispersion in polyol exhibited about 9% enhancement in terms of specific compressive strength due to better dispersion and cell nucleation properties of this system [9]. Furthermore, based on additional experiments, the PURNCF system which was prepared from 0.05 wt% dispersion of raw CNT in PMDI displayed about 9% improvement in terms of specific compressive strength due to better dispersion of raw CNTs within the PMDI dispersion which has much less viscosity and reactivity compared to that of polyol dispersion. However, on the other hand, the PURNCF system which was made from dispersion of 0.1 wt% functionalized CNT in PMDI exhibited the best performance such as almost 13% elevation in terms of specific compressive strength. Thus, the reason behind this finding was explained such that the carboxylic acid functional groups of CNTs helped to disperse CNTs much more effectively both in polyol and PMDI components, and led to the increase of mechanical properties more than the PURNCFs consisting of raw CNT type of carbon nanotubes [9].

In agreement with results from previously published works [1, 2, 9], usually, using surface modified CNTs rather than using unmodified CNTs improves closed-cellular morphology as well as mechanical properties of PURNCFs in comparison with those of pure PURF. Also, these types of property improvements can be observed not only in surface modified CNT-filled PURNCFs but also in PURNCF systems consisting of other types of surface-modified nanoparticles [7–9, 11, 25, 30, 32, 33, 38–40]. Specifically, the fabrication of advanced PURNCFs with improved mechanical properties is strongly correlated with generation of closed-cellular morphology having uniform distribution of cells with small sizes and larger cell densities. The formation of closed-cellular morphology in PURFs directly influence their mechanical properties. Moreover, the fabrication of advanced PURNCFs with improved mechanical properties is strongly affected by favorable interactions between surfaces of nanofillers and PU chains. Thus, in this context, surface modified nanoparticles have lots of benefits over unmodified nanoparticles in increasing number of favorable interactions due to establishment of beneficial secondary type of intermolecular bondings between surface functional groups of nanoparticles and polymer chains. By this way, mechanical properties of PURNCFs consisting of these types of surface-modified nanoparticles can be maximized over PURNCFs containing unmodified nanoparticles.

Previously, comparative study of PURNCFs containing CNTs and graphene nanosheets (GNSs) was published [10]. Based on the rheological measurements, it was shown that the dispersion of 0.3 wt% GNS in polyol was the optimum condition for the reactive foaming of polyol with polyisocyanate. Based on mechanical properties, it was highlighted that incorporation of 0.3 wt% CNTs enhanced compressive modulus values of PURNCFs. Moreover, it was reported that glass transition temperature elevated by 14 °C with incorporation of 0.3 wt% CNTs to PURFs. Based on these results, it was concluded that GNSs are more effective nanofillers compared to CNTs for PURFs in enhancing mechanical properties and thermal stability. The reason behind superior performance of GNSs over CNTs was explained such that characteristic two-dimensional geometry, surface structure, and higher specific surface area of GNSs pave the way for the establishment of more solid intermolecular

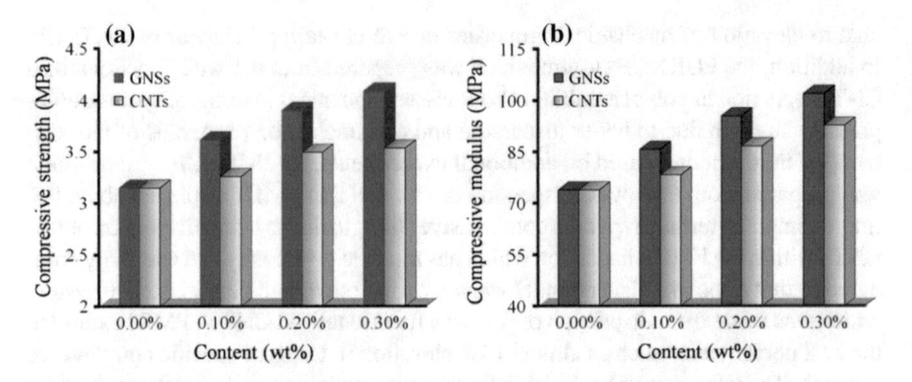

Fig. 4.25 Compressive **a** strength and **b** modulus values of pure PURF and PURNCFs consisting of GNSs and CNTs with different contents. Reproduced with permission from Ref. [10]. Copyright 2012 Society of Chemical Industry

interactions and reduction in the mobility of polymer chains at the interface between GNSs and PURNCF matrix [10].

Figure 4.25 shows compressive (a) strength and (b) modulus values of pure PURF and PURNCFs consisting of GNSs and CNTs with different contents. Based on the results, it was shown that compressive properties such as strength and modulus rise sufficiently with incorporation of GNSs and CNTs [10]. Specifically, it was shown that compressive strength and modulus values increased about 16 and 25% with inclusion of 0.3 wt% CNT to pure PURF, respectively. Furthermore, inclusion of 0.3 wt% GNS to pure PURF also enhanced compressive mechanical properties. Based on these results, it was suggested that CNTs or GNSs acted as very efficient nanofillers in terms of enhancing compressive mechanical properties in PURNCFs [10].

4.4 Thermal and Thermomechanical Properties

PURNCFs consisting of very small amount of carbon nanofibers (CNFs) were formed by creating a network structure of CNFs in dispersions of polyisocyanate and polyol components of PURFs [14]. Authors benefited from the reactive foaming process of these CNF dispersions in PURF components to produce PURFs with improved properties. According to results, it was pointed out that the conditions of polymerization kinetics and reactive foaming process in PURFs were not negatively influenced with incorporation of fillers. Furthermore, thermal conductivity values of PURNCFs were improved with incorporation of very low amount of filler which was less than 0.5 wt% [14].

Table 4.3 shows thermal conductivity values and other properties of pure PURF and PURNCFs consisting of CNFs prepared with different dispersion media. Based on the results, it was shown that incorporation of 1% CNF which was dispersed in

Table 4.3 Thermal conductivity values and other properties of pure PURF and PURNCFs consisting of CNFs prepared with different dispersion media

Sample	Cell number density (#/cm^3)	Open cell content ± 2 (%)	$k \pm 0.0001$ (W/mK)	Weight retention ± 0.2 (%)	Normalized compressive modulus ± 0.3 (MPa)
Conventional	1.97×10^4	87	0.0162	19.8	31.2
1% CNF-polyol	1.21×10^4	86	0.0157	20.5	35.3
1% CNF-MDI	1.01×10^4	86	0.0156	21.9	36.2

MDI component reduced thermal conductivity by 5.4%, thus effectively increases thermal insulating performance of PURFs [14]. The reason behind the reduction in thermal conductivity coefficient was explained such that the increase of nucleation of cells in the reaction medium and the reduction in radiative heat transfer by the presence of CNTs are the two main reasons that improve the insulation performance. In addition, it was previously [61, 62] shown that the convective heat transfer of the blowing gas is the major component of thermal conductivity in PURFs [14].

Recently, fabrication of cellulose nanocrystal (CNC) filled PURNCFs which were prepared from a 80:20 mixture of polyether and bio-based polyester polyols was performed to improve mechanical properties and thermal insulation characteristics of PURFs [41]. Based on the results, it was shown that thermal conductivity values of PURNCFs diminished by 2.4% with inclusion of 0.4 wt% CNC. In addition, the same authors also previously observed that 0.4 wt% introduction of acid-hydrolysed CNC into PURFs which had the same type of polyether polyol reduced thermal conductivity values by 5% due to improved nucleation property of modified CNC. In addition to this finding, specific compressive strength values diminished compared to the system having synthetic polyols due to unfavorable intermolecular interactions between CNC and bio-based polyester polyols [41].

Figure 4.26 shows DMA storage modulus and tan δ values of pure PURF and CNC filled PURNCFs based on hybrid polyol system with respect to temperature. In accordance with this data, it was shown that pure PURF sample has the highest storage modulus due to the separation of hard blocks of palm kernel oil polyol component in PURF [41]. In addition, it was reported that T_g peaks of all samples in Fig. 4.26 did not exhibit any shift since the degree of crosslinking is the same in pure PURF and CNC filled PURNCF. However, as shown in tan δ versus temperature graph in Fig. 4.26, a smaller peak with lower tan δ value was located at around 45 °C which is basically formed due to the effects of ultrasonication and the incorporation of CNC to PURFs. The reason for the occurrence of this small tan δ peak was previously associated with the existence of palm based polyol rich phase which is not miscible with highly crosslinked PURF matrix [63]. In Fig. 4.26, the small tan δ peak is not

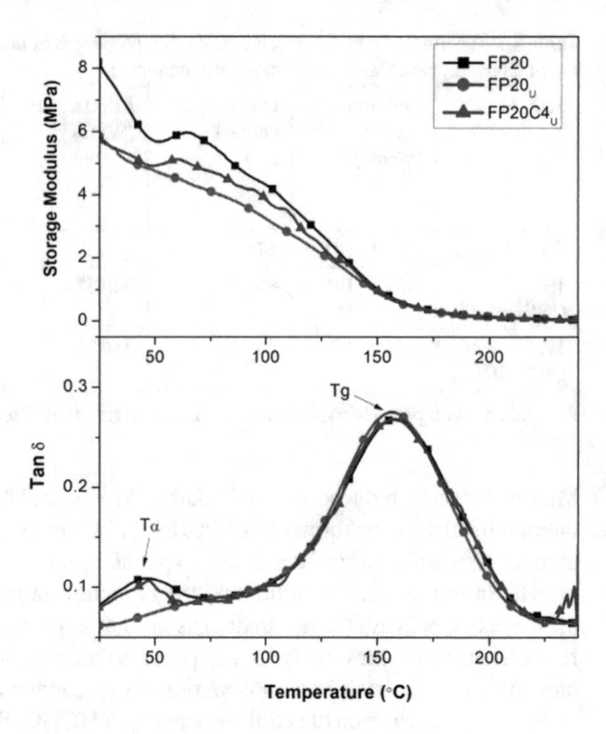

Fig. 4.26 DMA data of pure PURF and CNC filled PURNCFs based on hybrid polyol system with respect to temperature. Reproduced with permission from Ref. [41]. Copyright 2017 Elsevier B.V.

noticeable in the sample of FP20U in which application of ultrasonication during the sample preparation disrupted the polyester clusters that are responsible for the formation of this small tan δ peak. However, when the CNC was incorporated to hybrid polyol based PURFs, as shown in tan δ curve of Fig. 4.26, the smaller tan δ peak at 45 °C is also visible in sample FP20C4U since preferential association of CNC with ester groups of palm based polyol leads to the segregation of polyester polyol into hard segments [41].

It is mainly assumed that glass transition is a very vital concept to understand whether polymer matrix in polymer nanocomposites is mechanically reinforced or not. Thus, in polymer nanocomposites, the elevation in numerical value of T_g should be correlated with reduction of moving ability of polymer chains in presence of nanoparticles. Thus, mostly in thermoplastic based polymer nanocomposites, it is expected that T_g should increase with the addition of nanofiller to the system. However, in Fig. 4.26, the increase in T_g was not observed since the degree of crosslinking which is associated with T_g here in this system remained the same for all the samples.

Previously, cheap and continuous method was used to improve thermal insulation properties of PURFs [42]. The work was focused on an effective process of improving mechanical properties and decreasing thermal conductivity of PURFs with no change in foam density via inclusion of very low amounts of cellulose nanocrystals (CNCs) into PURFs by using an optimized solvent-free ultrasonication method. Based on the results, it was shown that the thermal conductivity of PURFs was reduced by 5% compared to pure PURF with incorporation of 0.4 wt% CNC which was not gone through any specific method of surface modification. The rationale behind this reduction in thermal conductivity was explained such that the favorable interaction between CNC and the polyol component increased the functioning of CNC as the nucleation agent, and thus caused the formation of smaller cell size and eventually the decrease of thermal conductivity [42].

Figure 4.27 shows influences of using CNC on (a) beginning and (b) aged thermal conductivity values of pure PURF and PURNCFs with and without ultrasonication. Based on the results, it was shown that initial thermal conductivity was reduced with inclusion of 0.4 wt% CNC due to smaller cell sizes and higher amount of closed cells in PURNCFs [42]. However, on the other hand, in Fig. 4.27a, it is clearly seen that incorporation of 0.8 wt% CNC elevates thermal conductivity since at this CNC concentration, the aggregation behavior of CNC leads to the generation of lower amount of closed cellular morphology and larger cell sizes [42].

Effects of using CNC on thermal conductivity aging of pure PURF and PURNCFs with and without ultrasonication after certain period of time is shown in Fig. 4.27b. Based on the results, it was shown that PURNCF filled with 0.4 wt% CNC displayed

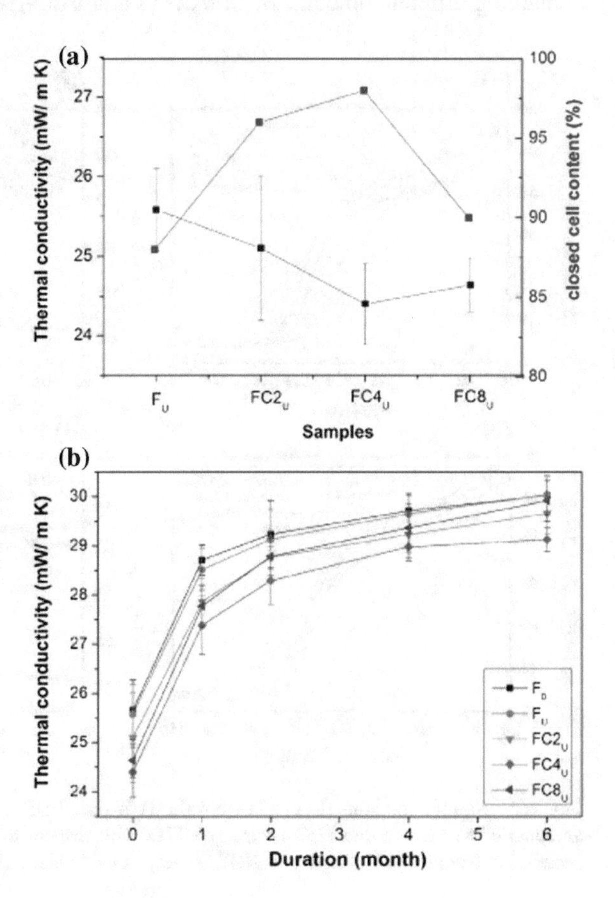

Fig. 4.27 Impact of using CNC on **a** beginning and **b** aged thermal conductivity values of pure PURF and PURNCFs with and without ultrasonication. Reproduced with permission from Ref. [42]. Copyright 2017 Elsevier B.V.

more superior thermal conductivity preservation compared to pure PURF and other PURNCFs. This positive finding based on the aged thermal conductivity data was also confirmed with other favorable properties of PURNCF consisting of 0.4 wt% CNC such as smaller cell size, higher closed cell content, stronger cell walls which also lead to improvement of mechanical properties in this particular sample [42].

Recently, MWCNTs which were grafted with poly(ethylene oxide) PEO-*graft*-MWCNTs (grafted MWCNTs) were prepared via ring-opening polymerization of ethylene oxide [1]. In addition, authors fabricated PURNCFs containing MWCNTs and grafted MWCNTs to explore influences of PEO molecules on morphological and thermomechanical properties and reactive foaming process of PURNCFs [1]. Based on the results, it was shown that using grafted MWCNTs had a little amount of impact on viscosity of polyol. Furthermore, it was revealed that incorporation of nanofiller increased the glass transition temperature and mechanical properties of PURNCFs due to favorable interactions between grafted MWCNTs and PU chains [1].

Figure 4.28 shows storage modulus (a, c) and tan δ (b, d) of pure PURF, PURNCFs containing different amounts of MWCNTs and PEO-*graft*-MWCNTs with respect

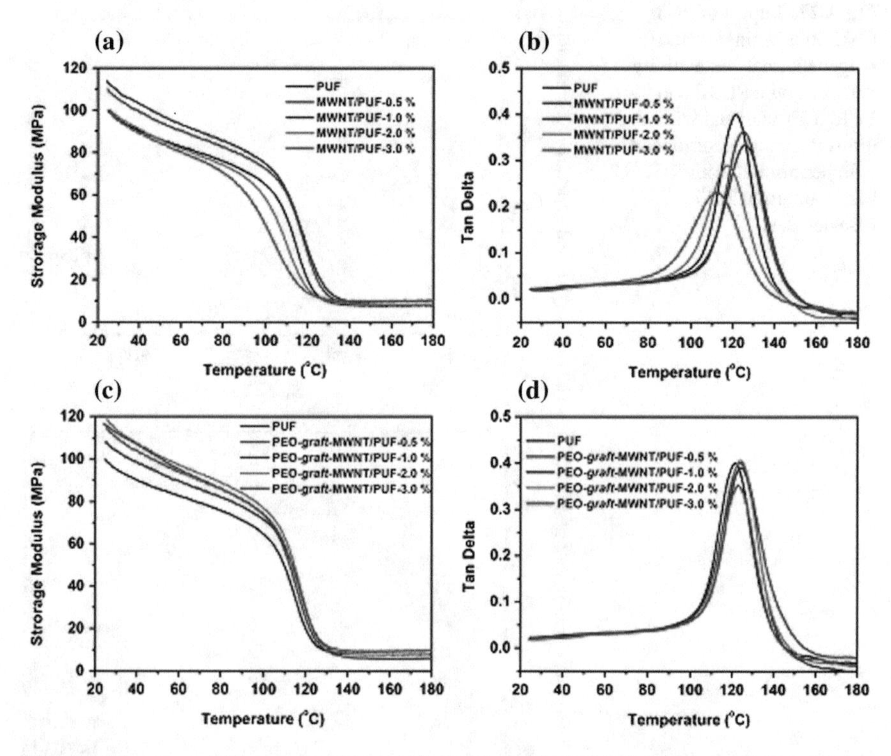

Fig. 4.28 Storage modulus (**a**, **c**) and tan δ (**b**, **d**) of pure PURF, PURNCFs containing different amounts of MWCNTs and PEO-*graft*-MWCNTs with respect to temperature. Reproduced with permission from Ref. [1]. Copyright 2018 Society of Chemical Industry

to temperature. Based on the results, in agreement with other works [4, 10, 11, 64], it was found out that inclusion of MWCNTs or PEO-*graft*-MWCNTs into PURFs increases storage modulus values, and also increasing nanofiller content moderately improves storage modulus [1].

In addition, based on glass transition temperature with respect to filler content results, the glass transition temperature of PURNCFs consisting of MWCNTs was found out to be increasing with filler concentration and the maximum T_g was obtained with incorporation of 1.0 wt% MWCNT. Moreover, glass transition temperature decreased with further addition of MWCNT [1]. The reason behind the decrease of T_g at higher MWCNT additions was explained such that excessive filler causes the aggregation of MWCNTs, which leads to the decrease of favorable interactions between MWCNTs and PURF matrix, and thus leads to the increase in mobility of polymer chains [1].

As shown in Fig. 4.28, the glass transition increases moderately with incorporation of PEO-*graft*-MWCNTs to PURFs. It was reported that the glass transition increase in PURNCFs consisting of PEO-*graft*-MWCNTs is much larger than that of PURNCFs consisting of MWCNTs. The reason behind this finding was explained such that PEO-*graft*-MWCNTs are much more effective nanofillers compared to bare MWCNTs in terms of enhancing favorable interactions between nanofillers and PU chains due to grafted-PEO groups on their surfaces [1, 44].

In agreement with previously published major PURCNF works containing different surface properties of MWCNTs [1, 2, 9], using surface modified MWCNTs basically improves mechanical and thermomechanical properties of PURNCFs via creation of favorable interfacial interactions between surfaces of nanofillers and PU chains. The reason behind the increase of storage modulus and glass transition temperature in PURNCFs consisting of surface modified MWCNTs in comparison with unmodified MWCNTs can be explained such that aggregation behavior of MWCNTs is basically reduced with introduction of compatible surface functional groups on surfaces of nanofillers via performing various surface modification techniques. Thus, the decrease of surface modified MWCNT aggregation in the PURF system, the homogenous dispersion of surface modified MWCNT cylinders with much higher surface areas compared to the aggregated state of bare MWCNTs in either polyol or isocyanates, and then the homogenous mixing of all components during the reactive foaming process lead to increase in the number of beneficial interactions between surface modified MWCNTs and PU chains. The higher number of intermolecular interactions between surface modified MWCNTs and PU chains basically decreases the moving ability of polymer chains due to reduction in the aggregation behavior of MWCNTs with higher surface areas. Thus, this physical mechanism provides suitable conditions for the increase of T_g and storage modulus of PURNCFs that were prepared with surface modified MWCNTs since molecular motions of PU chains become more restricted and the polymer matrix is converted into a much more rigid matrix due to the presence of unaggregated state of surface modified MWCNTs with higher surface areas [1, 2, 9].

Previously, thermal conductivity and fire resistance properties of PURNCFs containing MWCNT concentration range of 0.1–2 wt% were explored in detail [4].

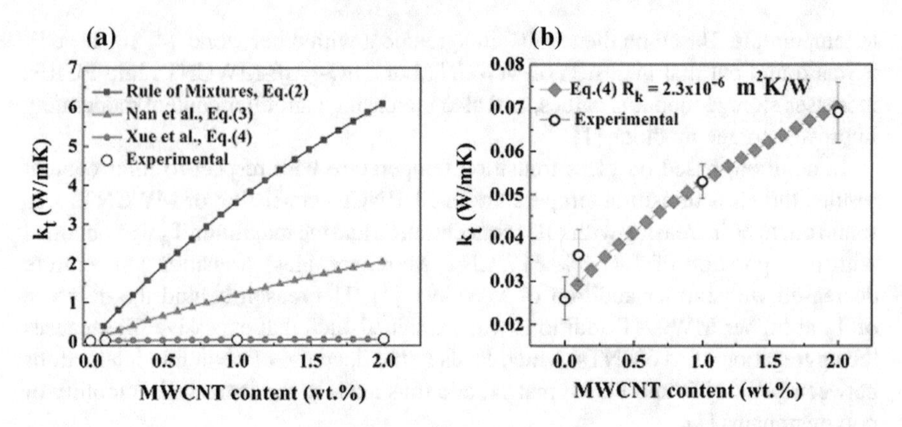

Fig. 4.29 The comparison of predicted and experimental thermal conductivity data of PURFs with respect to MWCNT content. **a** Both measured and predicted data, **b** equation fit to measured data. Reproduced with permission from Ref. [4]. Copyright The Author(s) 2016

Based on thermal conductivity measurement results, it was shown that increasing amount of MWCNTs elevated thermal conductivity values of PURNCFs moderately due to the presence of interfacial thermal resistance in these systems [4].

Figure 4.29 shows the comparison of predicted and experimental thermal conductivity data of PURFs with respect to MWCNT content. More specifically, Fig. 4.29 displays (a) both measured and predicted data, (b) equation fit to measured data [4]. In accordance with data in Fig. 4.29a, b, thermal conductivity values of PURNCFs consisting of MWCNTs increase gradually with increasing concentrations of MWCNTs. Previously, it was shown that the vibration ability of MWCNT structure via phonons mainly causes higher thermal conductivity values of MWCNTs [65]. Thus, increasing MWCNT amounts in PURNCFs elevates the intensity of vibrations related to phonons, and eventually leads to thermal conductivity increase in PURNCFs consisting of MWCNTs [4]. As observed in Fig. 4.29b, the prediction results overvalue the thermal conductivity measurement results of PURNCFs [4]. Based on the results of these models, it was reported that heat transfer is mainly controlled by the interfacial thermal resistance between MWCNTs and PU chains since it reduces the intensity of vibrations at the interface due to phonons and thus decreases the thermal conductivity of PURNCFs consisting of MWCNTs [4].

Previously, a comparative study of PURNCFs containing CNTs and graphene nanosheets (GNSs) was published [10]. Based on dynamic mechanical properties, it was highlighted that the glass transition temperature elevated by 14 °C with incorporation of 0.3 wt% CNTs to PURFs. Based on these results, it was concluded that GNSs are more effective nanofillers compared to CNTs for PURFs in improving thermal stability and mechanical properties. The reason behind superior performance of GNSs over CNTs was explained such that characteristic two-dimensional geometry, surface structure, and higher specific surface area of GNSs pave the way for the establishment of more solid intermolecular interactions and reduction in the mobility of polymer chains at the interface between GNSs and PURNCF matrix. Furthermore,

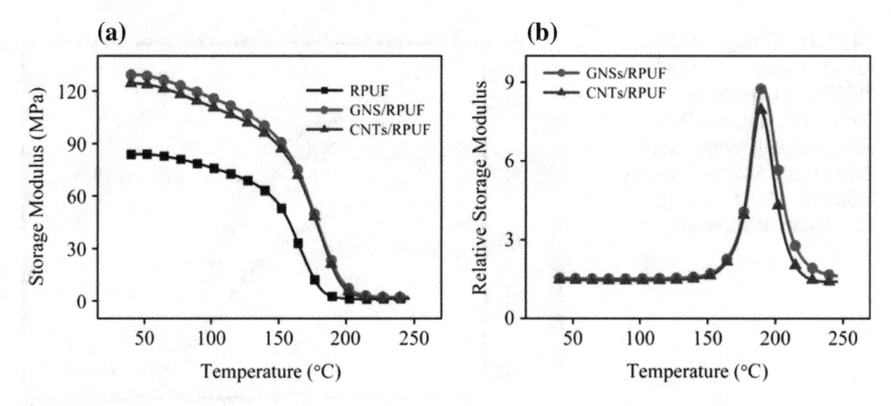

Fig. 4.30 DMA data of pure PURF and PURNCFs containing 0.3 wt% GNSs and CNTs with respect to temperature. Reproduced with permission from Ref. [10]. Copyright 2012 Society of Chemical Industry

based on thermal conductivity results, it was shown that thermal conductivity values of PURNCFs did not change that much with the incorporation of GNSs and CNTs [10].

Figure 4.30 shows DMA data of pure PURF and PURNCFs containing 0.3 wt% GNSs and CNTs with respect to temperature. In accordance with data, storage modulus values of pure PURF and PURNCFs decrease gradually and continuously in initial temperatures, then storage modulus decreases sharply between 180 and 210 °C which is associated with glass transition temperature of PURFs [10]. In addition, it is clearly seen that storage modulus values of PURNCFs consisting of GNSs are higher than those of PURNCFs filled with CNTs. Based on the relative storage modulus versus temperature graph in Fig. 4.30b, it was stated that incorporation of 0.3 wt% CNT or GNS to PURFs led to slight improvement of storage modulus at about 40 °C [10]. In contrast, at glass transition temperature (around 190 °C), the relative storage modulus values of PURNCFs consisting of GNSs and CNTs rise drastically. In agreement with previously published works [1, 11, 41], the reason behind storage modulus elevation with inclusion of CNTs or GNSs was explained such that favourable interactions are established between rigid nanofillers and PU chains which reduce the moving ability of PU chains near surfaces of nanofillers, thus leading to the increase of both T_g and storage modulus in PURNCFs consisting of these types of nanofillers [10].

In addition, in accordance with T_g and tan δ data, incorporation of GNS and CNT to PURFs definitely increases T_g and decreases the value of tan δ which proves increase of rigidity in PURNCFs by barrier effects of CNTs and GNSs [10]. Thus, it was previously shown that inclusion of GNSs and CNTs into PURFs highly decreases mobility of PU chains via favourable intermolecular bondings, and these powerful secondary bonding forces particularly act as crosslinks at the glass transition temperature [10, 66, 67].

Previously, PURNCFs consisting of low amounts of CNTs (about 1.2 wt%) were prepared with the establishment of uniform distribution CNTs throughout

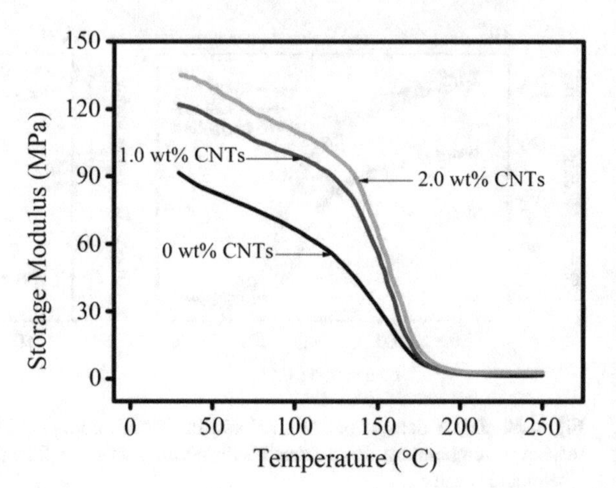

Fig. 4.31 Storage modulus values of pure PURF and PURNCFs containing various amounts of CNTs with changing temperature. Reproduced with permission from Ref. [11]. Copyright 2011 Wiley Periodicals, Inc.

closed-cellular morphology of PURFs [11]. Based on the results, it was shown that PURNCFs with conductive properties exhibited superior electrical stability within temperature range of 20–180 °C which proves the use these PURNCFs in suitable applications for many years. According to DMA and mechanical property test data, it was pointed out that 2.0 wt% CNT addition improved compression properties by 31% and increased storage modulus by 50% at room temperature [11].

Figure 4.31 shows storage modulus values of pure PURF and PURNCFs containing various amounts of CNTs with respect to temperature. In accordance with data, storage modulus values of pure PURF and PURNCFs consisting of different amount of CNTs decrease initially with increasing temperature, and then display a drastical decrease at glass transition temperature of PURFs between 140 and 160 °C [10, 11]. In addition, in the glassy region, storage modulus of PURNCF consisting of 2.0 wt% CNTs increases in comparison with pure PURF [11]. The reason behind this improvement in storage modulus was explained such that PURF matrix effectively transfers the load to the highly dispersed nanofillers consisting of very high strengths compared to PURF matrix [11, 55, 56, 67]. However, on the other hand, it was shown that above the glass transition temperature, there are no big differences between the storage modulus values of pure PURF and PURCNFs consisting of CNTs due to the decline of systematic stress shift between PURFs and CNTs in the rubbery region [10, 11].

4.5 Thermal Degradation and Flammability

The influences of using sonication technique consisting of various technical conditions were investigated on the preparation and properties of PURNCFs that were reinforced with carbon nanofibers (CNFs) [25]. Specifically, authors studied influences of various experimental parameters such as sonication temperature, selection of either polyol or polyisocyanate as the sonication medium, sonication time and power, concentration of nanoparticles and total amount of foam on morphological and thermal decomposition properties of PURNCFs filled with CNFs. Based on the results, it was found out that the application of ultrasound to CNF dispersions in the polyisocyanate part leads to improved properties. In addition, it was also reported that there is an optimum time limit for the sonication, and this time limit is closely dependent on the sonication power, concentration of CNFs in the dispersions and the foam amount that is used during the experiments [25].

Figure 4.32 shows the decomposition temperature plot of PURNCFs consisting of 0.5 wt% CNTs for different times of sonication such as 0, 10, 20, 30 and 40 min. Three samples were taken from each PURNCF sample that was sonicated for different times of sonication for TGA analysis and the average of thermal decomposition values from TGA tests with proper statistical analysis was displayed in Fig. 4.32. According to this data, the average value of decomposition temperatures for each PURNCF consisting of 0.5 wt% CNT was almost similar in all the sonication times. However, it was also reported that for 0 sonication time, the range of the error bar was found to be the highest while for 20 and 30 min of sonication times, the range of the error bar for the decomposition temperature was found to be the lowest [25].

Thus, based on these findings, the optimum sonication times for PURNCFs consisting of 0.5 wt% can be stated as 20 and 30 min due to the observation of a very narrow range for the error bar of decomposition temperature [25]. Thus, in these samples, application of 20 or 30 min of sonication times resulted in homogenous dispersion of CNTs within the PURF matrix, thus leading to very close decomposition temperature values from these samples based on their TGA analysis. However, on the other hand, application of no ultrasonication resulted in very diverse decomposition temperatures since homogenous dispersion of CNTs could not be obtained without any ultrasonication. In addition, application of higher amounts of sonication times again led to the observation of diverse decomposition temperatures since too

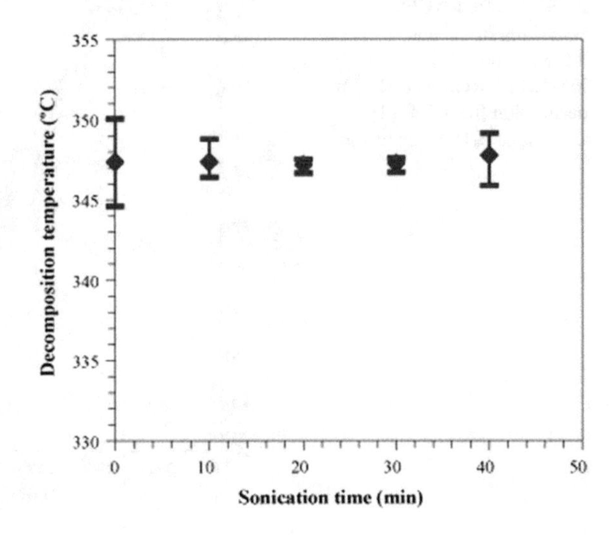

Fig. 4.32 Decomposition temperature plot of PURNCFs consisting of 0.5 wt% CNTs for different times of sonication. Reproduced with permission from Ref. [25]. Copyright 2007 Elsevier B.V.

much sonication energy was given to CNT dispersions at these higher sonication times, and this excess amount of energy was spent for the adhesion of CNTs rather than for their separation and well-dispersion within the PURF dispersions [25].

Previously, multi-walled carbon nanotubes consisting of hydroxyl functional groups were surface modified with 3-aminopropyltriethoxysilane (Si) and dipodal silane [2]. Authors prepared PURNCFs consisting of silanized MWCNTs to explore influences of these silane functionalized MWCNTs on thermal, closed-cellular morphological and mechanical properties of PURNCFs [2]. Based on morphological results, it was shown that incorporation of silanized MWCNTs into PURFs enhances the cell density in PURNCFs. More specifically, according to SEM and mechanical testing data, it was reported that mechanical properties of PURNCFs consisting of 1.5 wt% DSi-MWCNT were improved much more in comparison with those of PURNCFs consisting of 1.5 wt% Si-MWCNT (APTS-MWCNT) [2].

Figure 4.33 shows TGA data of pure PURF and PURNCFs consisting of 1.5 and 3.0 wt% MWCNTs that were surface modified with silanized functional groups. In accordance with results, it was shown that the decomposition temperature associated with 50% weight loss (T_{50}) increases by 17 and 12 °C with incorporation of 1.5 wt% Si-MWNT and MWCNTs modified with dipodal silane to pure PURF, respectively [2]. The reason behind this kind of enhancement in thermal decomposition temperature with inclusion of silanized MWCNTs was explained as in the following. The incorporation of small amounts of CNTs (less than 2 wt%) to PURFs led to uniform distribution of CNTs in PURF matrix, and thus favorable interactions between the well-dispersed CNT particles and polymer chains increased the decomposition temperature of PURNCF matrix due to the penetration of polymer chains within highly thermal stable CNT particles [2].

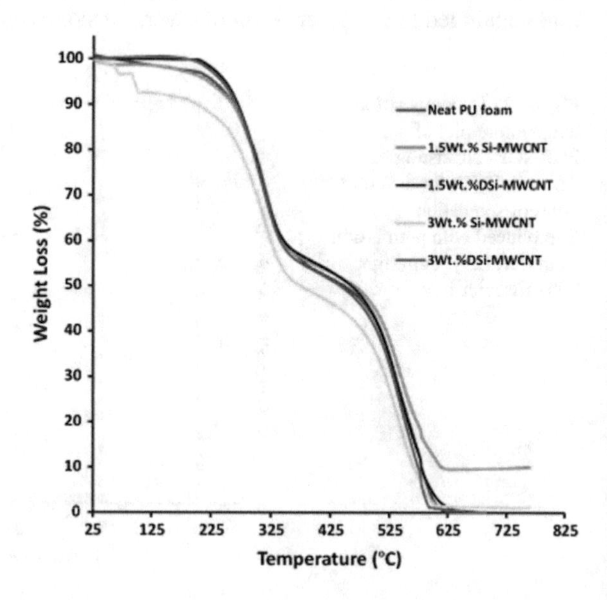

Fig. 4.33 TGA data of pure PURF and PURNCFs consisting of 1.5 and 3.0 wt% silanized MWCNTs. Reproduced with permission from Ref. [2]. Copyright 2018 Elsevier Ltd.

In Fig. 4.33, based on the TGA profiles of PURNCFs consisting of 1.5 wt% MWCNTs which were surface modified with APTS and dipodal silane molecules, respectively. It was shown that PURNCF containing 1.5 wt% MWCNT modified with APTS molecules displayed enhanced thermal stability compared to PURNCF sample containing 1.5 wt% MWCNT modified with dipodal silane since amine groups of APTS molecules on the Si-MWCNT surface could form more powerful and increased number of hydrogen bonds with urethane groups in comparison with those of DSi-MWCNTs [2]. On the other hand, it was also reported that thermal stability values of PURNCFs containing 3 wt% of two types of silanized MWCNTs diminished compared to that of pure PURF because of the aggregation behavior of silanized MWCNTs at higher concentrations. The heterogenous dispersion of silanized MWCNTs within PURNCFs leads to formation of strong secondary type of bonding interactions between MWCNTs. Moreover, this mechanism establishes basis for the aggregation behavior of MWCNTs within the PURF matrix, and the reduction of thermal stability in PURNCFs consisting of higher amounts of MWCNTs [2, 68].

In agreement with other previously published works in the literature [2, 4–6, 8, 9, 11], using cylindrical nanoparticles such as CNTs in PURNCFs improves thermal decomposition temperatures and thermal stability much more better compared to that of pure PURF. The reason behind the improvement of thermal stability via inclusion of small amounts of CNTs was explained such that the mobility of polymers chains with temperature and the diffusion of volatile thermal decomposition gases were obstructed by the evenly well-distributed CNTs and favorable interfacial secondary bonding forces between CNTs and the PURF matrix [11, 69]. Thus, this specific property of CNTs leads the way to improved thermal decomposition temperatures in PURNCFs filled with CNT fillers.

More enhancement in thermal stability of CNT filled PURNCFs can be provided if CNT nanocylinders are surface modified with surface functional groups which are physically and chemically similar to polymer chains in PURNCFs. Main reasons behind structure-property improvements in PURNCFs filled with surface-modified nanoparticles in comparison with those of PURNCFs consisting of unmodified nanoparticles were explained very clearly in previously published works [7–9, 11, 25, 30, 32, 33, 38–40]. The fabrication of advanced CNT filled PURNCFs with improved physical, thermal and mechanical properties depends strongly on the generation of closed-cellular morphology with uniform distribution of cell sizes and higher cell densities since these morphological parameters in rigid foams directly affect macroscopic performance of PURNCFs. Furthermore, the generation of this kind of perfect closed cellular morphology in CNT filled PURNCFs is strongly correlated with establishment of favorable interactions between surface of nanocylinders and PU chains. If carbon nanotubes are not surface modified, then there is a highly potential risk for aggregation of CNTs which ultimately reduces favorable CNT-PU interactions while increasing CNT-CNT type of interactions. On the other hand, if carbon nanotubes are surface modified with specific and compatible functional groups, the aggregation behavior of CNTs could be eliminated due to establishment of beneficial secondary type of intermolecular bondings between surface functional groups of CNTs and PU chains. In that case, PU-CNT type of interactions could be dominant over unfavorable CNT-CNT type of interactions. By this way, the properties of PURNCFs consisting of these surface-modified carbon nanotubes can be maximized over PURNCFs containing unmodified CNTs. Thus, for this reason, various surface modification methods should be employed to chemically attach surface functional groups onto CNTs which are physically and chemically compatible with PU matrix [1, 2, 30, 32, 33]. By this way, the number of favorable interactions between surface modified CNTs and PURNCF matrix could be maximized to produce PURNCFs with improved properties such as reduced cell sizes and higher cell densities. Using silane molecules as surface modifiers for CNTs is a specific and useful method for the creation of compatible nanocylinder surfaces with PU matrix [2]. More specifically, if the silane molecules have amine functionalities within their chemical structures, then the surface modification which is performed with these types of amino silanes can be more efficient in the generation of active and compatible surfaces. Thus, using CNTs which are surface modified with amino silanes in PURFs leads to the establishment of more H-bonding interactions between the surface of CNTs and PU chains. Also, using MWCNTs that are surface modified with amino silanes as nanocylindrical fillers in PURFs provides the basis for enhanced thermal stability and thermal decomposition temperatures compared to those of MWCNTs that are surface modified with other types of silanes [2, 57, 58]. Furthermore, using both silane modified CNT nanocylinders and chemically compatible flame retardant agents in PURNCFs is also another alternative method in terms of using these types of surface modified CNT nanofillers in the fabrication of PURNCFs with extra enhanced thermal decomposition and stability properties. The establishment of favorable interactions between silane modified CNTs and compatible flame retardants in PURNCFs might not only improve thermal stability and thermal decomposition temperatures of the system but also reduce the flammability properties of PURNCFs consisting of these hybrid silanized CNTs and compatible flame retardants.

Recently, thermal conductivity and fire resistance properties of PURNCFs containing MWCNT concentration range of 0.1–2 wt% were explored in detail [4]. Based on thermal conductivity measurement results, it was shown that increasing amount of MWCNTs elevated thermal conductivity values of PURNCFs moderately due to occurrence of interfacial thermal resistance in these systems. Moreover, it was observed that the content of MWCNTs in PURNCFs strongly affected fire resistant properties and closed-cellular morphology. More specifically, it was revealed that incorporation of 0.1 wt% MWCNTs to PURFs increases the flame propagation speed. However, in contrast, the flame propagation speed was observed to decrease while the amount of MWCNTs was further increased in PURNCFs. Furthermore, it was also pointed out that inclusion of MWCNTs to PURFs reduced the mass loss of PURNCFs after flame extinction. Thus, this observation suggest that the existence of MWCNTs increases resistance against fire in PURNCFs [4].

Figure 4.34 shows the flammability measurements of pure PURF and PURNCFs containing different amounts of MWCNTs. Basically, Fig. 4.34 shows the data of (a) time associated with remaining light after burning (t_f) and (b) mass loss (P_m) after burning. Based on the results, it was shown that the afterglow time first decreased with inclusion of 0.1 wt% MWCNTs and then it increases for PURNCFs consisting

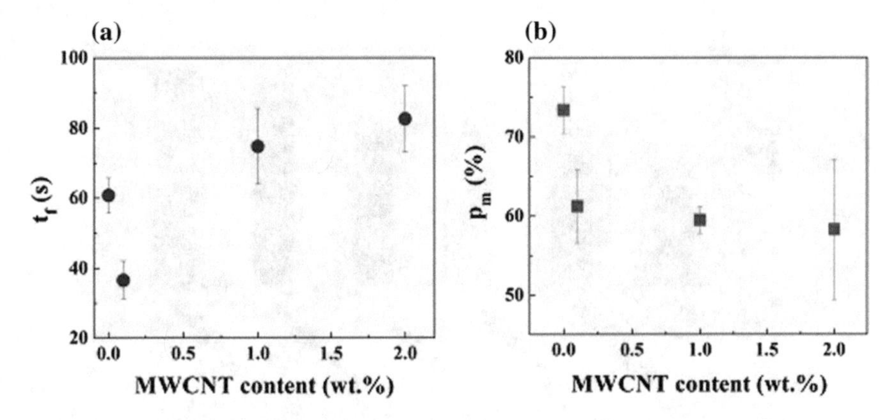

Fig. 4.34 The flammability measurements of pure PURF and PURNCFs containing different amounts of MWCNTs. **a** The time associated with remaining light after burning (t_f) and **b** mass loss (P_m) after burning. Reproduced with permission from Ref. [4]. Copyright The Author(s) 2016

of 1 and 2 wt% MWCNTs [4]. The reason behind this finding was explained such that in PURNCF consisting of 0.1 wt% MWCNT, due to smaller sizes of cells, oxygen gas within cells was homogeneously distributed throughout the closed-cellular morphology, causing the quick expending of flame. In contrast, MWCNTs formed more effective carbonaceous protective layers on foams and increased the value of afterglow time as the content of MWCNTs increased to 1 and 2 wt% [4]. According to data in Fig. 4.34b, the mass loss decreased gradually with inclusion of higher amounts of MWCNT. Specifically, it was reported that pure PURF lost about 75% of its mass after the burning process, but the mass loss for PURNCFs consisting of 0.1 wt% MWCNT decreased to 60%. The reason for this observation was explained such that the closed-cellular morphology with higher cellular density in PURNCFs consisting of 0.1 wt% MWCNTs provided the basis for the nanocomposite to become more thermal stable and more resistant to flame consumption [4].

Formerly, the reinforcement of PURFs was performed by using carbon nanotubes (CNT) or graphite as nano/micron-sized additives to enhance fire resistance thermal and mechanical properties [5]. In accordance with experimental data, it was pointed out that mechanical properties of PURNCFs improved by almost 20% with small additions of CNTs with maximum of 0.05 wt%. The reason behind this improvement in mechanical strength was attributed such that the strengthening effect in PURNCFs was mainly coming from CNTs and also incorporation of CNTs leads to generation of smaller cell sizes with much narrower distributions in comparison with those of pure PURF. Based on thermal stability results, it was reported that the flammability reduced and thermal stability increased with incorporation of CNTs because of protective carbon coating generation on the exterior of PURNCFs [5].

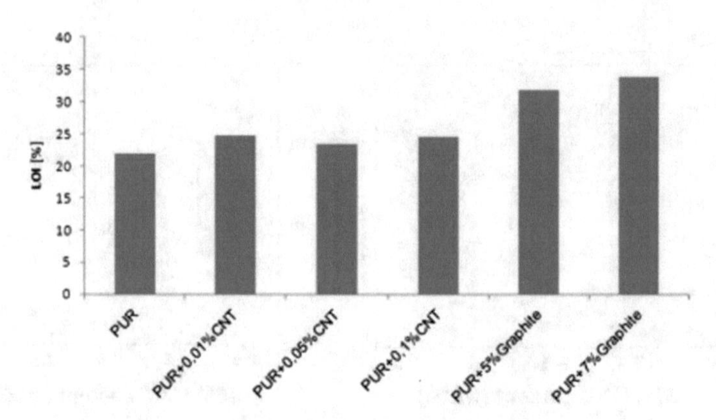

Fig. 4.35 LOI results of pure PURF and PURNCFs containing CNTs and PURCFs consisting of graphite with different concentrations. Reproduced with permission from Ref. [5]. Copyright 2016 Elsevier Ltd.

Figure 4.35 shows the LOI results of pure PURF and PURNCFs containing CNTs and PURCFs consisting of graphite with different concentrations. In accordance with experimental data, limiting oxygen index values enhanced with incorporation of CNTs and graphite containing different concentrations into pure PURF. Furthermore, it was emphasized that the inclusion of CNTs to pure PURF improved the LOI much less than the addition of graphite [5]. The reason behind the lower LOI values in CNT filled PURFs was explained such that CNTs distribute the heat flux throughout the sample evenly due to their higher thermal conductivity, thus this behavior leads to the decrease of decomposition temperatures in CNT filled PURFs. More specifically, it was also shown that PURCFs containing 5 and 7% of EG had LOI values of 31.8 and 33.9%, respectively. The substantial improvement in LOI values of EG filled PURCFs was explained such that transport of flammable and vaporous decomposed compounds was obstructed due to the expansion of graphite flakes via the generation of tightly packed, firm and protective char layers around PURCF matrix against the heat flux [5].

Earlier on, effects of using various types of carbon fillers such as MWCNTs and graphite sheets on flammability and thermal decomposition properties of PURFs were explored [6]. Authors used TGA and TG combined with infrared spectroscopy (TG–IR) to comprehend thermal degradation behavior of PURFs and also analyzed the products that were produced as a result of thermal decomposition process. The activation energy of PURFs was also specifically measured based on the results from TGA analysis. Furthermore, authors performed specific tests in order to measure LOI and smoke density values of PURFs. Based on the results, it was found out that the thermal stability of PURNCFs did not change that much compared to that of pure PURF. The activation energy of PURNCFs consisting of CNTs was increased compared to pure PURF. Furthermore, based on TG–IR results, it was revealed that no differences in the nature of volatile products were observed in PURNCFs and pure PURF during the thermal degradation process. The flame retardancy behavior

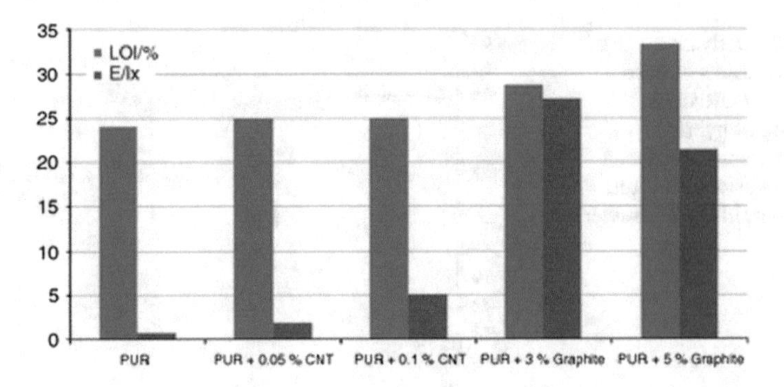

Fig. 4.36 Flammability results of pure PURF and PURNCFs consisting of CNTs and PURCFs consisting of graphite with various concentrations. Reproduced with permission from Ref. [6]. Copyright The Author(s) 2015

of PURNCFs consisting of CNTs was improved, however PURCFs consisting of graphite flakes exhibited better flame retardancy properties compared to CNT filled PURNCFs [6].

Figure 4.36 shows the flammability results of pure PURF and PURNCFs consisting of CNTs and PURCFs consisting of graphite with different contents. In agreement with other works [5], it was shown that inclusion of CNT moderately increased LOI values of PURFs [6]. However, on the other hand, it was reported that the PURCFs consisting of graphite had the highest LOI value in which the fire resistance was improved with increasing EG concentration. Specifically, it was highlighted that the LOI value of PURCF consisting of 5 wt% graphite was bigger than that of 3 wt% graphite [6]. Furthermore, it was found out that incorporation of CNTs to PURFs increased the average light intensity after it was burned for 4 min which was confirmed by the reduction of gas intensity from TG-IR results. As observed from PURCFs containing EGs, the generation of a stable char layer on the surface of PURNCF could be associated with the slight increase in the fire resistance of PURNCFs consisting of CNTs which leads to the reduction of flammability [6, 70, 71].

Previously, the effects of using nanohybrid particles of montmorillonite (MMT) and carbon nanotube (CNT) on the performance of PURNCFs were investigated [8]. Authors used chemical vapour deposition for the synthesis of hybrids, and then these MMT-CNT nanohybrid particles were incorporated into PURFs via in situ polymerization process [8]. Authors used SEM, TEM and optical microscopy and TGA in order to evaluate the closed cellular morphology and thermal decomposition behaviour of PURNCFs. According to experimental data, it was shown that smaller cell sizes and higher values of cell density were observed in PURNCFs in comparison with pure PURF. According to TGA data, it was revealed that thermal properties in PURNCFs were improved with the addition of nanoparticle hybrids [8].

Figure 4.37 shows the derivative TGA results of pure PURF and PURNCFs consisting of hybrid nanoparticles. Based on the results, it was shown that the thermal

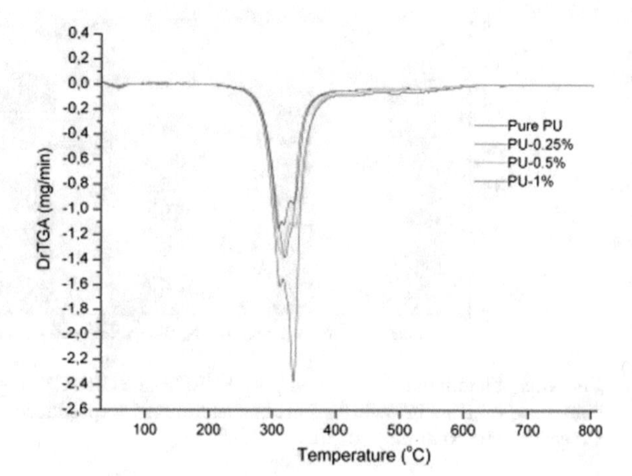

Fig. 4.37 Derivative TGA (DrTGA) results of pure PURF and PURNCFs consisting of hybrid nanoparticles. Reproduced with permission from Ref. [8]. Copyright 2012 Elsevier Ltd.

stability of PURNCFs consisting of hybrid nanoparticles displayed enhancements with the addition of nanohybrids such that PURNCFs consisting of nanohybrid particles started to degrade at slightly higher temperatures compared to pure PURF [8]. In addition, it was shown that PURNCFs consisting hybrid nanoparticles had the highest degrees of degradation which provided a route for the increase of thermal properties in PURNCFs. The rationale behind increase of degradation rate with incorporation of hybrid nanoparticles was explained such that the existence of metal particles within the hybrid nanoparticles led to the increase of degradation rates [8, 72–74]. Also, in addition to the increase of degradation rates in PURNCFs, it was shown that the residual mass at 650 °C decreased much more in the nanocomposites containing hybrid nanoparticles in comparison with that of pure PURF [8].

Recently, effects of homogeneous spreading of CNTs over foam matrix on the foamability, closed-cellular morphology and mechanical properties of PURFs were studied [9]. Authors incorporated raw and functional CNTs into PURFs up to 0.2 wt%, and analyzed the morphology, thermal stability and mechanical properties of pure PURF and PURNCFs consisting of CNTs [9]. Based on the results, it was shown that the dispersion of CNTs in PMDI with low degree of thickness and density led to greater distribution of CNTs, and also resulted in higher value of compressive strength in comparison with that of pure PURF. Based on TGA data, the mechanism behind thermal degradation behaviour of PURNCFs was explained by taking into account different dispersion states of raw and functionalized CNTs in both polyol and PMDI components [9].

Figure 4.38 shows TGA data of (a) pure PURF and (b) PURNCF filled with 0.1 wt% raw CNT in isocyanate component under nitrogen atmosphere. Based on the results, it was shown that pure PURF had two different kinds of degradation events [9]. The first one occurring after 200 °C corresponds to the thermal pyrolysis of urethane linkage that led to the formation of polyol and PMDI. The second degradation event takes place in the region of 200–350 °C which is related to the decomposition of polyol and, also PMDI component decomposes after 350 °C [9, 75]. In addition, it was

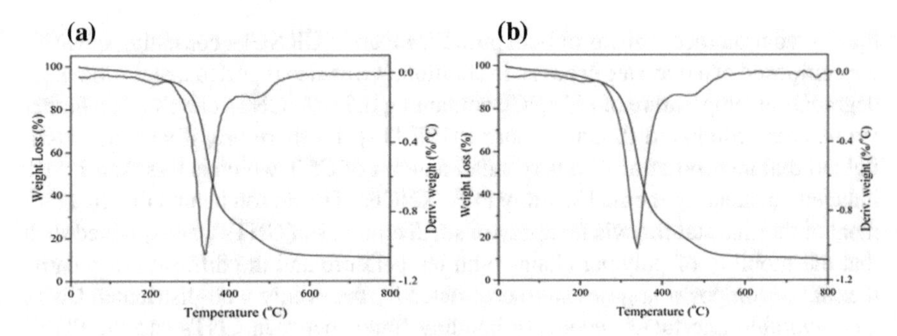

Fig. 4.38 TGA curves of **a** pure PURF and **b** PURNCF filled with 0.1 wt% raw CNT in isocyanate component under nitrogen gas. Reproduced with permission from Ref. [9]. Copyright 2018 Elsevier Ltd.

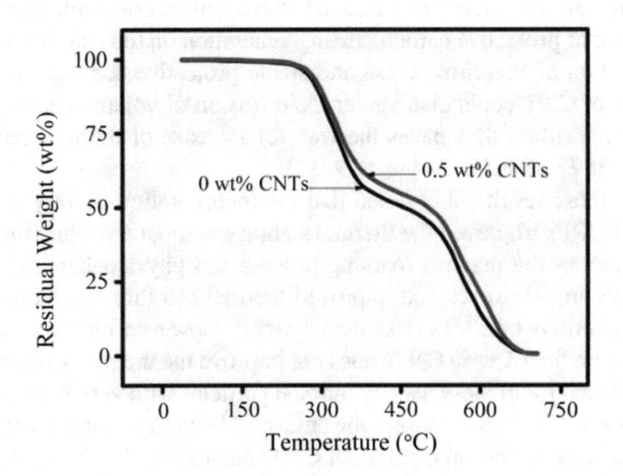

Fig. 4.39 TGA curves of pure PURF and PURNCF consisting of 0.5 wt% CNTs. Reproduced with permission from Ref. [11]. Copyright 2011 Wiley Periodicals, Inc.

highlighted that incorporation of CNTs to PURFs did not change the thermal stability behavior since the maximal degradation temperatures of PURNCFs consisting of CNTs were reported as almost equal with that of pure PURF [9].

Formerly, PURNCFs containing low amounts of CNTs (about 1.2 wt%) were prepared with the establishment of uniform distribution CNTs throughout closed-cellular morphology of PURFs [11]. Based on the results, it was shown that PURNCFs with conductive properties exhibited superior electrical stability within temperature range of 20–180 °C which proves the successful use of these PURNCFs in suitable applications. Furthermore, only 0.5 wt% CNT inclusion into PURFs improved thermal stability in such a way that thermal decomposition temperature at 50% weight loss rose from 450 to 499 °C [11].

Figure 4.39 shows TGA results of pure PURF and PURNCFs consisting of 0.5 wt% CNTs. In agreement with the results from other works [9], it was shown that the degradation mechanisms of both pure PURF and PURNCFs consisting of CNTs are composed of a two-step process. In addition, it was also reported that the maximal degradation temperature of PURNCF containing 0.5 wt% CNTs elevated to a higher temperature compared to that of pure PURF [11]. Furthermore, it was also highlighted that incorporation of a very small amount of CNT which is less than 1 wt% induced outstanding thermal stability in PURNCFs. The reason behind the improvement of thermal stability via inclusion of small amounts of CNTs was explained such that the mobility of polymer chains with temperature and the diffusion of volatile thermal decomposition gases were obstructed by the evenly well-distributed CNTs and favorable interfacial secondary bonding forces between CNTs and the PURF matrix [11, 69]. Also, other studies of PURNCFs consisting low amounts of CNTs revealed the very basic reasons behind the reduction in flammability properties and enhancement in thermal stability. Based on thermal stability results, it was reported that the flammability reduces and thermal stability increases with incorporation of CNTs because of protective carbon coating generation on the exterior of PURNCFs [5]. The creation of this firm, dense and stable protective coating which is a special property of CNT could also hinder the diffusion of volatile decomposed gases during decomposition, thus paves the way for increase of thermal decomposition temperatures in PURNCFs [2, 4–6, 8, 9, 11].

Based on these results, it is noted that the incorporation of very small amount of CNTs in PURFs improved the thermal stability without changing the main characteristics such as the reactive foaming process and physicochemical structure of the PURF system. However, the improved thermal stability can be obtained with very small additions of CNTs (less than 1 wt%). However, on the other hand, the addition of more than 1 wt% CNTs does not improve the thermal stability due to the aggregation behavior of these nanocylindrical particles with very high surface areas. Thus, at higher CNT concentrations, the unfavorable interactions between CNTs and PURF matrix leads to the aggregation of CNTs and eventually does not improve the thermal stability of PURNCFs.

However, in PURNCFs consisting of surface modified CNTs, it was previously shown that thermal stability could be improved with incorporation of 1.5 wt% silanized MWCNTs [2]. Thus, this specific finding shows that it is possible to add higher amounts of surface-modified nanoparticles in comparison with unmodified nanoparticles and obtain higher improved properties such as thermal stability and mechanical properties at higher nanoparticle concentrations. The reason behind this specific finding can be explained such that higher number of favorable interactions can be formed in PURNCF systems containing surface modified CNTs in comparison with those consisting of unmodified CNTs. Also, at the same time, the aggregation behavior of CNTs can be reduced much more by using surface modified CNTs in PURNCFs. Thus, this specific outcome leads to the property that higher weight fractions of surface modified CNTs in PURNCFs can be exploited without inducing too much aggregation in the system in comparison with using much lower concentrations of unmodified CNTs in PURNCFs.

Very recently, Burgaz et al. prepared and studied PURNCFs containing different concentrations of carboxylic acid functionalized MWCNTs and hydrophilic fumed nanosilica particle mixtures [76]. Authors investigated intermolecular secondary bonding interactions, closed-cellular morphology, thermomechanical and mechanical properties, thermal conductivity and thermal stability behavior of PURNCFs in detail. In accordance with experimental data, it was basically shown that using nanoparticle mixtures having concentrations such as less than 0.5 wt% carboxylic acid functionalized MWCNTs and less than 0.2 wt% hydrophilic fumed nanosilica in PURFs enhanced mechanical, thermomechanical and thermal stability values compared to PURNCFs consisting of only either surface modified MWCNTs or hydrophilic fumed nanosilica particles. Specifically, it was unveiled that PURNCF filled with 0.4 wt% modified-MWCNTs and 0.1 wt% silica nanoparticles exhibited enhanced thermal stability and reduced compressive strength and much higher values in connection with thermomechanical properties such as glass transition temperature and elastic response values compared to PURCNF filled with 0.5 wt% modified MWCNTs. In accordance with FT-IR data, the principal mechanism behind enhancement in these properties was explained such that suitable and improved Hydrogen bonding secondary forces of urethane linkages in PURNCFs with hydroxyl functionalities of fumed nanosilica and carboxylic acid groups on MWCNT cylindrical nanoparticles were basically generated in PURNCFs consisting of suitable concentrations of these two diverse nanoparticle blends [76].

Figure 4.40 displays TGA curves and decomposition temperatures of pure PURF and PURNCFs containing different amounts of CNT and nanosilica. In Fig. 4.40a, it is plainly noticed that incorporation of 0.5 wt% fumed silica nanoparticles into PURFs enhances all of degradation temperatures compared to pure PURF. Moreover, it was also pointed out that including same amount of modified-MWCNTs elevated thermal decomposition temperatures of PURNCFs compared to pure PURF. However, it was revealed that inclusion of only nanosilica to PURFs enhanced thermal stability of PURNCFs much more better in comparison with adding same amount of modified-MWCNTs to PURFs [76]. Furthermore, based on data in Fig. 4.40b, it was emphasized that degradation temperatures of PURNCFs consisting of both modified-MWCNTs and nanosilica improved when nanosilica concentration was less than 0.25 wt%. Thus, as a consequence of this finding, it was unveiled that maximum thermal stability was noted in PURNCF which contains 0.4 wt% modified-MWCNTs and 0.1 wt% hydrophilic fumed silica nanoparticles [76]. Moreover, the basic reason behind reduction in thermal stability and thermal decomposition temperatures of PURNCF sample consisting of higher amount of nanosilica (0.4 wt%) and lower amounts of modified-MWCNTS (0.1 wt%) can be explained due to generation of higher amounts of silica-silica type of interactions rather than polymer-silica type of interactions. In contrast, based on experimental data in Fig. 4.40b, PURNCF sample consisting of both 0.1 wt% modified-MWNCTs and 0.4 wt% silica nanoparticles exhibited minimum degradation behaviour among all samples containing 0.5 wt% nanoparticle concentration in total [76].

Figure 4.41 exhibits simplified drawings of secondary intermolecular forces between (a) silica nanoparticles and PU chains in PURNCF/silica nanoparticle blend, (b) modified-MWCNTs and PU molecules in PURNCF/MWCNT blend and (c) silica nanoparticles, modified-MWCNTs and PU chains in PURNCF/modified-

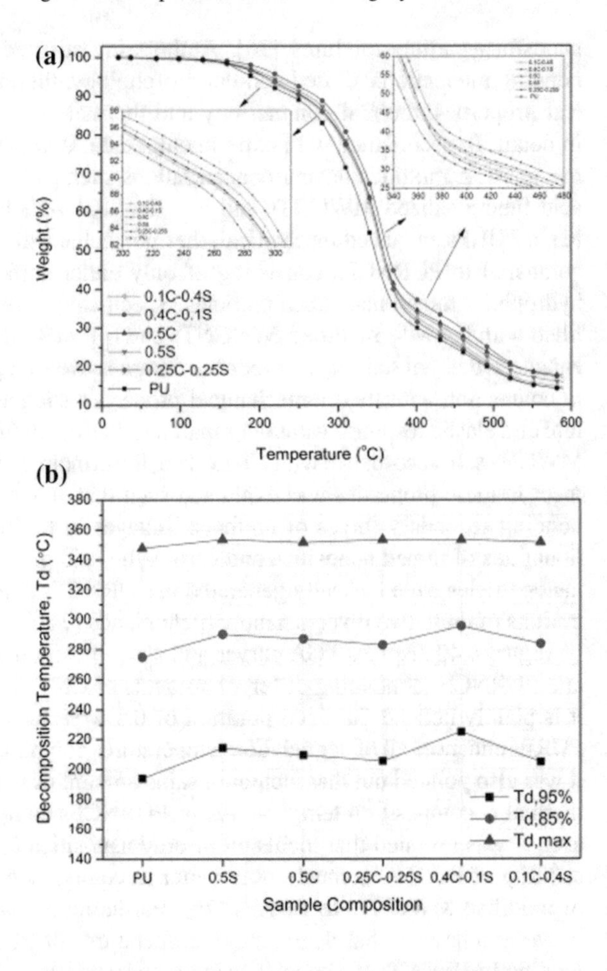

Fig. 4.40 **a** TGA curves and **b** decomposition temperatures of pure PURF and PURNCFs containing different amounts of CNT and nanosilica. Reproduced with permission from Ref. [76]. Copyright 2019 Elsevier Ltd.

MWCNT/silica nanoparticle blend. Based on specific drawings in Fig. 4.41a, b, the reason behind observation of higher thermal stability in PURNCF/nanosilica system in comparison with that of PURNCF/modified-MWCNTs can be understood much more better [76]. It is well-known that hydrophilic nanosilica particles have higher amounts of functionalities on their surfaces compared to modified-MWCNTs since nanosilica particles have larger surface areas and smaller particles sizes in comparison with those of modified-MWCNTs. Thus, for this reason, PURNCFs containing nanosilica particles exhibited much higher thermal stability compared to PURNCFs filled with same amount of modified-MWCNTs [76]. Based on specific drawing of Fig. 4.41c, the reason behind observation of higher thermal stability in PURNCF filled with 0.4 wt% modified-MWCNT and 0.1 wt% silica nanoparticles in comparison with that of PURNCF doped with 0.1 wt% modified-MWCNT and 0.4 wt% silica nanoparticles can be perceived in more detail. If the content of silica nanoparticles is about 0.1 wt% in PURCNFs filled with modified-MWCNTs and silica nanoparticles, then nanosilica additives could easily position themselves between modified

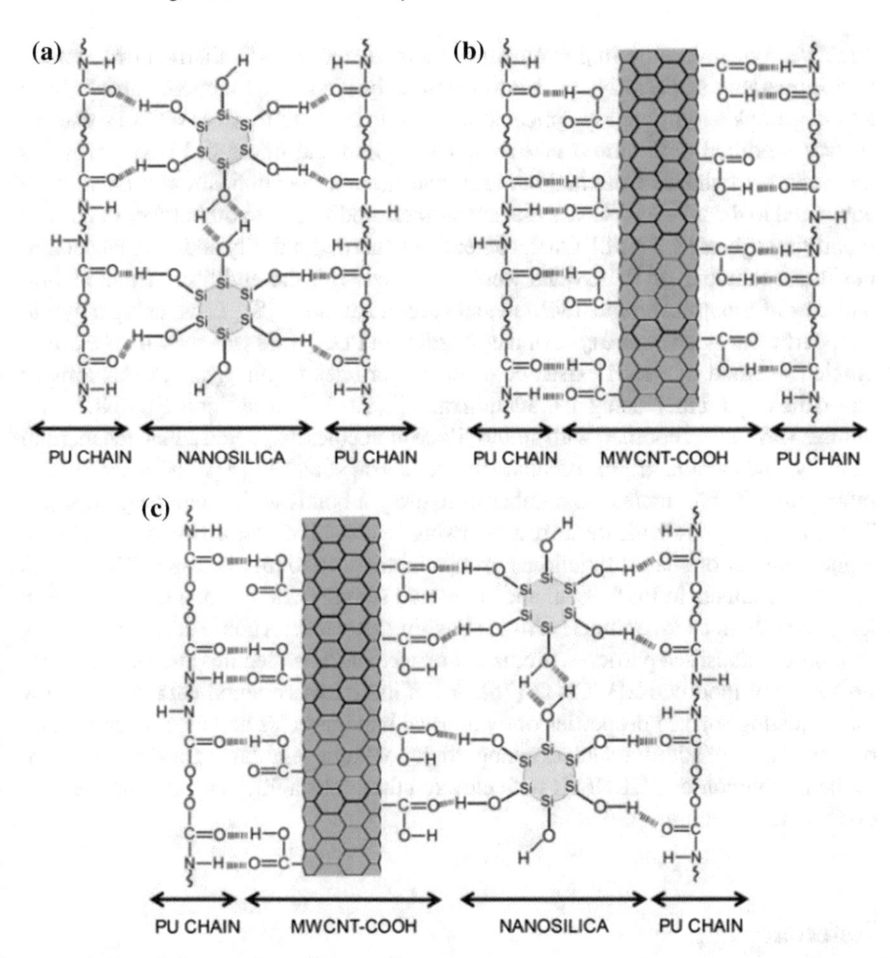

Fig. 4.41 Simplified drawings of secondary intermolecular forces between **a** silica nanoparticles and PU chains in PURNCF/silica nanoparticle blend, **b** modified-MWCNTs and PU molecules in PURNCF/MWCNT blend and **c** silica nanoparticles, modified-MWCNTs and PU chains in PURNCF/modified-MWCNT/silica nanoparticle blend. Reproduced with permission from Ref. [76]. Copyright 2019 Elsevier Ltd.

MWCNTs and PU chains due to their favorable Hydrogen bonding interactions with both modified-MWCNTs and urethane linkages of PU chains. But, as an alternative, if the concentration of silica nanoparticles is about 0.4 wt% in PURCNFs consisting of both types of nanoparticles, then nanosilica additives could not form suitable H-bonding interactions with both modified-MWCNTs and urethane linkages of PU chains since strong Hydrogen bonding forces between nanosilica particles dominate over other types of interactions at this very high nanosilica concentration [76].

In accordance with previously published works [2, 8, 57, 58], performing surface modification of CNTs is a specific and useful method for the creation of compatible nanocylinder surfaces with PU matrix, and also this method was shown to be very effective concerning the improvement of thermal stability and thermal decomposition temperatures in PURNCFs. More specifically, if surface functional groups have amino groups within their chemical structures, then using these MWCNTs that are surface modified with amino groups as nanocylindrical fillers in PURFs provides the basis for enhanced thermal stability and thermal decomposition temperatures compared to those of MWCNTs that are surface modified with other types of surface modifying agents [2, 57, 58]. On the other hand, thermal stability and thermal decomposition behaviour of PURNCFs were improved with the addition of nanohybrid particles of montmorillonite (MMT) and carbon nanotube [8]. Thus, using a hybrid nanoparticle instead of one type of nanoparticle in PURNCFs was shown to be a very efficient method in which existence of metal particles within hybrid nanoparticles paved the way for increasing degradation rates [8, 72–74]. In addition, PURNCF containing MWCNTs modified with amino silane molecules displayed enhanced thermal stability since amine groups of silane molecules on silanized MWCNTs could form more powerful and increased number of hydrogen bonds with urethane groups [2]. Thus, in analogous with these results, using hydrophilic nanosilica particles with higher number of silanol functional groups in addition to modified-MWCNTs with optimum contents in PURFs enhanced thermal stability and thermal decomposition temperatures since favorable Hydrogen bonding interactions could take place among silanols of nanosilica particles, urethane linkages of PU molecules and surface functionalities of modified-MWCNTs [76]. All of these experimental data clearly show that adjusting surface properties of cylindrical nanoparticles and also incorporation of second type of additive such as nanoparticles or flame retardants are very important in the production of PURNCFs with elevated thermal stability and thermal decomposition temperatures.

References

1. Chen, K.P., Cao, F., Liang, S.E., Wang, J.H., Tian, C.R.: Preparation of poly (ethylene oxide) brush-grafted multiwall carbon nanotubes and their effect on morphology and mechanical properties of rigid polyurethane foam. Polym. Int. **67**(11), 1545–1554 (2018)
2. Yaghoubi, A., Nikje, M.M.A.: Silanization of multi-walled carbon nanotubes and the study of its effects on the properties of polyurethane rigid foam nanocomposites. Compos. Part A Appl. Sci. Manuf. **109**, 338–344 (2018)
3. Wang, X., Kalali, E.N., Wan, J.T., Wang, D.Y.: Carbon-family materials for flame retardant polymeric materials. Prog. Polym. Sci. **69**, 22–46 (2017)
4. Espadas-Escalante, J.J., Aviles, F., Gonzalez-Chi, P.I., Oliva, A.: Thermal conductivity and flammability of multiwall carbon nanotube/polyurethane foam composites. J. Cell. Plast. **53**(2), 215–230 (2017)
5. Ciecierska, E., Jurczyk-Kowalska, M., Bazarnik, P., Gloc, M., Kulesza, M., Kowalski, M., Krauze, S., Lewandowska, M.: Flammability, mechanical properties and structure of rigid polyurethane foams with different types of carbon reinforcing materials. Compos. Struct. **140**, 67–76 (2016)
6. Ciecierska, E., Jurczyk-Kowalska, M., Bazarnik, P., Kowalski, M., Krauze, S., Lewandowska, M.: The influence of carbon fillers on the thermal properties of polyurethane foam. J. Therm. Anal. Calorim. **123**(1), 283–291 (2016)

7. Espadas-Escalante, J.J., Aviles, F.: Anisotropic compressive properties of multiwall carbon nanotube/polyurethane foams. Mech. Mater. **91**, 167–176 (2015)
8. Madaleno, L., Pyrz, R., Crosky, A., Jensen, L.R., Rauhe, J.C.M., Dolomanova, V., Timmons, A.M.M.V.D., Pinto, J.J.C., Norman, J.: Processing and characterization of polyurethane nanocomposite foam reinforced with montmorillonite-carbon nanotube hybrids. Compos. Part A Appl. Sci. Manuf. **44**, 1–7 (2013)
9. Caglayan, C., Gurkan, I., Gungor, S., Cebeci, H.: The effect of CNT-reinforced polyurethane foam cores to flexural properties of sandwich composites. Compos. Part A Appl. Sci. Manuf. **115**, 187–195 (2018)
10. Yan, D.X., Xu, L., Chen, C., Tang, J.H., Ji, X., Li, Z.M.: Enhanced mechanical and thermal properties of rigid polyurethane foam composites containing graphene nanosheets and carbon nanotubes. Polym. Int. **61**(7), 1107–1114 (2012)
11. Yan, D.X., Dai, K., Xiang, Z.D., Li, Z.M., Ji, X., Zhang, W.Q.: Electrical conductivity and major mechanical and thermal properties of carbon nanotube-filled polyurethane foams. J. Appl. Polym. Sci. **120**(5), 3014–3019 (2011)
12. Faruk, O., Sain, M., Farnood, R., Pan, Y.F., Xiao, H.N.: Development of lignin and nanocellulose enhanced bio PU foams for automotive parts. J. Polym. Environ. **22**(3), 279–288 (2014)
13. Saha, M.C., Kabir, M.E., Jeelani, S.: Enhancement in thermal and mechanical properties of polyurethane foam infused with nanoparticles. Mater. Sci. Eng. A **479**(1–2), 213–222 (2008)
14. Harikrishnan, G., Singh, S.N., Kiesel, E., Macosko, C.W.: Nanodispersions of carbon nanofiber for polyurethane foaming. Polymer **51**(15), 3349–3353 (2010)
15. Li, H., Zhong, J., Meng, J., Xian, G.J.: The reinforcement efficiency of carbon nanotubes/shape memory polymer nanocomposites. Compos. Part B Eng. **44**(1), 508–516 (2013)
16. Lopes, M.C., de Castro, V.G., Seara, L.M., Diniz, V.P.A., Lavall, R.L., Silva, G.G.: Thermosetting polyurethane-multiwalled carbon nanotube composites: thermomechanical properties and nanoindentation. J. Appl. Polym. Sci. **131**(23), 41207 (2014)
17. Saha, S., Saha, U., Singh, J.P., Goswami, T.H.: Thermal and mechanical properties of homogeneous ternary nanocomposites of regioregular poly(3-hexylthiophene)-wrapped multiwalled carbon nanotube dispersed in thermoplastic polyurethane: dynamic- and thermomechanical analysis. J. Appl. Polym. Sci. **128**(3), 2109–2120 (2013)
18. Coleman, J.N., Khan, U., Blau, W.J., Gun'ko, Y.K.: Small but strong: a review of the mechanical properties of carbon nanotube-polymer composites. Carbon **44**(9), 1624–1652 (2006)
19. Bafekrpour, E., Simon, G.P., Naebe, M., Habsuda, J., Yang, C.H., Fox, B.: Preparation and properties of composition-controlled carbon nanofiber/phenolic nanocomposites. Compos. Part B Eng. **52**, 120–126 (2013)
20. Zhu, J.H., Wei, S.Y., Ryu, J., Budhathoki, M., Liang, G., Guo, Z.H.: In situ stabilized carbon nanofiber (CNF) reinforced epoxy nanocomposites. J. Mater. Chem. **20**(23), 4937–4948 (2010)
21. Prolongo, S.G., Campo, M., Gude, M.R., Chaos-Moran, R., Urena, A.: Thermo-physical characterisation of epoxy resin reinforced by amino-functionalized carbon nanofibers. Compos. Sci. Technol. **69**(3–4), 349–357 (2009)
22. Burgaz, E.: Thermomechanical analysis of polymer nanocomposites. In: Huang, X., Zhi, C. (eds.) Polymer Nanocomposites. Springer, Cham (2016)
23. Zhu, Z.Z., Wang, Z., Li, H.L.: Functional multi-walled carbon nanotube/polyaniline composite films as supports of platinum for formic acid electrooxidation. Appl. Surf. Sci. **254**(10), 2934–2940 (2008)
24. Ma, P.C., Siddiqui, N.A., Marom, G., Kim, J.K.: Dispersion and functionalization of carbon nanotubes for polymer-based nanocomposites: a review. Compos. Part A Appl. Sci. Manuf. **41**(10), 1345–1367 (2010)
25. Kabir, M.E., Saha, M.C., Jeelani, S.: Effect of ultrasound sonication in carbon nanofibers/polyurethane foam composite. Mater. Sci. Eng. A **459**(1–2), 111–116 (2007)
26. Byrne, M.T., Gun'ko, Y.K.: Recent advances in research on carbon nanotube-polymer composites. Adv. Mater. **22**(15), 1672–1688 (2010)
27. Nikje, M.M.A., Yaghoubi, A.: Preparation and properties of polyurethane/functionalized multi-walled carbon nanotubes rigid foam nanocomposites. Polimery **59**(11–12), 776–782 (2014)

28. Velasco-Santos, C., Martinez-Hernandez, A.L., Lozada-Cassou, M., Alvarez-Castillo, A., Castano, V.M.: Chemical functionalization of carbon nanotubes through an organosilane. Nanotechnology **13**(4), 495–498 (2002)
29. Ma, P.C., Kim, J.K., Tang, B.Z.: Functionalization of carbon nanotubes using a silane coupling agent. Carbon **44**(15), 3232–3238 (2006)
30. Lee, J.H., Kathi, J., Rhee, K.Y., Lee, J.H.: Wear properties of 3-aminopropyltriethoxysilane-functionalized carbon nanotubes reinforced ultra high molecular weight polyethylene nanocomposites. Polym. Eng. Sci. **50**(7), 1433–1439 (2010)
31. Zhang, L.F., Yilmaz, E.D., Schjodt-Thomsen, J., Rauhe, J.C., Pyrz, R.: MWNT reinforced polyurethane foam: processing, characterization and modelling of mechanical properties. Compos. Sci. Technol. **71**(6), 877–884 (2011)
32. Ganesan, Y., Salahshoor, H., Peng, C., Khabashesku, V., Zhang, J.N., Cate, A., Rahbar, N., Lou, J.: Fracture toughness of the sidewall fluorinated carbon nanotube-epoxy interface. J. Appl. Phys. **115**(22), 224305 (2014)
33. Dolomanova, V., Rauhe, J.C.M., Jensen, L.R., Pyrz, R., Timmons, A.B.: Mechanical properties and morphology of nano-reinforced rigid PU foam. J. Cell. Plast. **47**(1), 81–93 (2011)
34. Scheibe, B., Borowiak-Palen, E., Kalenczuk, R.J.: Oxidation and reduction of multiwalled carbon nanotubes—preparation and characterization. Mater. Charact. **61**(2), 185–191 (2010)
35. Datsyuk, V., Kalyva, M., Papagelis, K., Parthenios, J., Tasis, D., Siokou, A., Kallitsis, I., Galiotis, C.: Chemical oxidation of multiwalled carbon nanotubes. Carbon **46**(6), 833–840 (2008)
36. Kim, Y.S., Davis, R.: Multi-walled carbon nanotube layer-by-layer coatings with a trilayer structure to reduce foam flammability. Thin Solid Films **550**, 184–189 (2014)
37. Wesolek, D., Gieparda, W.: Single- and multiwalled carbon nanotubes with phosphorus based flame retardants for textiles. J. Nanomater. **2014**, Article ID 727494 (2014)
38. Yeh, J.M., Chang, K.C., Peng, C.W., Chand, B.G., Chiou, S.C., Huang, H.H., Lin, C.Y., Yang, J.C., Lin, H.R., Chen, C.L.: Preparation and insulation property studies of thermoplastic PMMA-silica nanocomposite foams. Polym. Compos. **30**(6), 715–722 (2009)
39. Goods, S.H., Neuschwanger, C.L., Whinnery, L.L., Nix, W.D.: Mechanical properties of a particle-strengthened polyurethane foam. J. Appl. Polym. Sci. **74**(11), 2724–2736 (1999)
40. Zhou, X.J., Sethi, J., Geng, S.Y., Berglund, L., Frisk, N., Aitomaki, Y., Sain, M.M., Oksman, K.: Dispersion and reinforcing effect of carrot nanofibers on biopolyurethane foams. Mater Design **110**, 526–531 (2016)
41. Septevani, A.A., Evans, D.A.C., Martin, D.J., Annamalai, P.K.: Hybrid polyether-palm oil polyester polyol based rigid polyurethane foam reinforced with cellulose nanocrystal. Ind. Crop Prod. **112**, 378–388 (2018)
42. Septevani, A.A., Evans, D.A.C., Annamalai, P.K., Martin, D.J.: The use of cellulose nanocrystals to enhance the thermal insulation properties and sustainability of rigid polyurethane foam. Ind. Crop Prod. **107**, 114–121 (2017)
43. Shen, J., Han, X.M., Lee, L.J.: Nanoscaled reinforcement of polystyrene foams using carbon nanofibers. J. Cell. Plast. **42**(2), 105–126 (2006)
44. You, K.M., Park, S.S., Lee, C.S., Kim, J.M., Park, G.P., Kim, W.N.: Preparation and characterization of conductive carbon nanotube-polyurethane foam composites. J. Mater. Sci. **46**(21), 6850–6855 (2011)
45. Modesti, M., Lorenzetti, A., Besco, S.: Influence of nanofillers on thermal insulating properties of polyurethane nanocomposites foams. Polym. Eng. Sci. **47**(9), 1351–1358 (2007)
46. Cao, X., Lee, L.J., Widya, T., Macosko, C.: Polyurethane/clay nanocomposites foams: processing, structure and properties. Polymer **46**(3), 775–783 (2005)
47. Zeng, C.C., Hossieny, N., Zhang, C., Wang, B.: Synthesis and processing of PMMA carbon nanotube nanocomposite foams. Polymer **51**(3), 655–664 (2010)
48. Han, X.M., Zeng, C.C., Lee, L.J., Koelling, K.W., Tomasko, D.L.: Extrusion of polystyrene nanocomposite foams with supercritical CO_2. Polym. Eng. Sci. **43**(6), 1261–1275 (2003)
49. Wladyka-Przybylak, M., Wesolek, D., Gieparda, W., Boczkowska, A., Ciecierska, E.: The effect of the surface modification of carbon nanotubes on their dispersion in the epoxy matrix. Pol. J. Chem. Technol. **13**(2), 62–69 (2011)

50. Shen, J., Zeng, C.C., Lee, L.J.: Synthesis of poly styrene-carbon nanofibers nanocomposite foams. Polymer **46**(14), 5218–5224 (2005)
51. Chen, L.M., Ozisik, R., Schadler, L.S.: The influence of carbon nanotube aspect ratio on the foam morphology of MWNT/PMMA nanocomposite foams. Polymer **51**(11), 2368–2375 (2010)
52. Kvien, I., Oksman, K.: Orientation of cellulose nanowhiskers in polyvinyl alcohol. Appl. Phys. A Mater. **87**(4), 641–643 (2007)
53. Sehaqui, H., Mushi, N.E., Morimune, S., Salajkova, M., Nishino, T., Berglund, L.A.: Cellulose nanofiber orientation in nanopaper and nanocomposites by cold drawing. ACS Appl. Mater. Interfaces **4**(2), 1043–1049 (2012)
54. Xiong, J.W., Zheng, Z., Qin, X.M., Li, M., Li, H.Q., Wang, X.L.: The thermal and mechanical properties of a polyurethane/multi-walled carbon nanotube composite. Carbon **44**(13), 2701–2707 (2006)
55. Diez-Pascual, A.M., Naffakh, M., Gomez, M.A., Marco, C., Ellis, G., Martinez, M.T., Anson, A., Gonzalez-Dominguez, J.M., Martinez-Rubi, Y., Simard, B.: Development and characterization of PEEK/carbon nanotube composites. Carbon **47**(13), 3079–3090 (2009)
56. Chae, H.G., Minus, M.L., Kumar, S.: Oriented and exfoliated single wall carbon nanotubes in polyacrylonitrile. Polymer **47**(10), 3494–3504 (2006)
57. Sahoo, N.G., Jung, Y.C., Yoo, H.J., Cho, J.W.: Influence of carbon nanotubes and polypyrrole on the thermal, mechanical and electroactive shape-memory properties of polyurethane nanocomposites. Compos. Sci. Technol. **67**(9), 1920–1929 (2007)
58. Nikje, M.M.A., Tehrani, Z.M.: Novel modified nanosilica-based on synthesized dipodal silane and its effects on the physical properties of rigid polyurethane foams. Des. Monomers Polym. **13**(3), 249–260 (2010)
59. Hamilton, A.R., Thomsen, O.T., Madaleno, L.A.O., Jensen, L.R., Rauhe, J.C.M., Pyrz, R.: Evaluation of the anisotropic mechanical properties of reinforced polyurethane foams. Compos. Sci. Technol. **87**, 210–217 (2013)
60. Madaleno, L., Pyrz, R., Jensen, L.R., Pinto, J.J.C., Lopes, A.B., Dolomanova, V., Schjodt-Thomsen, J., Rauhe, J.C.M.: Synthesis of clay-carbon nanotube hybrids: growth of carbon nanotubes in different types of iron modified montmorillonite. Compos. Sci. Technol. **72**(3), 377–381 (2012)
61. Glicksman, L.R.: Heat transfer in foams. In: Hilyard, N.C., Cunningham, A. (eds.) Low Density Cellular Plastics: Physical Basis of Behavior. Chapmann & Hall, London (1994)
62. Modesti, M., Lorenzetti, A., Dall'Acqua, C.: New experimental method for determination of effective diffusion coefficient of blowing agents in polyurethane foams. Polym. Eng. Sci. **44**(12), 2229–2239 (2004)
63. Veronese, V.B., Menger, R.K., Forte, M.M.D., Petzhold, C.L.: Rigid polyurethane foam based on modified vegetable oil. J. Appl. Polym. Sci. **120**(1), 530–537 (2011)
64. Pei, A.H., Malho, J.M., Ruokolainen, J., Zhou, Q., Berglund, L.A.: Strong nanocomposite reinforcement effects in polyurethane elastomer with low volume fraction of cellulose nanocrystals. Macromolecules **44**(11), 4422–4427 (2011)
65. Xu, J., Fisher, T.S.: Enhancement of thermal interface materials with carbon nanotube arrays. Int. J. Heat Mass Transf. **49**(9–10), 1658–1666 (2006)
66. Verdejo, R., Barroso-Bujans, F., Rodriguez-Perez, M.A., de Saja, J.A., Lopez-Manchado, M.A.: Functionalized graphene sheet filled silicone foam nanocomposites. J. Mater. Chem. **18**(19), 2221–2226 (2008)
67. Xia, H.S., Song, M.: Preparation and characterization of polyurethane-carbon nanotube composites. Soft Matter **1**(5), 386–394 (2005)
68. Chen, W., Tao, X.M.: Self-organizing alignment of carbon nanotubes in thermoplastic polyurethane. Macromol. Rapid Commun. **26**(22), 1763–1767 (2005)
69. Huang, X.Y., Brittain, W.J.: Synthesis and characterization of PMMA nanocomposites by suspension and emulsion polymerization. Macromolecules **34**(10), 3255–3260 (2001)
70. Kashiwagi, T., Grulke, E., Hilding, J., Groth, K., Harris, R., Butler, K., Shields, J., Kharchenko, S., Douglas, J.: Thermal and flammability properties of polypropylene/carbon nanotube nanocomposites. Polymer **45**(12), 4227–4239 (2004)

71. Kashiwagi, T., Du, F.M., Winey, K.I., Groth, K.A., Shields, J.R., Bellayer, S.P., Kim, H., Douglas, J.F.: Flammability properties of polymer nanocomposites with single-walled carbon nanotubes: effects of nanotube dispersion and concentration. Polymer **46**(2), 471–481 (2005)
72. Guan, G.H., Li, C.C., Zhang, D., Jin, Y.: The effects of metallic derivatives released from montmorillonite on the thermal stability of poly(ethylene terephthalate)/montmorillonite nanocomposites. J. Appl. Polym. Sci. **101**(3), 1692–1699 (2006)
73. Liu, X.Y., Yu, W.D.: Evaluating the thermal stability of high performance fibers by TGA. J. Appl. Polym. Sci. **99**(3), 937–944 (2006)
74. Kumar, A.P., Depan, D., Tomer, N.S., Singh, R.P.: Nanoscale particles for polymer degradation and stabilization-trends and future perspectives. Prog. Polym. Sci. **34**(6), 479–515 (2009)
75. Jiao, L.L., Xiao, H.H., Wang, Q.S., Sun, J.H.: Thermal degradation characteristics of rigid polyurethane foam and the volatile products analysis with TG-FTIR-MS. Polym. Degrad. Stab. **98**(12), 2687–2696 (2013)
76. Burgaz, E., Kendirlioglu, C.: Thermomechanical behavior and thermal stability of polyurethane rigid nanocomposite foams containing binary nanoparticle mixtures. Polym. Testing **77**, 105930 (2019). https://doi.org/10.1016/j.polymertesting.2019.105930

Chapter 5
PU Rigid Nanocomposite Foams Containing Spherical Nanofillers

5.1 Introduction

In previously published major works in the literature, nanosilica, polyhedral oligomeric silsesquioxanes (POSS), TiO_2 and ZnO can be given as nanoadditive examples for spherical nano-sized fillers that have been used in PURNCFs [1–7]. However, among PURF/spherical nano-sized filler studies which were performed by researchers in academia, the most generally investigated nanoadditive is nanosilica due to its excellent characteristics in PURFs such as improved thermal solidity and decomposition behavior, mechanical, thermal and thermomechanical properties [3–7].

Many different types of nanosilica particles such as nanosilica surface-treated with oligomers [8], hydrophilic and hydrophobic nanosilica and nanosilica containing pores with diameters of 2–50 nm [9, 10], and nanosilica particles prepared via generation of colloidal solution and integrated network of separate particles [11] have been extensively exploited as nanoadditives consisting of both distinct physical and chemical surface characteristics in strengthening of structure, morphology, intermolecular interactions, mechanical and thermomechanical properties and thermal decomposition behaviour of polymer nanocomposites [12]. Under the category of nanosilica particles, pyrogenic silica which is also called as fumed silica has been extensively utilized in academic and industrial research projects as a filler which serves efficiently as a medium in fluids and powders and as a nanofiller for the property improvement of elastomers [13]. Silica (especially fumed silica) is an important and effective nanofiller in terms of enhancing the viscosity of organic solutions [13–15]. Especially, hydrophilic silica is a very effective and multipurpose type of silica which generates steady and thin solutions consisting of lower values of viscosity. Furthermore, another important characteristic of hydrophilic silica is that it can establish powerful hydrogen bonding interactions if it is dispersed in liquid media such as low molecular weight poly (ethylene glycol) and alcohols containing short chain structures [13]. In particular, in terms of geometry, fumed silica is

E. Burgaz, *Polyurethane Insulation Foams for Energy and Sustainability*,
Advanced Structured Materials 111,
https://doi.org/10.1007/978-3-030-19558-8_5

composed of amorphous silicon dioxide in which stable aggregates of approximately 100–250 nm are formed due to the fusion of primary particles with sizes of 1–3 nm [16]. Due to their large surface area (50–400 m^2/g), the formation of particle-particle interactions leads to higher degrees of aggregation, and consequently can form a three-dimensional network in polymer solutions, molten polymers and casted polymer nanocomposite films [16].

The effects of nanosilica addition to PURFs have been investigated by several researchers in terms of morphology, density, thermal stability, mechanical, thermal and thermomechanical properties [3–7]. However, the most important problems that should be solved in the preparation of PURNCFs consisting of nanosilica particles are aggregation and dispersion problems in the PURF matrix. The poor dispersion of nanosilica in the PURF matrix is mainly caused by their large aspect ratio and strong dynamic hydrogen bonding forces [3–7, 13, 16]. Previously, PURFs consisting of nanosilica with an average diameter of 12 nm were prepared and the content of nanosilica was varied in the range of 0–20 wt%, and the density of foams was about 30 kg/m^3. In accordance with results, incorporation of nanosilica enhanced the density of PURNCFs only at concentrations above 20 wt%. In addition, nanosilica lowered the compression strength of PURNCFs at all concentrations [3].

Previously, effects of using nanosilica as a nanofiller on the closed-cellular morphology, reactive foaming kinetics, thermal conductivity, viscoelastic properties, dynamic mechanical and compressive mechanical properties of PURNCFs were systematically studied [4]. Based on the reaction kinetics results from in situ FTIR spectroscopy, it was shown that PURNCFs consisting of nanosilica particles exhibit higher degrees of isocyanate conversion. In addition, it was pointed out that during the highly reactive foaming process, nano silica particles acted as extra nucleation sites, and thus paved the way for diminishing of average cell size and also uniform distribution of cell size. Moreover, it was found out that open cell character in cellular morphology increases with increasing amount of nanosilica in PURNCFs [4].

Hydrophilic nanosilica particles form hydrogen bonding interactions among themselves due to their surface hydroxyl functional groups. Thus, for this reason, hydrophilic nanosilica particles are usually found as aggregates. In order to reduce aggregation behavior of hydrophilic nanosilica, their surface consisting of silanol groups is typically surface modified with non-polar silyl groups. Thus, by this way, the hydrophilic nature of their surface is converted into hydrophobic character [4, 13]. In PURNCFs, in order to maintain a well-dispersion of nanosilica particles and provide stronger interfacial interactions between nanosilica and PURF matrix, usually different types of silanes are used in the surface modification of silica nanoparticles. Previously, various silanes such as *N*-(2-aminoethyl)-3-aminopropyltrimethoxysilane (AEAP) [7], gamma-glycidoxypropyltrimethoxysilane (GPTS), 3-Aminopropyltriethoxysilane (APTS) [5] and diethanol amine (DEA) [6] were exploited as surface modifying agents for nanosilica particles. Moreover, these surface modified nanosilica particles were used as nanospherical fillers to explore their influences on closed-cellular morphology, mechanical and thermomechanical properties and thermal decomposition behavior of PURNCFs. Recently, effects of using two different nanosilicas that were surface modified with APTS and AEAP on thermal stability, morphology, mechanical and dynamic mechanical properties of PURNCFs were studied in detail. It was found out that PURNCFs consisting of APTS modified nanosilica particles exhibit diminished mechanical properties and enhanced dynamic mechanical properties compared to those containing AEAP modified nanosilica particles. Thus, based on these results, it was concluded that in PURNCFs consisting of AEAP modified nanosilica particles, the interfacial area between nanoparticles and PU matrix becomes stronger via two functional groups of AEAP, however the strength of bulk PU matrix is diminished due to the reduction of covalent bonds in the system [5].

The influences of using modified nanosilica particles which were successively surface modified with silane coupling agents such as GPTS and DEA on the closed-cellular morphology, mechanical and thermomechanical properties and thermal and thermal decomposition behaviors of PURNCFs were investigated comprehensively [6]. According to results, it was revealed that mechanical properties and thermal decomposition behaviour were enhanced with the addition of doubly modified nanosilica. Moreover, storage modulus diminished due to reduction in the amount of hard segments since reactive functional groups on the surface of nanosilica changed the ratio of reactants during the reactive foaming process [6]. Recently, fumed nanosilica that was surface-modified with AEAP was incorporated as a nanofiller into PURFs, and it was found out that using higher amounts of modified nanosilica in PURNCFs elevated mechanical and thermal properties. In addition, the decrease of glass transition temperature in PURNCFs was observed since AEAP surface functional groups on silica nanoparticles changed stoichiometric ratio of reactants and reduced the amount of hard phase formation in PURF matrix [7].

Polyhedral oligosilsesquioxanes (POSS) are listed under silsesquioxane molecules. Here, the name "silsesquioxanes" are given to a group of molecules whose chemical and physical properties are identified with the chemical formula of $R_nSi_nO_{1.5n}$ [17, 18]. According to their geometrical structure, POSS are categorized under a class of molecules with isotropic structures and diameters of 1–3 nm. Also, the geometrical structure of POSS is closely correlated with quantity of silicon atoms in the central part and amount of outermost R groups of the molecule [19]. As an illustration, a cubic silsesquioxane having the chemical formula of Si_8O_{12} which has a central part with cubic geometry and is encircled by eight substitution cyclopentyl groups has a spherical geometry with an average diameter of 1.5 nm. By using various synthetic techniques, a functional group which goes through successful polymerization reaction can be switched with one of outermost substitution R groups in POSS. Due to this advantageous capability, POSS units can be efficiently surface modified with useful polymer chains which are chemically linked to their cage-like structures via strong covalent bonding interactions [17, 19, 20]. As a consequence of this specific property, POSS are identified both as a spherical nanofiller and a massive molecule consisting of a functional group which can be connected with polymer chains via successful polymerization reactions.

The performance of POSS molecules in various types of polymer matrices is strongly related to several factors such as molecular dimensions of POSS units, physicochemical properties of organic groups within POSS molecules, and important functionings of POSS molecules such as their concentration, solubility and compatibility when they are used as nanofillers inside polymer matrices [21]. Various types of methods such as copolymerization, grafting and blending can be exploited to integrate POSS nanoparticles efficiently as spherical nanofillers into different types of polymeric materials [22, 23]. As a consequence of this specific characteristics, POSS nanoparticles were extensively used as spherical nanofillers in various polymer matrices to enhance morphology, mechanical, thermal and thermomechanical properties and thermal decomposition behaviour of polymer nanocomposites. On the other hand, there are only a few studies about the use of POSS nanoparticles that were used as filler materials in PURFs [2, 24]. Previously, thermal degradation of PURNCFs consisting of two different types of functionalized POSS nanoparticles in which the first one consists of eight functional substitution groups and the second one contains only one functional substitution group was systematically explored by using multiple combinations of characterization techniques such as thermogravimetric analysis (TGA), FTIR spectroscopy and quadrupole mass spectrometry (QMS) [2]. Based on TG/FTIR/QMS data, it was revealed that POSS nanoparticles had positive thermal stabilization influences on the polymer matrix since the initial decomposition temperature was elevated in PURNCFs. Moreover, it was observed that thermal decomposition process of PURNCFs consisting of POSS nanoparticles started at 150 °C due to the breakage of urethane bonds [2].

5.2 Morphology

Previously, Nikje et al. performed a surface alteration method of fumed nanosilica by exploiting *N*-(2-aminoethyl)-3-aminopropyltrimethoxysilane molecule as a surface refining instrument, and incorporated this surface adjusted nanosilica into PURFs [7]. Authors used several characterization techniques such as FT-IR, TGA, SEM, tensile testing, DMA and TMA to explore physical, thermal, mechanical and thermomechanical properties of PURNCFs consisting of modified nanosilica. Based on the results, it was found out that increasing amount of modified nanosilica in PURNCFs paved the way for enhancing both thermal and mechanical properties. However, in contrast, it was pointed out that dynamic mechanical properties of PURNCFs consisting of modified nanosilica exhibited a different behavior compared to mechanical properties due to establishment of various types of interactions between nanosilica particles and PURNCF matrix. Furthermore, it was revealed that glass transition temperatures of PURNCFs consisting of modified nanosilica were diminished compared to that of pure PURF. The reason behind this observation was explained such that functional groups that are present on the surface of nanosilica particles influenced the reactant ratio of PU components during cross-linking reactions and reduced hard phase formation in PURNCF matrix [7].

Figure 5.1 shows SEM images of pure PURF and PURNCFs consisting of different percentages of modified nanosilica. Based on SEM results, it was shown that both pure PURF and PURNCFs consisting of modified nanosilica unveiled

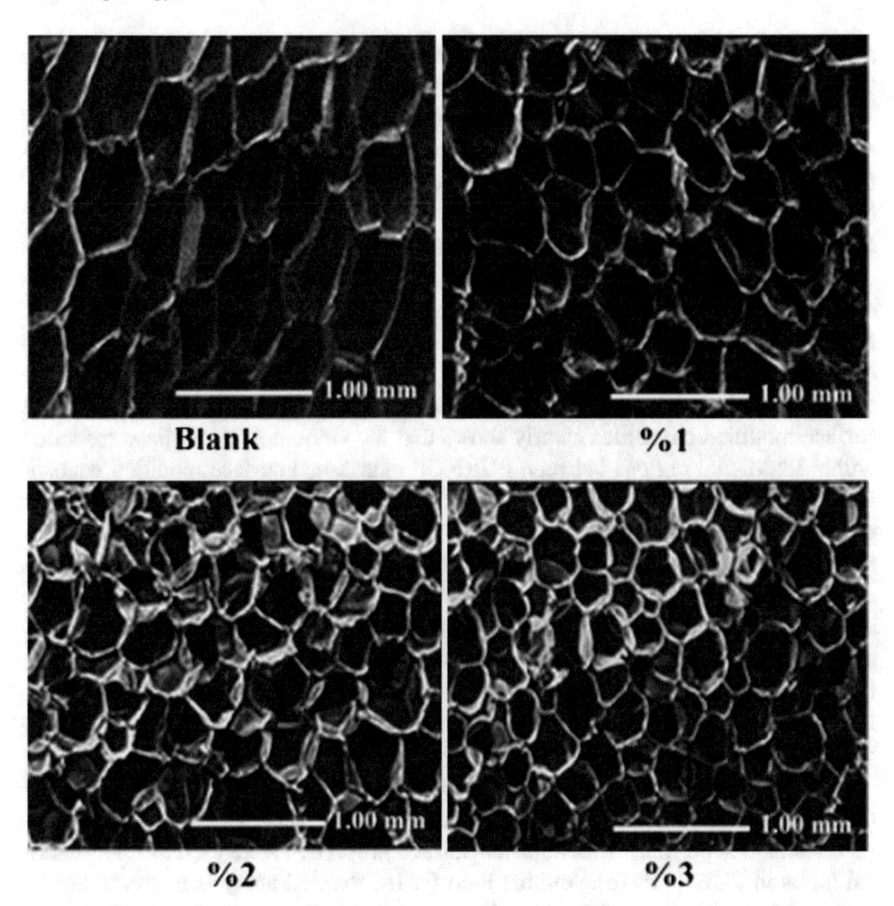

Fig. 5.1 SEM images of pure PURF and PURNCFs consisting of different amounts of surface modified nanosilica particles. Reproduced with permission from Ref. [7]. Copyright 2009 Society of Plastics Engineers

a closed-cellular morphology which was occupied with thermodynamically firm polygonal cell geometries [7]. It was also reported that increasing amount of modified nanosilica particles in PURNCFs elevated cell density and diminished cell size. The main mechanism behind this result was explained such that degree of nanosilica distribution throughout PURNCF matrix has a principal task on the refinement of cell size during reactive foaming event. Thus, homogenous dispersion of nanosilica within the PURNCF matrix could be shown as the major mechanism in elevation of cell density and shrink of cell size. Based on the results, it was also shown that ultimate foam density in PURNCFs was directed by contentious series of events in which cell nucleation and growth and merging of cells took place. Previously, it was shown that interfacial regions between PURNCF matrix and homogenously distributed modified nanosilica particles are the usual locations for nucleation process of cells [7, 25]. Moreover, previous experiments also revealed that nanoparticles took action as sites for non-homogeneous nucleation and cell formation during reactive foaming process in PURNCFs [7, 26].

The improved interactions between the surface functional groups of modified nanosilica and polymer chains leads to homogeneous distribution of nanosilica particles throughout PURNCF matrix. Thus, well-dispersion of nanoparticles dramatically increases the number of heterogenous nucleation sites per unit volume in the PURNCF matrix which leads to the formation of cells with smaller sizes and cell density compared to the case of using bare nanosilica particles in PURNCFs. Consequently, physical, thermal and mechanical properties of PURNCFs filled with surface modified nanosilica particles improve with reduction in cell size and elevation of cell density in comparison with those of pure PURF. However, on the other hand, the decrease of glass transition temperature or storage modulus with incorporation of surface modified nanosilica clearly shows that the amount of hard phase reduction within interfacial regions between PURNCF matrix and surface modified nanosilica particles is basically the main reason for the decrease of storage modulus and glass transition temperature which are associated with rigidity or stiffness of system. In addition, the uniform distribution of surface modified nanosilica particles and especially their active tasks in the formation and nucleation of cells within these interfacial regions clearly facilitate the reduction of hard domains in these interfacial regions between PURNCF matrix and surface modified nanosilica particles. Thus, thermomechanical properties such as glass transition temperature and storage modulus diminish since these parameters are closely connected with stiffness and rigidity values associated with interfacial regions [7, 25, 26].

Previously, surface properties of nanosilica were altered by using gamma-Glycidoxypropyltrimethoxysilane and diethanol amine, successively and these modified nanosilica particles with adjusted surface properties were used as nanospherical fillers in PURNCFs [6]. Authors used FT-IR, thermal analysis methods, tensile testing, TMA, DMA and SEM in order to systematically analyze mechanical properties, intermolecular interactions, closed-cellular morphology, thermomechanical properties and thermal stability behaviour of the system. According to results, it was shown that incorporation of modified nanosilica enhanced the values of thermal stability and mechanical properties in PURNCFs. However, on the other hand, it was revealed that storage modulus values of PURNCFs decreased since surface reactive functional groups associated with nanosilica particles altered relative reactant ratios during cross-linking reactions and also paved the way for reducing the amount of hard segments in PURNCFs [6]. Thus, the stiffness or the rigidity of PURNCF system consisting of modified nanosilica is greatly dependent on the surface properties of nanoparticles that are used as nanofillers in the system.

Figure 5.2 shows the SEM images of pure PURF and PURNCFs consisting of modified nanosilica with different contents. As shown in SEM micrographs, using higher amounts of surface modified nanosilica particles led to alterations in closed-cellular morphology in which sizes of cells diminished and the value of cell density increased. The reason behind the increased cell density and the reduction in cell size was explained such that the dispersion state of modified nanosilica has a major effect on the control of cell size and density during the reactive foaming process [6]. Also,

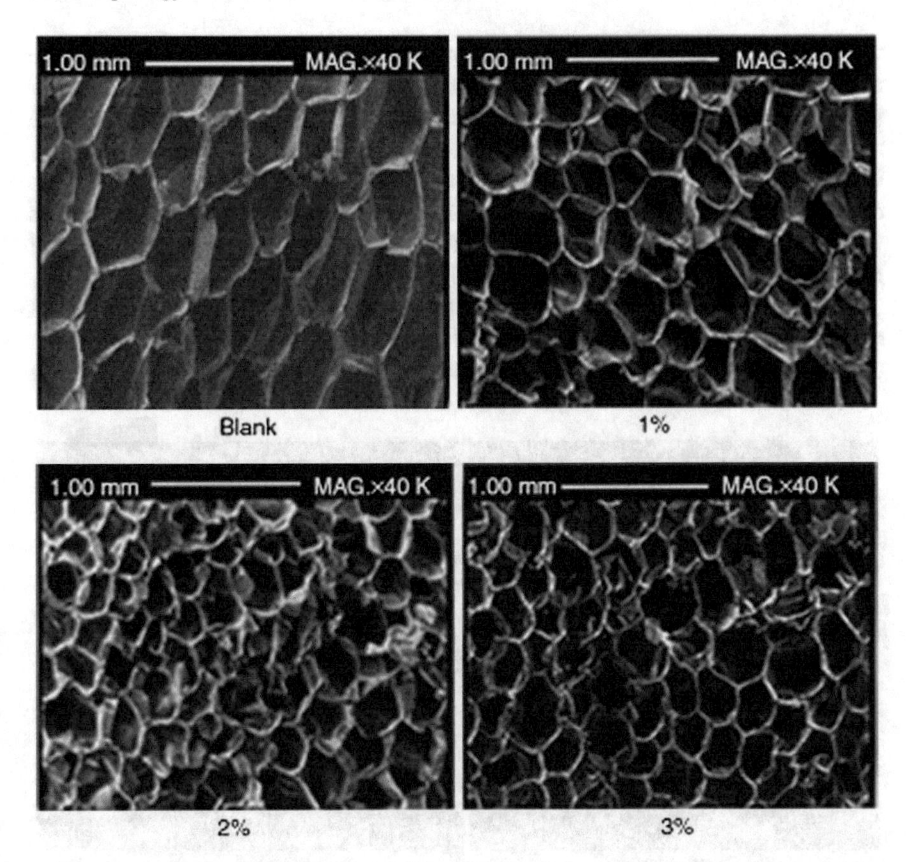

Fig. 5.2 SEM images of pure PURF and PURNCFs containing various amounts of modified nanosilica particles. Reproduced with permission from Ref. [6]. Copyright The Author(s), 2010

it was previously shown that the cell nucleation mostly occurs within interfacial regions between PURNCF matrix and well-distributed modified nanosilica particles [6, 25]. Thus, if the well-dispersion of nanosilica particles can be maintained within the polymer matrix, then there would be high number of nucleation sites in these interfacial regions due to the higher surface area of modified nanosilica in which aggregates could not be formed due to the presence of silane functional groups.

Previously, 3-Aminopropyltriethoxysilane (APTS) was used as a surface modifying molecule and this modified nanosilica was used as a nanofiller in PURNCFs [5]. Furthermore, the effects of using APTS on morphology, mechanical, thermomechanical, physical and thermal properties of PURNCFs were compared with those of PURNCFs filled with nanosilica particles which were surface modified with *n*-(2-aminoethyl)-3-aminopropyltrimethoxysilane (AEAP). Authors used the characterization tools such as FT-IR, TGA, SEM, DMA, tensile testing and TMA in order to comprehend the physical, thermal, mechanical and thermomechanical properties of PURNCFs consisting of modified nanosilica particles. In accordance with experimental results, it was pointed out that PURNCFs consisting of nanosilica modified with APTS exhibited better dynamic mechanical properties but inferior static mechanical properties in comparison with PURNCFs consisting of nanosilica modified by AEAP. The reason behind the poor dynamic mechanical properties that were observed in the AEAP system was explained such that the stoichiometric ratio of cross-linking reactions was altered by two surface functional groups associated with modified nanosilica, thus leading to the reduction of stiffness or the number of hard segments in PURNCF bulk matrix [5].

Figure 5.3 shows the SEM images of pure PURF and PURNCFs consisting of modified nanosilica with different contents. Based on the results, it was shown that using higher amounts of modified nanosilica in PURNCFs basically caused cell size

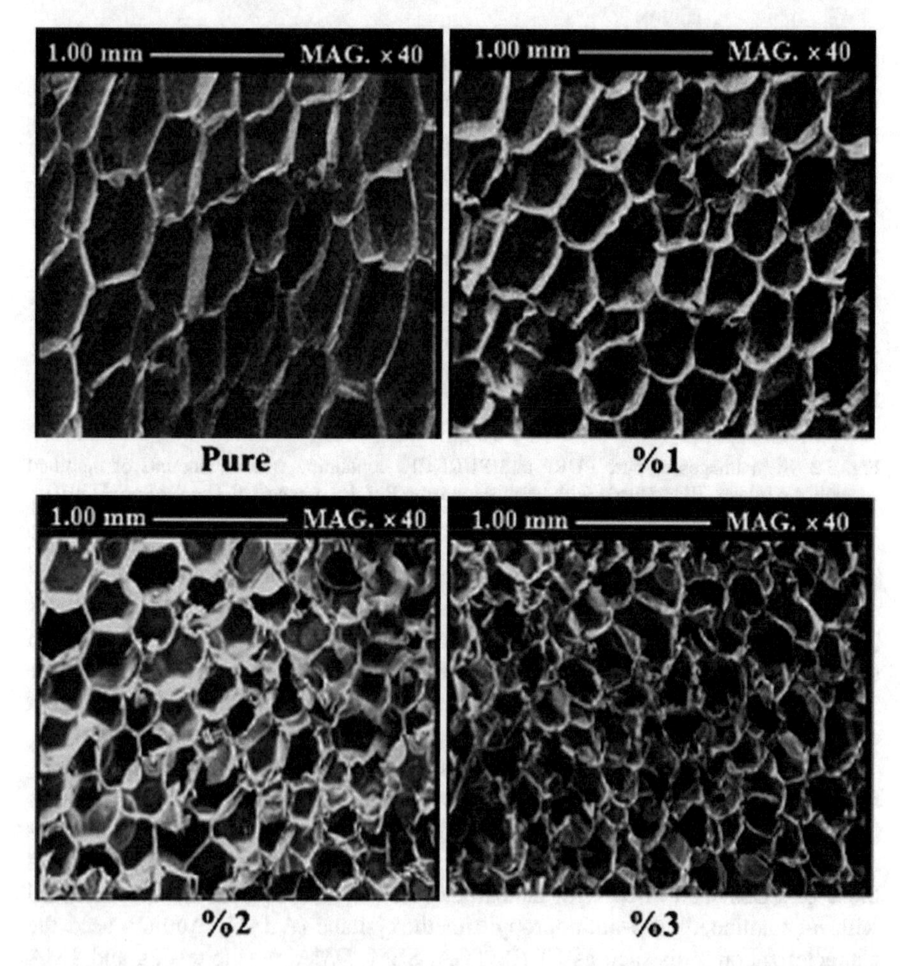

Fig. 5.3 SEM images of pure PURF and PURNCFs containing different amounts of modified nanosilica particles. Reproduced with permission from Ref. [5]. Copyright Koninklijke Brill NV, Leiden, 2011

reduction and elevation of cell density. The reason behind this finding was explained such that the smaller cell sizes with higher density was formed due to the presence of small amounts of nanoparticles in PURNCFs which act as nucleation sites for the bubble nucleation process [5, 27]. Also, it was reported that the cell density in the AEAP system was much higher compared to the APTS system as a result of much better dispersion of modified nanosilica particles due to the presence of stronger interactions in PURNCFs consisting of AEAP modified nanosilica [5].

In accordance with results from other previously published works [4, 6, 7, 25, 26, 28, 29], it is very well known that using surface modified nanosilica particles that were treated with various surface modification methods in PURFs generally leads to fabrication of PURNCFs with improved properties. Especially, using silane molecules consisting of more than one type of functional group which are also physically and chemically compatible with PURNCF matrix paves the way for enhancement of intermolecular interactions. By this way, the number of favorable interactions between surface modified nanosilica particles and PURNCF matrix could be maximized to produce PURNCFs with improved closed-cellular morphological properties such as reduced cell sizes and higher cell densities. Thus, based on the comparison of SEM results in Figs. 5.1, 5.2 and 5.3, it is clearly seen that using silane molecules consisting of more than one type of surface functional groups on nanosilica particles has many advantages over using bare nanosilica particles in PURNCFs since the detailed features of closed-cellular morphology (cell size, density and shape) are positively affected with incorporation of nanosilica particles modified with compatible surface functional groups to PURFs [4–7, 25, 26, 28, 29].

Earlier on, Santiago-Calvo et al. studied influences of using nanofillers on reactive foaming kinetics of PURFs by using different types of nanoclay and nanosilica which have different surface functional groups [4]. Authors used FT-IR to measure systematically the degree of isocyanate consumption with the addition of nanofillers to PURFs. Specifically, urea and urethane amounts from carbonyl absorption regions in FT-IR data were quantitatively determined to understand the detailed principles behind blowing and gelling reactions during reactive foaming event. In addition, authors also investigated the density, closed-cellular morphology, thermal conductivity and mechanical properties of PURNCFs consisting of nanosilica and nanoclay particles with the experimental data that was obtained from the reaction kinetics study of the system [4].

Figure 5.4 displays SEM micrographs of (a) pure PURF and PURNCFs consisting of 5 wt% (b) nanosilica without modifier, (c) nanosilica with hexamethyldisilazane, (d) nanosilica with dimethyldichlorosilane, (e) natural MMT and (f) MMT with modifier. Based on the results, it was shown that the cell size of PURNCFs consisting of nanosilica and nanoclay decreased by about 22 and 34%, respectively compared to that of pure PURF [4]. The decrease of cell size with the addition of nanosilica and nanoclay was caused by the mechanism in which the favorable interactions between surface functional groups of these nanoparticles and PU components in the dispersion leads to increase in the number of nucleation centers which form higher number of bubbles and consequently higher cells per unit volume in comparison with that of pure PURF. Based on the results in terms of cell size, it was also reported that nanoclays

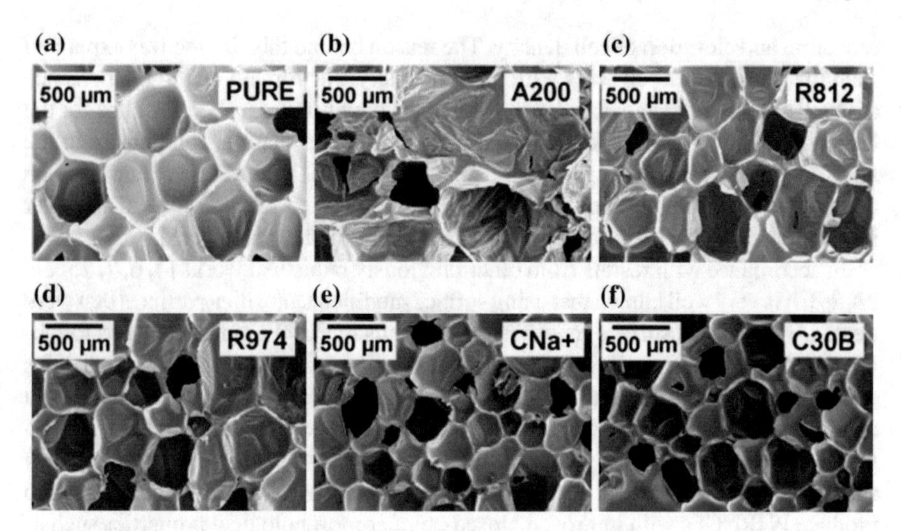

Fig. 5.4 SEM micrographs of **a** pure PURF and PURNCFs consisting of 5 wt% **b** nanosilica without modifier, **c** nanosilica with hexamethyldisilazane, **d** nanosilica with dimethyldichlorosilane, **e** natural MMT and **f** MMT with modifier. Reproduced with permission from Ref. [4]. Copyright 2018 Elsevier Ltd.

served as much more better nucleation centers than nanosilica particles due to higher surface areas of nanoclay particles. In addition, it was also revealed that both pure PURF and PURNCFs consisting of nanoparticles (both clay and nanosilica) have isotropic cell shapes and homogeneous cell structures based on cell size distribution results [4].

Formerly, Nikje et al. adjusted the surface properties of nanosilica particles by utilizing a specific surface modification technique in which an original dipodal silane consisting of 3-aminopropyltriethoxysilane and *gamma*-glycidoxypropyltrimethoxysilane was used as the surface modifier. Authors then used this surface modified nanosilica as a nanospherical filler in PURNCFs [28]. Basically, the polar surface functional groups such as hydroxyl and amine groups on the modified nanosilica were responsible for the well-dispersion of nanosilica particles within the PURNCF matrix. Authors used characterization tools such as FT-IR, TGA, SEM, proton nuclear magnetic resonance spectroscopy (1H-NMR), DMA, tensile testing and TMA in order to evaluate types of intermolecular interactions, closed-cellular morphology, physical, thermal, mechanical and thermomechanical properties of PURNCFs consisting of modified nanosilica particles. In accordance with results, it was pointed out that mechanical and thermal properties of PURNCFs consisting of modified nanosilica particles enhanced with incorporation of nanosilica particles. However, on the contrary, it was suggested that the addition of modified nanosilica decreased dynamic mechanical properties of the system due to the suppression of the hard domain formation within the bulk polymer of PURNCF system [28].

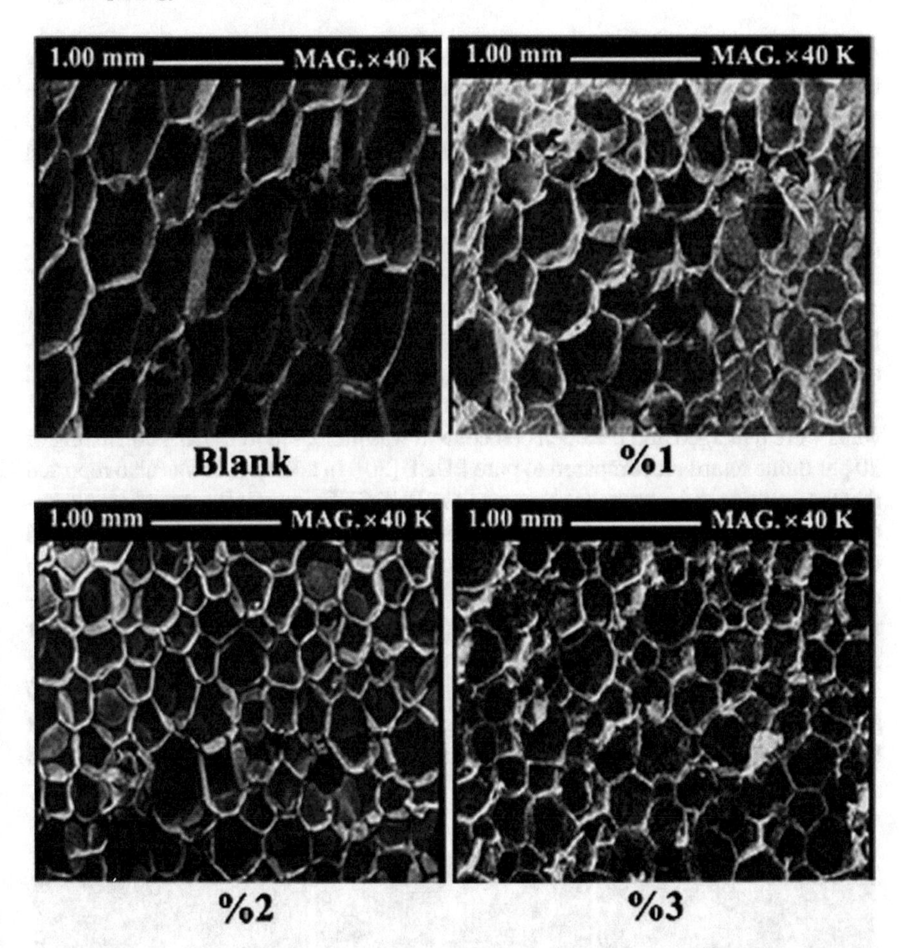

Fig. 5.5 SEM images of pure PURF and PURNCFs consisting of modified nanosilica with different concentrations. Reproduced with permission from Ref. [28]. Copyright Koninklijke Brill NV, Leiden, 2010

Figure 5.5 shows the SEM images of pure PURF and PURNCFs consisting of modified nanosilica with different concentrations. Based on the results, it was shown that the cell size decreased and cell density increased with increasing amount of modified nanosilica in PURNCFs [28]. The well-dispersion of modified nanosilica within the PURNCF matrix was shown to be the main reason for the decrease of cell size and increase of cell density. Also, small amount of modified nanosilica particles which are well-dispersed within the polymer matrix could act as extra nucleation centers which lead to the formation of higher number of cells per unit volume in comparison with pure PURF [28, 29].

Previously, Wang et al. prepared halogen-free flame-retardant PURNCFs by incorporating nanohybrid particles of graphene oxide and nanospherical silica and a flame retardant (dimethyl methylphosphonate) in various contents [30]. Authors investigated mechanical, flame retardant and thermal properties of PURNCFs containing various contents of dimethyl methylphosphonate and nanosilica/graphene oxide hybrid. Based on the results, it was shown that flame retardant and mechanical properties of PURNCFs extensively improved by incorporating both flame retardant and nanohybrid particle compared to pure PURF and PURFs containing only dimethyl methylphosphonate. In addition, it was also reported that much more smaller and uniform cell sizes were obtained in PURNCFs consisting of both dimethyl methylphosphonate and nanosilica/graphene oxide nanohybrid in comparison with PURFs containing only dimethyl methylphosphonate [30].

Figure 5.6 shows SEM images of (a) pure PURF, (b) PURF consisting of 20 phr flame retardant and PURNCFs consisting of (c) 2 phr GO and 18 phr flame retardant, (d) 2 phr nanohybrid particle and 18 phr flame retardant, (e) 4 phr nanohybrid particle and 16 phr flame retardant. Based on SEM results, it was shown that the cell walls were damaged and the open cell content was increased in PURFs consisting of 20 phr flame retardant compared to pure PURF [30]. In addition, it was also reported that the cell size did not change that much in PURCNFs containing only flame retardant and GO compared to pure PURF. However, on the other hand, it was revealed that the cell sizes and shapes became much more smaller and more uniform when nanohybrid particles were added into PURFs containing dimethyl methylphosphonate. Furthermore, it was pointed out that the cell size of PURNCF containing 2 phr nanohybrid particle and 18 phr flame retardant was apparently smaller than those of PURF consisting of 20 phr flame retardant and PURNCF consisting of 2 phr GO and 18 phr flame retardant [30]. The cell size reduction in PURNCFs consisting of both nanohybrid particle and flame retardant was due to different physicochemical

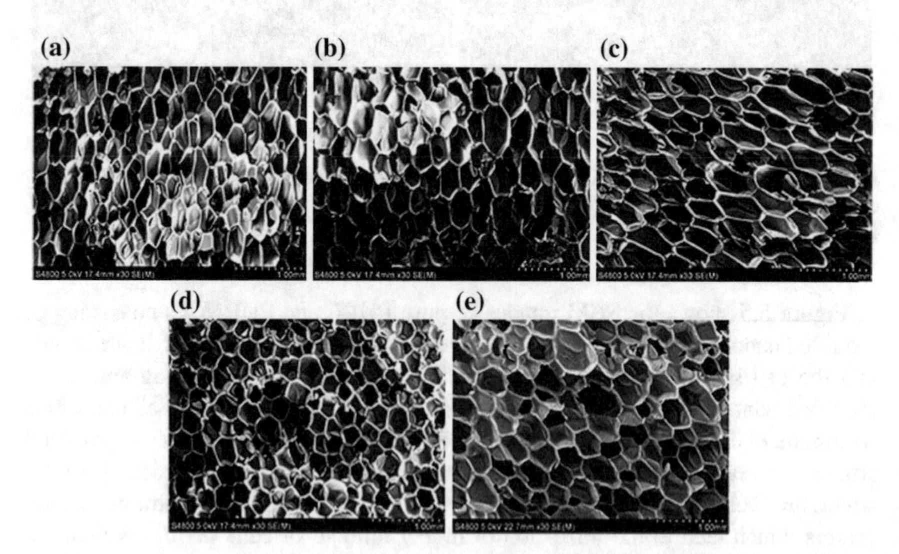

Fig. 5.6 SEM micrographs of **a** pure PURF, **b** PURF consisting of 20 phr flame retardant and PURNCFs consisting of **c** 2 phr GO and 18 phr flame retardant, **d** 2 phr nanohybrid particle and 18 phr flame retardant, **e** 4 phr nanohybrid particle and 16 phr flame retardant. Reproduced with permission from Ref. [30]. Copyright 2018 Taylor & Francis

properties of nanohybrid particle consisting of different surface functional groups on its surface. Also, nanohybrid particle was shown to be behaving as a much better nucleation medium than GO during reactive foaming event due to the decoration of GO surface by nanosilica particles [30, 31].

Using one type of nanofillers such as spherical, cylindrical or plate-like nanoparticles in PURNCFs definitely improves closed-cellular morphological properties. Furthermore, introducing compatible surface functional groups to these nanoparticles enhances closed-cellular morphology much more compared to the case of using bare nanoparticles in PURNCF systems. However, besides using one type of nanofiller or surface modified nanofiller in PURNCFs, hybrid nanofiller systems consisting of at least two types of nanofillers with different properties and geometries can be used in order to fabricate PURNCFs with more superior properties compared to those consisting of only one type of nanofiller. Thus, using nanohybrid particles of GO and nanosilica in PURNCFs resulted in more enhancement of closed-cellular morphology due to favorable interactions between nanosilica and GO nanoparticles [30, 31]. Basically, compatible surface functional groups on GO nanoparticles increased in number due to attachment of nanosilica particles onto the surfaces of GO nanaoparticles. Thus, as a result of this process, the number of compatible functional groups on GO-silica nanohybrid particles became much higher compared to those of either GO or silica nanoparticles. Thus, consequently, cell sizes of foams reduced and their cell density values increased with the incorporation of hybrid nanofillers into PURFs. As a result, the enhancement in the overall properties was observed because of the fact that dispersion of GO nanoparticles in PURNCFs was facilitated by the attachment of nanosilica particles and positive synergy was provided by favourable interactions between GO-nanosilica hybrid nanoparticles and PU chains in PURNCF matrix [30, 31].

Previously, superparamagnetic Fe_3O_4/silica nanoparticles were incorporated into PURFs in order to prepare new magnetic PURNCFs by using a one-step synthesis route [32]. The process of producing nanoparticles on substrates from a colloidal suspension or gel state was used to synthesize nanoparticles consisting of both core and shell layers, and the synthesized nanoparticles were characterized by using TEM, XRD, FT-IR to comprehend the detailed morphology of nanoparticles. Authors prepared PURNCFs consisting of maximum 3% of magnetic nanoparticles, and they also used various characterization tools such as FT-IR, DMA and TGA on PURNCFs in order to systematically evaluate closed-cellular structure and intermolecular interactions, thermomechanical, and thermal decomposition properties of nanocomposite foams. In accordance with results, it was revealed that thermal, mechanical, and magnetic properties of PURNCFs were improved compared to pure PURF [32].

Figure 5.7 shows the schematical drawing for hydrogen bonding interactions between magnetic nanoparticles and urethane groups of PURNCF matrix. As shown in Fig. 5.7, hydrogen bonding interactions could be formed between hydroxyl groups of nanosilica particles and amide groups of urethane functional groups with the addition of Fe_3O_4/silica nanoparticles into PURFs [32]. The incorporation of nanoparticles definitely increased the mechanical properties since nanoparticles were attached to the polymer backbone with strong H-bonding forces. However, the

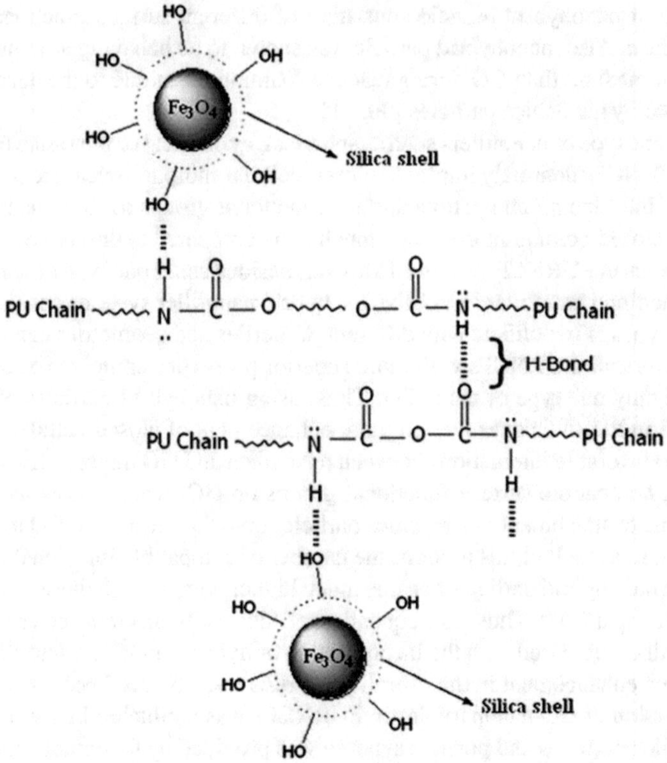

Fig. 5.7 Schematical drawing for hydrogen bonding interactions between nanoparticles and urethane groups of PURNCF matrix. Reproduced with permission from Ref. [32]. Copyright Springer-Verlag 2012

presence of nanoparticles which are responsible for H-bonding with the polymer matrix might decrease the stiffness or storage modulus since nanosilica particles could reduce the number of crosslinking points or the concentration of hard domains within the PURNCF matrix during the reactive foaming process. The decrease in stiffness values of PURNCFs filled with nanosilica particles clearly shows that hydroxyl functional groups of nanosilica particles established effective hydrogen bonding interactions with urethane linkages of PU chains, thus leading to the decrease of crosslinking density in PURNCFs [5–7, 32].

Previously, Nazeran et al. outlined synthesis of PURNCFs containing silica aerogel nanoparticles with efficient thermal insulation properties [33]. In the study, PURNCFs were prepared in two methods in which silica aerogel was added to MDI component in the first method, and in the second procedure silica aerogel was added to the polyol component by including various concentrations of silica aerogel nanoparticles in the range of 1–5 wt%. Authors used many characterization techniques to explore influences of incorporating these nanoparticles on thermal, mechanical, morphological and surface properties of PURNCFs. Based on FT-IR results, the covalent bonding of nanoparticles with MDI component in PURNCF matrix was confirmed, however no chemical reaction occurred reaction between silica aerogel nanoparticles and polyol component of PURFs [33]. In accordance with water contact angle and thermal conductivity data, it was shown that thermal conductivity values diminished and also values of hydrophobicity elevated in cyclopentane blown PURNCFs with incorporation of 5 wt% silica aerogel nanoparticles compared to those of pure PURF. According to mechanical properties testing data, it was revealed that yield stress and modulus values of PURNCFs enhanced considerably with incorporation of 3 wt% silica aerogel nanoparticles to PURFs. However, it was concluded that mechanical properties of PURNCFs decreased with inclusion of more than 3 wt% silica aerogel nanoparticles [33].

Figure 5.8 shows SEM images along free foam growth direction of pure PURF and PURNCFs that were prepared in which silica aerogel was added to MDI component. Basically, PURF samples in Fig. 5.8 consist of silica aerogel nanoparticles such as (a) 0%, (b) 1%, (c) 3% and (d) 5%. As observed from SEM micrographs in Fig. 5.8,

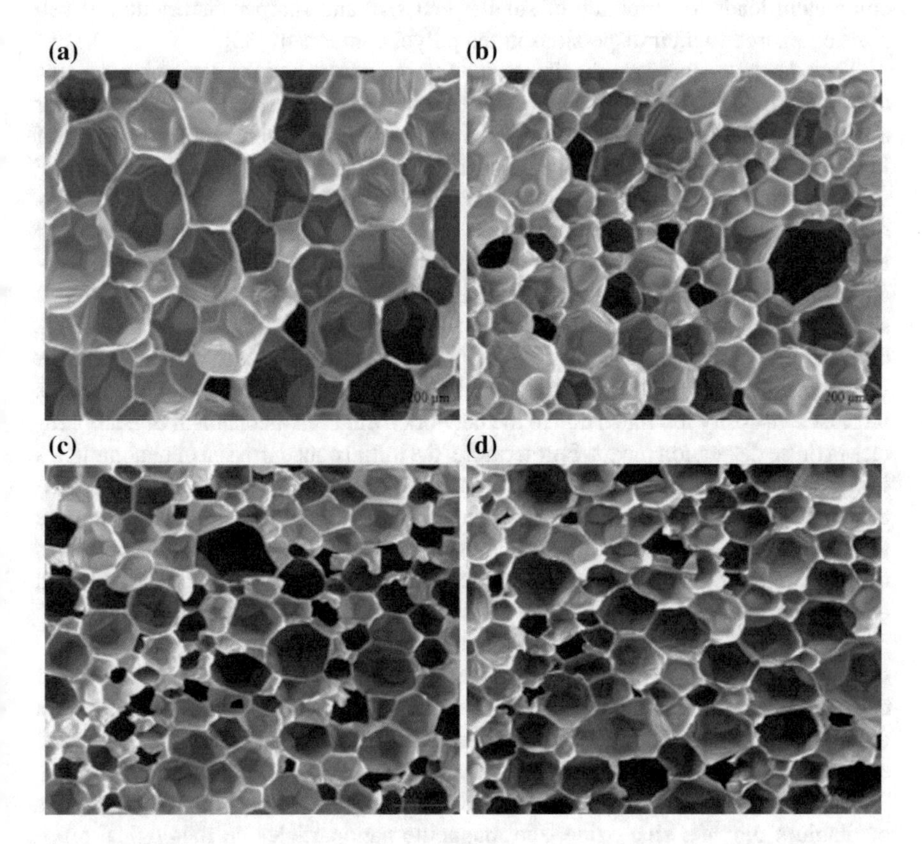

Fig. 5.8 SEM images along free foam growth direction of pure PURF and PURNCFs that were prepared in which silica aerogel was added to MDI component consisting of silica aerogel nanoparticles such as **a** 0%, **b** 1%, **c** 3%, and **d** 5%. Reproduced with permission from Ref. [33]. Copyright 2017 Elsevier B.V.

it was shown that the closed-cellular morphology containing polyhedral cell walls is clearly visible in PURNCFs [33]. In addition, it was reported that cell sizes of both pure PURF and PURNCFs along free foam growth direction are larger than their crosswise directions. This finding was explained such that smaller cells are formed in the crosswise direction since local cell density in crosswise direction is much higher than that of free foam growth direction [33]. Furthermore, it was pointed out that a larger reduction in cell size and much sharper distribution of cell sizes were observed in PURNCFs by including 3 wt% silica aerogel nanoparticles to either MDI or polyol components. In contrast, it was also shown that the mean cell size of PURNCFs elevated with incorporation of 5 wt%. silica aerogel nanoparticles. The reason behind elevation of cell size with inclusion of 5 wt% silica aerogel was explained such that presence of 5 wt% aerogel silica nanoparticles leads to viscosity elevation and decrease in homogenous dispersion of nanoparticles, and thus this behavior leads to the increase of cell size in PURNCFs [33]. Also, based on SEM results, it was confirmed that the dispersion of aerogel silica nanoparticles in MDI component leads to formation of smaller cell size and sharper distribution of cell sizes compared to their dispersions in the polyol component [33].

Thus, based on these results, it is noted that the nature and the dispersion degrees of nanoparticles within the PURNCF matrix has a great effect on closed-cellular morphology, thermal, physical, mechanical and thermomechanical properties of PURNCFs. Also, choosing the right dispersion medium for each nanoparticle/PURF system during sample preparation of PURNCFs is very critical since each nanoparticle forms different types of interactions in PURF components such as polyol and PMDI due to distinct physicochemical properties of nanoparticles. Mainly, technical parameters of closed-cellular morphology such as cell size, cell density and cell shapes in PURNCFs are greatly affected by the specific conditions of nanoparticle dispersions. Most of the time, the viscosity of PURF components such as polyol or PMDI plays the critical role because adding nanoparticles into these components increases viscosity too much due to the network formation mechanism of nanoparticles in these dispersion media. Furthermore, the initial concentration of nanoparticles in polyol or PMDI should be kept at optimum level since higher amounts of nanoparticles in these dispersion media increase the viscosity of nanoparticle solution mixture to very high levels such that effective crosslinking reactions and optimum conditions for the formation of well-defined closed-cellular morphology cannot be established during the reactive foaming event in PURNCFs.

Consequently, before the reactive foaming process, mixing and dispersion conditions are very important to end up with positive PURNCF properties such as decreased cell size and increase of cell density. Also, these two important rigid foam criteria directly control final PURNCF characteristics such as thermal conductivity, thermal stability, thermomechanical and mechanical properties.

Akkoyun et al. investigated the effects of using various types and geometries of titanium dioxide, zinc oxide, and magnetite nanoparticles on rheological properties of polyol/nanoparticle suspensions, closed-cellular morphology and reactive foaming process of PURFs [1]. Based on the results, it was shown that rheological formation of long range connectivity in polyol/nanoparticle suspensions diminished with increasing the ratio of dimensions associated with nanoparticles whose shapes are not spherical. Based on the reaction kinetics results, it was reported that reaction rate enhanced with respect to elevation of surface area above a critical value of 30 m^2. Furthermore, it was revealed that introduction of oxide surfaces with critical surface areas of 30 m^2 diminished ultimate cell sizes. However, above these well-reported critical surface areas, it was pointed out that the reaction rate increased and the exothermic nature of polymerization reaction led to the formation of uncontrolled foam structure [1].

Table 5.1 shows the cell diameters that were obtained for pure PURF and PURNCFs consisting of various nanoparticles. Based on the results, it was shown that the cell diameter decreased by about 40% for PURNCFs consisting of nanoparticles compared to that of pure PURF [1]. More specifically, it was revealed that the decrease in cell size is more apparent in PURNCFs consisting of nanofiller concentrations of 10 wt% in the polyol component. Furthermore, it was shown that incorporation of 10 wt% nanoparticles to PURFs diminished the average cell size by about 20% compared to that of pure PURF. As a consequence, based on these results, it was confirmed that incorporation of nanofillers with optimum contents led to fabrication of PURNCFs consisting of smaller cell sizes and higher cell densities [1].

Previously, Lorusso et al. investigated influences of utilizing commercial nanoparticles on closed-cellular morphology, thermal, mechanical and thermomechanical properties of PURFs [34]. Authors used various sorts of nanoparticles such as titanium dioxide nanoparticles with spherical geometries and halloysite clay nanotubes with rod-like geometries as nanofillers to produce PURNCFs consisting of titania nanocrystals and nanohalloysite with improved properties. According to experimental findings of the study, it was pointed out that PURNCFs consisting of 10% of these commercial nanoparticles had improved thermal and mechanical properties. In addition, it was pointed out that the thermal properties improved by adding titania and halloysite nanoparticles in the range of 4–8%. On the contrary, it was revealed

Table 5.1 Cell sizes of pure PURF and PURNCFs consisting of various nanoparticles with different concentrations

	Nanofiller (wt%)			
	Cell diameter, d_{50} (μm)[a]		Cell diameter, d_{90} (μm)[a]	
PU/Nanofiller systems	1	10	1	10
PU/Unfilled	227		236	
PU/TiO_2 nanospheres	208	181	200	180
PU/TiO_2 nanorods	175	145	179	141
PU/ZnO nanospheres	191	164	200	160
PU/ZnO nanoplatelets	200	145	188	155
PU/Fe_3O_4 nanoplatelets	220	168	208	167

[a]SEM

that the mechanical properties of PURNCFs started to improve with the addition of 6% nanoparticle content. Based on SEM results, it was shown that incorporation of nanoparticles to PURFs resulted in reduction of cell size and elevation of cell density [34].

Figure 5.9 shows SEM micrographs of (a) pure PURF, PURNCFs consisting of (b) 2%, (d) 4%, (f) 6%, (h) 8% and (j) 10% titanium dioxide nanoparticles and PURNCFs consisting of (c) 2%, (e) 4%, (g) 6%, (i) 8% and (k) 10% halloysite clay nanotubes. Based on the results, it was shown that both pure PURF and PURNCFs exhibited closed-cellular morphology consisting of energetically solid cells with polygonal geometrical shapes [34]. In addition, it was reported that PURNCFs consisting of various concentration of these nanoparticles displayed higher cell densities and smaller cell sizes compared to pure PURF. Thus, this result shows that the well-dispersion state of nanoparticles led nanoparticles to behave as heterogenous nucleation centers during the reactive foaming and cell formation process. Moreover, it was also shown that larger amount of broken cells were formed, the overall cell structure and cell walls became less uniform and thicker in PURNCFs consisting of 10% of nanoparticles [34].

Previously, Saha et al. investigated effects of using various types of nanoparticles such as titanium dioxide nanoparticles with spherical geometries, nanoclay particles with plate-like geometries and carbon nanofibers (CNFs) with rod-like shapes on closed-cellular morphology, thermal and mechanical properties and thermal stability of PURFs [35]. Authors used the well-known sonication method for homogenously distribution of nanoparticles in PMDI component of PURF which was then immediately mixed with the polyol component by using a mechanical stirrer. Afterwards, the complete polyol/nanoparticle/PMDI mixture was poured into an aluminum mold which was heated previously. Finally, the polyol/nanoparticle/PMDI mixture was completely cross-linked inside an oven. Based on the experimental data, it was revealed that mechanical and thermal properties of PURNCFs consisting of only 1 wt% nanoparticles exhibited significant enhancement compared to pure PURF. In addition, it was also reported that the maximum improvement in terms of mechanical and thermal properties was found for PURNCFs consisting of CNFs whereas TiO_2 filled PURNCFs displayed the minimum performance [35].

Figure 5.10 shows the SEM micrographs of (a) pure PURF and PURNCFs consisting of 1% (b) TiO_2, (c) nanoclay and (d) CNFs. Based on SEM results, it was shown that cell sizes and shapes were observed to be quite uniform throughout all foams. In addition, it was pointed out that cell shapes changed from spherical to elliptical geometry with the addition of CNFs to pure PURF [35]. The reason behind the change of cell geometry was explained such that CNFs with larger aspect ratio elongated cell dimensions along the longest dimension of nanoparticles because of favorable interactions between PURNCF matrix and nanoparticles during reactive foaming event. In addition, it was revealed that the cell size was reduced and the cell density was increased with the addition of nanoparticles to PURFs [35].

Previously, Hebda et al. studied synthesis of PURNCFs filled with disilanolisobutyl polyhedral oligomeric silsesquioxane nanoparticles [36]. During synthesis, authors chemically incorporated disilanolisobutyl POSS nanoparticles with two

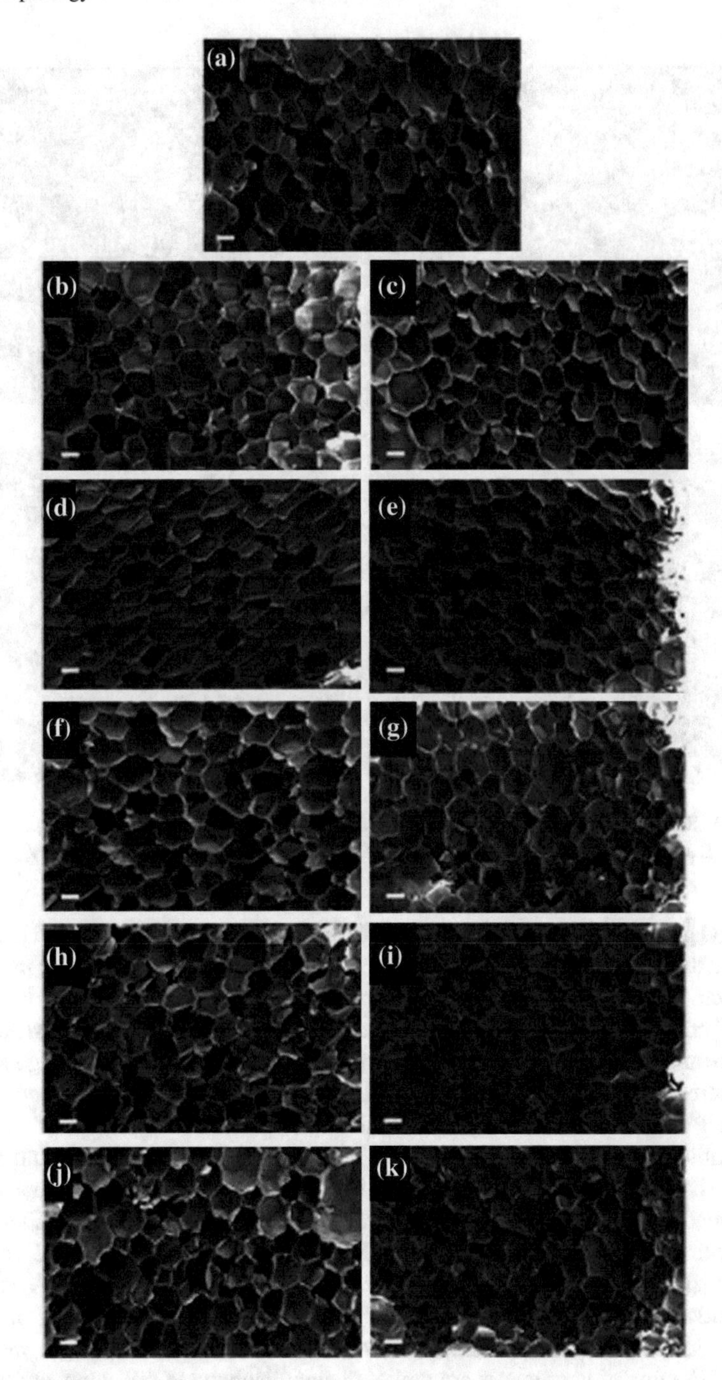

Fig. 5.9 SEM micrographs of **a** pure PURF, PURNCFs consisting of **b** 2%, **d** 4%, **f** 6%, **h** 8% and **j** 10% titanium dioxide nanoparticles and PURNCFs consisting of **c** 2%, **e** 4%, **g** 6%, **i** 8% and **k** 10% halloysite clay nanotubes. Reproduced with permission from Ref. [34]. Copyright The Author(s) 2017

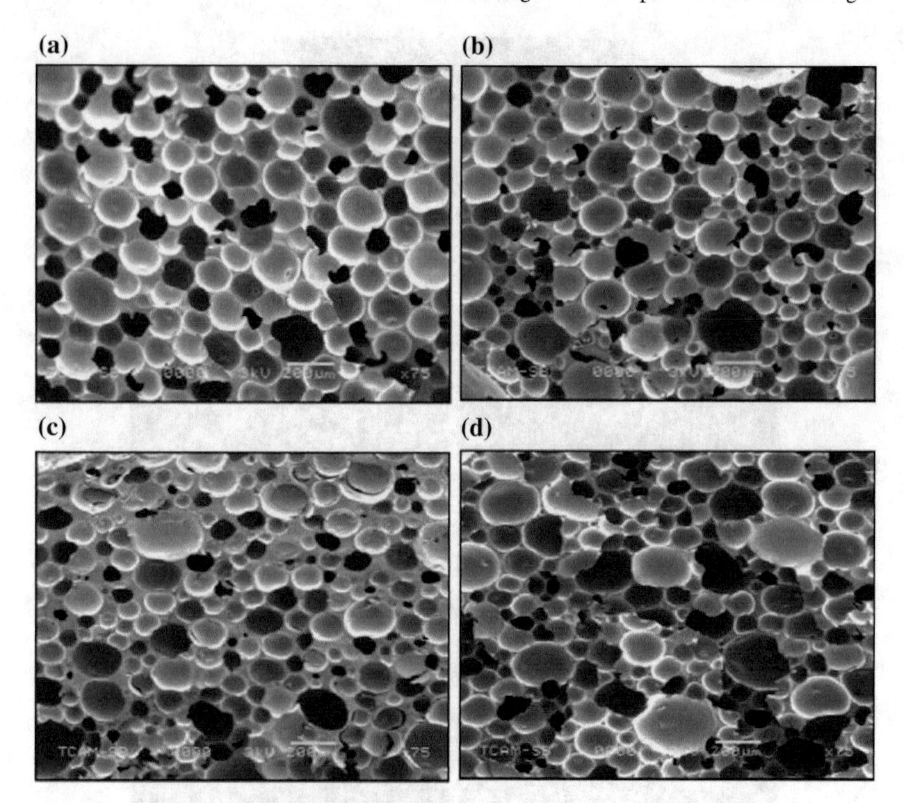

Fig. 5.10 SEM micrographs of **a** pure PURF and PURNCFs consisting of 1% **b** TiO_2, **c** nanoclay and **d** CNFs. Reproduced with permission from Ref. [35]. Copyright 2007 Elsevier B.V.

hydroxyl groups directly on PU backbone to produce hybrid POSS-based PURNCFs. Authors examined effects of using POSS nanoparticles on closed cellular morphology such as amount of closed cells, specific geometry of cells, foam density and thermal conductivity and compressive mechanical properties of PURNCFs. According to results, it was shown that population of cells diminished and average area of cells increased by incorporating 1.5 and 2.0 wt% disilanolisobutyl POSS compared to pure PURF. On the contrary, it was also shown that incorporation of 0.5 wt% disilanolisobutyl POSS to PURFs led to elevation in amount of cells and reduction in size of cells compared to pure PURF. Based on X-ray tomography results, the reduction in surface area of pores was observed. The uniform distribution of disilanolisobutyl POSS nanoparticles in PURNCFs was confirmed based on SEM and energy-dispersive X-ray spectroscopy results. In accordance with TGA data, the degradation process of PURFs did not change that much with inclusion of POSS nanoparticles into PURFs. It was also revealed that compressive strength values of PURNCFs consisting of lowest amount of disilanolisobutyl POSS along parallel and perpendicular to foam growth direction were much larger than those of pure PURF [36].

Figure 5.11 shows SEM micrographs of (a) pure PURF and (b–e) PURNCFs including different amounts of disilanolisobutyl POSS and (e) EDS mapping results for PURNCF consisting of 2.0% disilanolisobutyl POSS. In accordance with experimental data, it was pointed out that all PURNCFs consisting of disilanolisobutyl POSS nanoparticles have slightly open cells with various cell sizes [36]. In addition, it was also reported that cell sizes of PURNCFs consisting of 1.5 and 2.0 wt% disilanolisobutyl POSS nanoparticles were much larger compared to those of pure PURF and PURNCF consisting of 0.5 wt% disilanolisobutyl POSS. Based on EDS elemental mapping results of PURNCF containing 2.0 wt% disilanolisobutyl POSS, the uniform distribution of silicon was observed within spherical enclosures of POSS nanoparticles which were located on surface of cells [36].

Previously, modified POSS nanoparticles as suspended groups and POSS nanoparticles as chemical crosslinks were used in the fabrication of PURNCFs [24]. Thermal conductivity and water absorption behaviours, closed-cellular morphology and mechanical properties of PU rigid nanocomposite foams consisting of 0–15 wt% POSS were characterized thoroughly. Based on FT-IR results, it was shown that there is a good amount of interaction between POSS nanoparticles and PURNCF matrix. It was also revealed that cell size elevated and cell density diminished with incorporation of POSS nanoparticles as chemical crosslinks to pure PURF. However, on the other hand, it was reported that the addition of suspended POSS nanoparticles decreased cell sizes and elevated cell density compared to those of pure PURF [24]. Based on SEM-EDS results, it was revealed that various sizes of lamellar crystals were generated by POSS nanoparticles in which suspended POSS nanoparticles were uniformly dispersed in bulk while POSS nanoparticles as chemical crosslinks were situated close to surfaces of cells. Based on the compression tests, it was found out that compressive strength values of PURNCFs containing POSS nanoparticles along parallel and perpendicular to foam growth direction are larger than those of pure PURF. In terms of compressive strength, it was concluded that the incorporation of

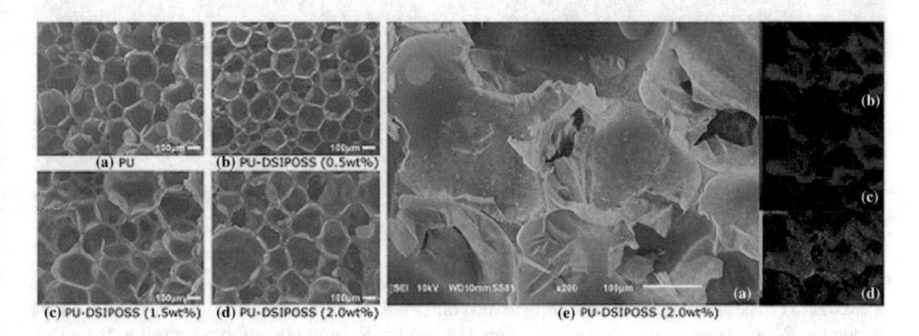

Fig. 5.11 SEM micrographs of **a** pure PURF and **b–e** PURNCFs including different amounts of disilanolisobutyl POSS and **e** EDS mapping results for PURNCF consisting of 2.0% disilanolisobutyl POSS nanoparticles. Reproduced with permission from Ref. [36]. Copyright 2018 John Wiley & Sons, Ltd.

suspended POSS nanoparticles to PURFs gives better results compared to that of PURNCF filled with POSS nanoparticles as chemical crosslinks [24].

Figure 5.12 shows SEM micrographs of PURNCFs consisting of (a) 5%, (b) 10%, (c) 15% POSS nanoparticles as chemical crosslinks with EDS mapping results and PURNCFs consisting of (d) 5%, (e) 10%, (f) 15% suspended POSS nanoparticles with EDS mapping results. In accordance with experimental data, it was shown that there is some aggregation of POSS nanoparticles with sizes of 1–2 μm in PURNCF consisting of 5 wt% POSS nanoparticles as chemical crosslinks [24]. In addition, it was reported that POSS crystals had much larger sizes which were on the order of 10 μm in PURNCFs filled with 10 and 15 wt% of POSS nanoparticles as chemical crosslinks. Furthermore, it was revealed that POSS lamellar crystals consisting of thickness values less than 1 μm were uniformly dispersed in PURF matrix. Moreover, it was shown that POSS crystals in the PURNCF containing 15 wt% suspended POSS nanoparticles have the same sizes with those of PURNCF consisting of 15 wt% POSS nanoparticles as chemical crosslinks. During the reactive foaming process, stresses could be formed due to the evaporation of blowing agents. Thus, these stresses could stretch and further crack POSS crystals. Consequently, this mechanism leads to the damage of foam cells whose closed cell amounts were observed to be lower based on SEM results [24].

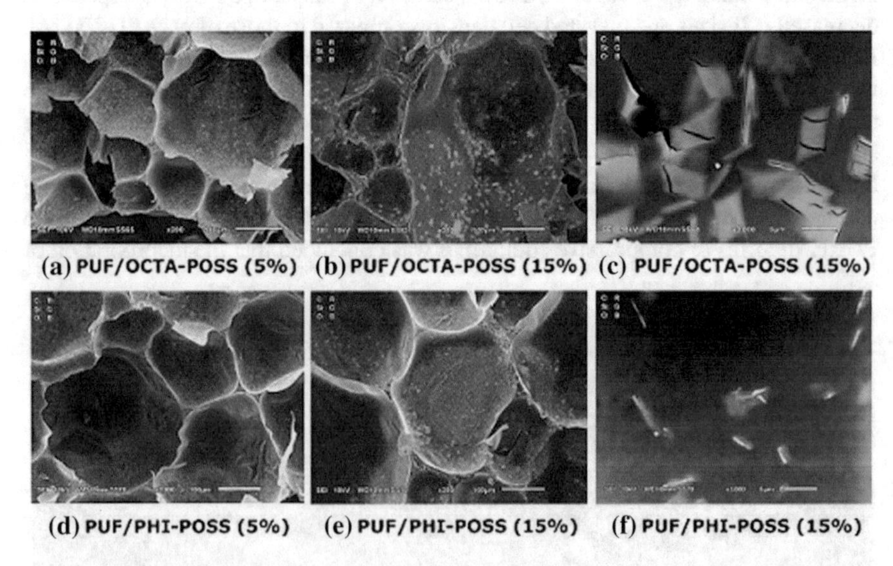

Fig. 5.12 SEM micrographs of PURNCFs consisting of **a** 5%, **b** 10%, **c** 15% POSS nanoparticles as chemical crosslinks with EDS mapping results and PURNCFs consisting of **d** 5%, **e** 10%, **f** 15% suspended POSS nanoparticles with EDS mapping results. Reproduced with permission from Ref. [24]. Copyright 2015 John Wiley & Sons, Ltd.

5.3 Mechanical Properties

Formerly, the effects of using nanofillers on the reaction kinetics of PURFs were investigated by using different types of nanoclay and nanosilicas which have different surface functional groups [4]. Authors used FT-IR to measure systematically the degree of isocyanate consumption with the addition of nanofillers to PURFs. The experimental conditions associated with blowing and gelling phenomena during reactive foaming event were evaluated based on the quantitative determination of urea and urethane amounts from carbonyl absorption regions in FT-IR data. In addition, authors also investigated the density, closed-cellular morphology, thermal conductivity and mechanical properties of PURNCFs consisting of nanosilica and nanoclay particles with the experimental data that was obtained from the reaction kinetics study of the system [4].

Figure 5.13 shows relative (a) compressive modulus and (b) collapse stress values of pure PURF and PURNCFs consisting of different nanoparticles. In accordance with experimental data, it was pointed out that mechanical properties of PURNCFs mainly depend on closed cellular morphology, foam density and specific interac-

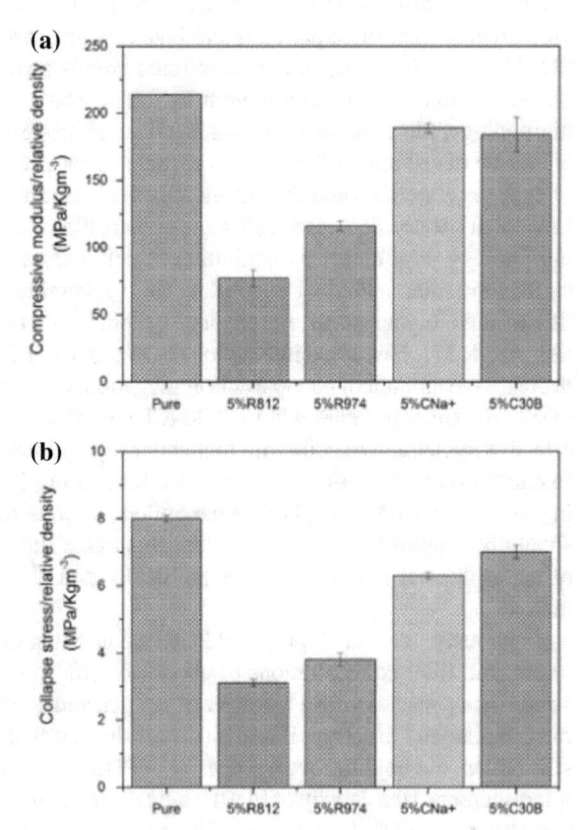

Fig. 5.13 Relative **a** compressive modulus and **b** collapse stress values for pure PURF and PURNCFs consisting of different nanoparticles. Reproduced with permission from Ref. [4]. Copyright 2018 Elsevier Ltd.

tions between nanoparticles and PURNCF matrix [4]. In addition, it was reported that the compression properties decreased extensively with the addition of nanosilica to PURFs. Specifically, it was shown that compressive modulus values of PURNCFs containing nanosilicas decreased while their collapse stress values were reduced in comparison with pure PURF. The reason behind the reduction of mechanical properties as a function of nanosilica loading was explained such that the presence of nanosilica particles in PURNCFs increased the formation of open cells within the closed-cellular morphology, thus leading to reduction in mechanical properties. On the other hand, PURNCFs consisting of nanosilica with dimethyldichlorosilane exhibited better compression properties compared to those containing other nanosilica due to the higher amounts of urea groups in this PURNCF system [4, 37].

Thus, based on these results, it is well noted that uniform distribution of nanosilica particles in PURNCF matrix, and also the types of nanosilica particles (surface modified or not) that are used in nanocomposite foams are the two most important parameters that affect mechanical properties of PURNCFs. Physicochemical properties of nanosilica particles have direct effects on the nanocomposite foams such that they can change the reactivity of cross-linking reactions during the reactive foaming process and consequently alter the mechanical properties of PURNCFs. In addition, the dispersion of nanosilica particles is directly related with the surface properties of nanosilica particles and favorable interactions between nanosilica and PURNCF matrix. Thus, surface functional groups on nanosilica particles which are physicochemically compatible with PURNCF matrix pave the way for enhancing morphology, physical, thermal and mechanical properties because of their positive effects on distribution of nanosilica particles throughout PURNCF matrix.

It is very well known that using spherical nanofillers with compatible surface functional groups decreases cell sizes and enables the formation of closed-cellular morphology with uniform distribution of cell sizes and higher cell densities since surface-modified spherical nanofillers form favorable interactions with PU chains due to their higher surface areas and beneficial surface physicochemical properties [4, 24, 37]. Favorable interactions between spherical nanofillers and PU chains decrease the amount of nanoparticle aggregation. Consequently, uniform distribution of spherical nanoparticles within PURNCF components paves the way for generating effective nucleation of cells with higher density and smaller sizes during the reactive foaming event. Because of this reason, for the fabrication of PURNCFs consisting of surface-modified spherical nanofillers, various surface modification methods should be employed to chemically attach surface functional groups on the surface of nanospherical fillers which are physically and chemically compatible with PU matrix.

Previously, Javni et al. prepared PURFs consisting of two different types of silica fillers with filler concentrations of up to 20% [3]. In their study, they used micron-sized silica particles with a diameter of 1.5 μm and nano-sized silica particles with a median diameter of 12 nm. Based on the results, it was shown that using micron-sized silica fillers did not improve the density of PURFs that much while nano-sized silica fillers increased the density of PURFs only at concentration above 20%. In addition, it was also reported that nano-sized silica particles decreased the compression strength of PURFs. According to wide angle X-ray scattering (WAXS) results, it was revealed that the PURNCFs consisting of nano-sized silica particles exhibited the amorphous morphology [3].

Figure 5.14 shows impacts of utilizing micron and nano-sized silica particles on compressive mechanical properties of PURCFs. In accordance with experimental findings, it was emphasized that increasing filler concentrations led to reduction in compression strength values of PURCFs [3]. Specifically, it was revealed that the addition of 5 and 10% micron-sized silica showed a slight reinforcing effect but the compression strength reduced at higher concentrations. In addition, it was shown that the compression strength reduced with the addition of nano-sized silica much more compared to the use of micron-sized silica as fillers. The reason behind the decrease of compression strength as a function of nano and micron-sized silica addition was explained such that the closed cellular morphology was disrupted with the addition of silica particles at high concentrations which lead to deterioration of mechanical properties [3].

Recently, the synthesis of PURNCFs containing silica aerogel nanoparticles with efficient thermal insulation properties was outlined in detail [33]. In the study, PURNCFs were prepared in two methods in which silica aerogel was added to MDI component in the first method, and in the second procedure silica aerogel was added to the polyol component by including various concentrations of silica aerogel nanoparticles in the range of 1–5 wt%. Authors used many characterization techniques to explore influences of incorporating these nanoparticles on thermal, mechanical, morphological and surface properties of PURNCFs. According to mechanical properties testing data, it was revealed that yield stress and modulus values of PURNCFs enhanced considerably with incorporation of 3 wt% silica aerogel nanoparticles to PURFs. However, it was concluded that mechanical properties of

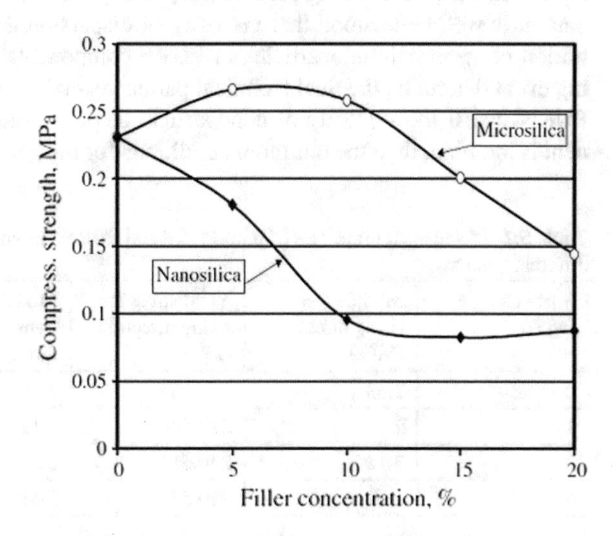

Fig. 5.14 Impacts of using micron and nano-sized silica particles on compressive mechanical properties with respect to filler amount. Reproduced with permission from Ref. [3]. Copyright 2002 Sage Publications

PURNCFs decreased with inclusion of more than 3 wt% silica aerogel nanoparticles [33].

Table 5.2 shows the mechanical properties of pure PURF and PURNCFs consisting of silica aerogels with different contents. In accordance with experimental data, it was shown that incorporation of silica aerogel nanoparticles enhanced mechanical properties of PURFs in both free rising and perpendicular directions in comparison with those of pure PURF [33]. In addition, it was also reported that using lower amounts of silica aerogel nanoparticles within range of 1–3 wt% in PURNCFs led to enhanced modulus and yield strength values. The reason behind the behavior of modulus and yield strength as a function of silica content was explained such that the addition of silica aerogel has various effects such as changes in the hydrogen bonding, polymerization reactivity, and cellular morphology in the system [33].

The positive effects of using aerogel silica nanoparticles can be stated as reinforcement of PURNCF matrix and enhancement of closed-cellular morphology while their negative impacts can be identified as disruptions of hydrogen bonding interactions and the ratio of reactants during polymerization reaction [26, 33]. In PURNCFs containing silica aerogel nanoparticles up to 3 wt%, positive effects of silica aerogel nanoparticles dominated over their negative effects in the performance of PURNCFs. However, at 5 wt% or higher concentrations of silica aerogel nanoparticles, the negative effects such as the decrease of hydrogen bonds and homogenous dispersion of silica aerogels due to the aggregation and viscosity problems lead to the deterioration of closed cellular morphology and mechanical properties [33].

Thus, the performance of PURNCFs consisting of silica aerogels strictly depends on the concentration of silica within the PURNCF matrix. If the silica aerogel concentration is too high above the optimum concentration, morphology and mechanical properties decrease due to the dispersion and viscosity problems that the system faces during the reactive foaming process [33]. Consequently, based on this experimental data, it is well understood that viscosity of dispersion medium and initial concentration of spherical nanoparticles in PURF components before the reactive foaming event determine the final technical parameters of closed-cellular morphology in PURNCFs. If the viscosity of nanoparticle dispersion in polyol or PMDI component is too high, then the optimum conditions for the generation of well-established

Table 5.2 Mechanical properties of pure PURF and PURNCFs consisting of silica aerogels with different contents

Silica aerogel (wt%)	Module free rising direction (MPa)	Yield stress free rising direction (kPa)	Module transverse direction (MPa)	Stress at 10% strain transverse direction (kPa)
0	2.63	211.36	2.08	107.74
1	2.78	229.58	2.15	119.82
3	3.12	240.61	2.22	121.65
5	2.61	219.54	2.11	111.98

closed-cellular morphology cannot be achieved during the reactive foaming event. However, on the other hand, if the viscosity of nanoparticle dispersion or nanoparticle concentration is at the optimum level, then the fabrication of PURNCFs consisting of preferred closed-cellular morphology can be realized. Hence, since closed-cellular morphology directly controls the final mechanical properties of PURNCFs consisting of spherical nanofillers, it is very important to adjust and determine the optimum conditions such as concentration of nanoparticles and viscosity of reaction medium during the reactive foaming event to obtain PURNCFs with enhanced closed-cellular morphology and mechanical properties.

Earlier on, influences of utilizing commercial nanoparticles on closed-cellular morphology, thermal, mechanical and thermomechanical properties of PURFs were investigated [34]. Authors used various sorts of nanoparticles such as titanium dioxide nanoparticles with spherical geometries and halloysite clay nanotubes with rod-like geometries as nanofillers to produce PURNCFs consisting of titania nanocrystals and nanohalloysite with improved properties. According to experimental findings of the study, it was pointed out that PURNCFs consisting of 10% of these commercial nanoparticles had improved thermal and mechanical properties. On the contrary, it was revealed that the mechanical properties of PURNCFs started to improve with the addition of 6% nanoparticle content [34].

Figure 5.15 shows the compression stress-strain plots of pure PURF and PURNCFs consisting of TiO_2 with different contents. Based on the results, it was shown that compressive mechanical property values of PURNCFs containing TiO_2 higher than 6 wt% enhanced in comparison with those of pure PURF [34]. However, on the other hand, it was reported that PURNCFs consisting of TiO_2 amounts lower than 4% exhibited lower values in terms of compressive strength and modulus compared to pure PURF. In accordance with compressive testing data, it was pointed out that the lower slope and the shape of plateau region in the compression stress-strain curves are closely related with the morphological behavior of PURFs [34]. Thus, the parameters of closed cellular morphology such as cell size, cell density, uni-

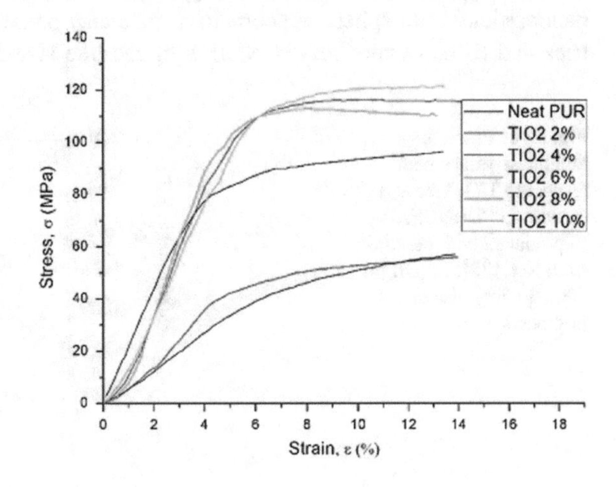

Fig. 5.15 Compression stress–strain plots of pure PURF and PURNCFs containing different amounts of TiO_2. Reproduced with permission from Ref. [34]. Copyright The Author(s) 2017

form dispersion of cells and also homogeneous distribution of nanoparticles within PURNCF matrix can be adjusted such that in the overall the mechanical properties of PURNCFs should be designed for the use of these materials in specific applications.

Previously, Saha et al. investigated fracture toughness properties of PURNCFs consisting of nano-sized fillers by performing three-point bending experiments [38]. Authors prepared PURNCFs which have foam density values on the order of 260 kg/m^3. In the study, nano-sized titanium dioxide, nanoclay, carbon nanofibers (CNFs) and multiwall carbon nanotubes (MWNTs) were incorporated into PURFs as nano-sized fillers with various geometries. The PURNCFs consisting of nanoparticles in the range of 0.5–2 wt% were prepared by infusion of nanoparticles into PURF liquid components via applying sonochemical method [38]. Authors also applied linear elastic fracture mechanics modelling to find out fracture toughness behaviour of PURNCFs. Based on the results, it was shown that the highest enhancement in fracture toughness was observed in PURNCF sample consisting of 0.5 wt% CNF. In addition, it was reported that among different sized TiO_2 nanoparticles, 5 nm nanoparticles with the addition of 1.0 wt% displayed the highest improvement in terms of fracture toughness by about 16% compared to pure PURF [38].

Figure 5.16 shows effects of using TiO_2 nanoparticle with different sizes and concentrations on fracture toughness properties of PURNCFs. Based on the results, it was shown that enhancement in fracture toughness properties of PURNCFs is strongly correlated with size of nanoparticles [38]. In addition, it was reported that the optimum concentration of nanoparticles was found to be about 1 wt% for TiO_2 nanoparticles which have diameters of 5 and 10 nm. Furthermore, it was also pointed out that inclusion of 1 wt% TiO_2 nanoparticles having average size of 5 nm led to the best enhancement in terms of fracture toughness compared to pure PURF. The reason behind this finding was explained such that the addition of TiO_2 nanoparticles with smaller sizes affects the toughening mechanism positively via the formation of multiple crack branchings. Thus, this mechanism which was induced by smaller nanoparticles led to the increase of fracture toughness in PURNCFs [38].

Recently, effects of using various types of nanoparticles such as titanium dioxide nanoparticles with spherical geometries, nanoclay particles with plate-like geometries and carbon nanofibers (CNFs) with rod-like shapes on closed-cellular mor-

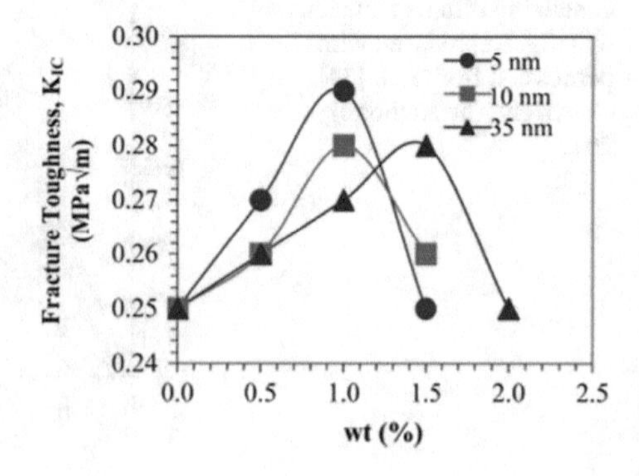

Fig. 5.16 Fracture toughness values with respect to TiO_2 size and amount in PURNCFs. Reproduced with permission from Ref. [38]. Copyright 2008 Society of Plastics Engineers

phology, thermal and mechanical properties and thermal stability of PURFs were investigated thoroughly [35]. Authors used the well-known sonication method for homogenously distribution of nanoparticles in PMDI component of PURF which was then immediately mixed with the polyol component by using a mechanical stirrer. Based on the experimental data, it was revealed that mechanical and thermal properties of PURNCFs consisting of only 1 wt% nanoparticles exhibited significant enhancement compared to pure PURF. In addition, it was also reported that the maximum improvement in terms of mechanical and thermal properties was found for PURNCFs consisting of CNFs whereas TiO_2 filled PURNCFs displayed the minimum performance [35].

Figure 5.17 shows the compression stress–strain plots of pure PURF and PURCNFs with 1% of nanoparticles. According to experimental data, it was pointed out that compressive strength and modulus values increased with incorporation of nanoparticles to PURFs. In addition, it was reported that PURNCFs consisting of TiO_2 and CNF exhibit the lowest and highest increase, respectively in terms of compression properties [35]. More specifically, it was stated that only 1 wt% inclusion into PURF enhanced compressive mechanical properties substantially. Moreover, it was pointed out that slopes associated with stress-strain curves in PURNCFs were much larger compared to that of pure PURF. In addition, the CNF filled PURNCFs have the highest slope of linear plateau region while the lowest slope corresponds to PURNCFs consisting of TiO_2 [35].

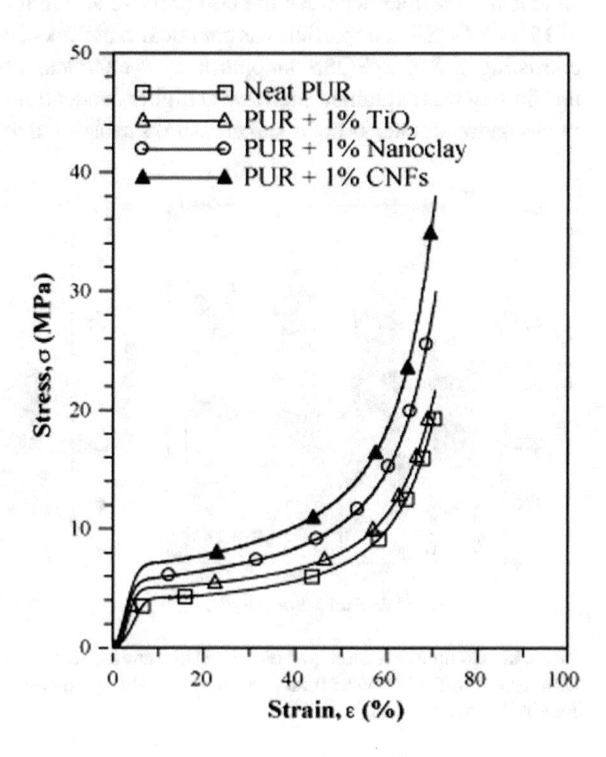

Fig. 5.17 Compression stress–strain plots of pure PURF and PURCNFs with 1% of nanoparticles. Reproduced with permission from Ref. [35]. Copyright 2007 Elsevier B.V.

Formerly, modified POSS nanoparticles as suspended groups and POSS nanoparticles as chemical crosslinks were used in the fabrication of PURNCFs [24]. Thermal conductivity and water absorption behaviours, closed-cellular morphology and mechanical properties of PU rigid nanocomposite foams consisting of 0–15 wt% POSS were characterized thoroughly. Based on FT-IR results, it was shown that there is a good amount of interaction between POSS nanoparticles and PURNCF matrix. Based on compression tests, it was found out that compressive strength values of PURNCFs containing POSS nanoparticles along parallel and perpendicular to foam growth direction are larger than those of pure PURF. In terms of compressive strength, it was concluded that incorporation of suspended POSS nanoparticles to PURFs gives better results compared to that of PURNCF filled with POSS nanoparticles as chemical crosslinks [24].

Figure 5.18 shows the compressive strengths of pure PURF and PURNCFs consisting of POSS nanoparticles. In conformity with experimental data, it was found out that mechanical properties of PURNCFs are closely related to closed-cellular morphology, foam density and types of fillers that are used during sample preparation [24]. In addition, it was reported that compressive strength values of pure PURF and PURNCFs consisting of two types of POSS nanoparticles that were tested in parallel and perpendicular directions of foam growth direction are greater than those of pure PURF material. Specifically, the largest increase in compressive strength compared to pure PURF was found in PURNCF consisting of 5 wt% POSS nanoparticles as chemical crosslinks whereas the compressive strength decreased with the addition of 15 wt% POSS nanoparticles as chemical crosslinks compared to that of PURNCF consisting of 5 wt% POSS nanoparticles as chemical crosslinks. The reason behind this finding was explained such that at higher concentrations, the crystalline nature of POSS nanoparticles as chemical crosslinks causes the destruction of closed-cellular

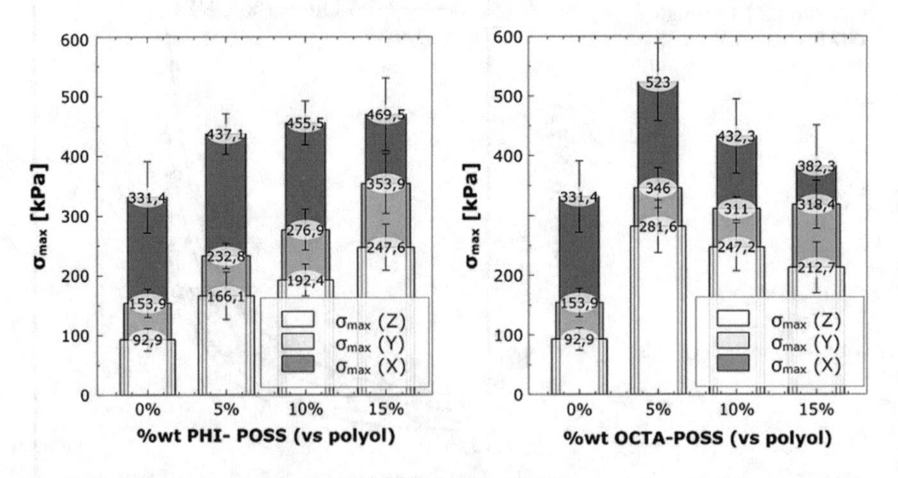

Fig. 5.18 Compressive strengths of pure PURF and PURNCFs consisting of POSS nanoparticles with respect to POSS concentration. Reproduced with permission from Ref. [24]. Copyright 2015 John Wiley & Sons, Ltd.

morphology of PURFs during reactive foaming event [24]. In accordance with these experimental results, it is well-understood that introducing higher amounts of spherical nanoparticles consisting of higher amount of crystallinity leads to disruption of closed-cellular morphology in PURNCFs. The reason behind this finding can be explained such that spherical nanoparticles consisting of higher crystallinity might function as rigid obstacles against the nucleation and growth of closed cells by damaging the closed-cellular geometries and transforming them into more like open cell geometries.

However, on the other hand, PURNCFs consisting of suspended POSS nanoparticles exhibited totally different behavior in terms of compression properties compared to system containing POSS nanoparticles as chemical crosslinks. It was shown that compressive strength values elevated with increasing concentrations of suspended POSS nanoparticles in all three directions within the PURNCF matrix compared to pure PURF. The reason behind the increase of compressive strength values in PURNCFs consisting of suspended POSS nanoparticles was explained such that randomly distributed suspended POSS crystals within PURNCF matrix did not cause any cell destruction, thus they behaved as reinforcing media for enhancement of mechanical properties in POSS filled PURNCF system [24]. In accordance with these experimental results, it is well-understood that introducing suspended spherical nanoparticles consisting of some degree of crystallinity into PURNCFs leads to disorganized distribution of these spherical nanoparticles throughout PURNCF matrix. Thus, because of this reason, spherical nanoparticles are not very concentrated locally in PURNCFs. Thus, this specific property of suspended nanoparticles makes them efficient nanoadditives for the enhancement of mechanical properties of PURNCFs.

5.4 Thermal and Thermomechanical Properties

Formerly, a surface alteration method of fumed nanosilica particles was performed by exploiting *N*-(2-aminoethyl)-3-aminopropyltrimethoxysilane molecule as a surface refining instrument, and these surface adjusted nanosilica particles were incorporated into PURFs [7]. Authors used several characterization techniques such as FT-IR, TGA, SEM, tensile testing, DMA and TMA in order to understand the physical, thermal, mechanical and thermomechanical properties of PURNCFs consisting of modified nanosilica particles. Based on the results, it was reported that dynamic mechanical properties of PURNCFs consisting of modified nanosilica exhibited a different behavior compared to mechanical properties due to establishment of various types of interactions between nanosilica particles and PURNCF matrix. Furthermore, it was revealed that glass transition temperatures of PURNCFs consisting of modified nanosilica were diminished compared to that of pure PURF. The reason behind this observation was explained such that functional groups that are present on the surface of nanosilica particles influenced the reactant ratio of PU components during cross-linking reactions and reduced hard phase formation in PURNCF matrix [7].

Figure 5.19 shows the storage modulus of pure PURF and PURNCFs consisting of surface modified nanosilica with different contents. According to experimental data, it was revealed that storage modulus values of PURNCFs consisting of modified nanosilica particles decreased in comparison with that of pure PURF [7]. The reason behind the decrease of storage modulus with incorporation of modified nanosilica was explained such that the hard phase formation was reduced as a result of the disruption of the stoichiometric ratio because of reactions between surface functional groups of modified nanosilica and isocyanates of PURNCF matrix [7]. As pointed out in Fig. 5.19, storage modulus of PURNCF consisting of 1% nanosilica has the lowest values at all temperatures compared to pure PURF and other PURNCFs containing 2 and 3% of nanosilica. However, on the contrary, it can be easily seen that at all temperatures, the storage modulus values of PURNCFs containing 2 and 3% modified nanosilica particles are much higher than those of PURNCF consisting of 1% nanosilica, but much lower than those of pure PURF. This result basically shows that incorporation of 2 and 3% nanosilica to PURFs leads to aggregation problem and this property decreases the uniform distribution of modified nanosilica particles throughout PURNCF matrix. Then, this dispersion problem decreases favorable interactions between nanosilica particles and PU chains. Thus, stoichiometric ratio changes much less compared to the system of 1% nanosilica. Consequently, the hard domain formation and the values of storage modulus reduce much less in comparison with the PURNCF containing 1% nanosilica [7].

Previously, surface properties of nanosilica were altered by using gamma-Glycidoxypropyltrimethoxysilane and diethanol amine, successively and these modified nanosilica particles with adjusted surface properties were used as nanospherical fillers in PURNCFs [6]. Authors used FT-IR, thermal analysis methods, tensile testing, TMA, DMA and SEM in order to systematically analyze intermolecular

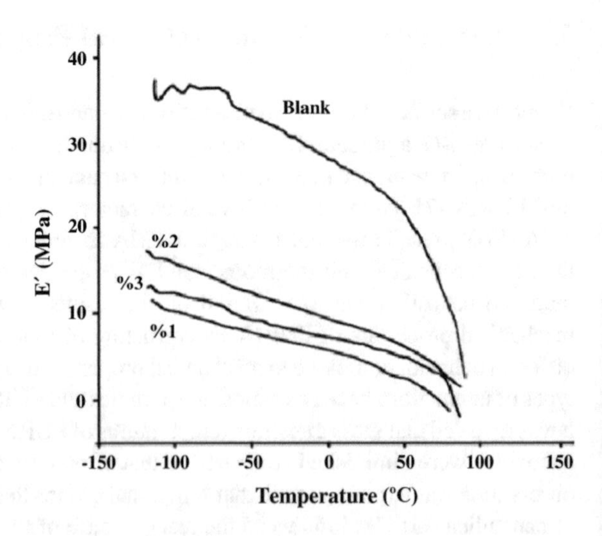

Fig. 5.19 Storage modulus of pure PURF and PURNCFs consisting of surface modified nanosilica with different contents. Reproduced with permission from Ref. [7]. Copyright 2009 Society of Plastics Engineers

interactions, closed-cellular morphology, mechanical and thermomechanical properties and thermal decomposition behavior of the system. According to results, it was revealed that storage modulus values of PURNCFs decreased since surface reactive functional groups associated with nanosilica particles altered relative reactant ratios during cross-linking reactions and also paved the way for reducing the amount of hard segments in PURNCFs [6]. Thus, the stiffness or the rigidity of PURNCF system consisting of modified nanosilica is greatly dependent on the surface properties of nanoparticles that are used as nanofillers in the system.

Figure 5.20 shows the storage modulus of pure PURF and PURNCFs consisting of modified nanosilica with different contents. Based on the results, it was shown that PURNCFs consisting of modified nanosilica exhibited lower storage modulus values at all temperatures compared to pure PURF [6]. In agreement with the results from previously published other works [7], the reactions between surface functional groups on modified nanosilica and isocyanate functional groups in PURNCFs led to the disruption of stoichiometric ratio and thus decreased the crosslink density and the ratio of hard domains in the system [6].

The influences of using nanofillers on reactive foaming kinetics of PURFs were investigated by using different types of nanoclay and nanosilica which have different surface functional groups [4]. Authors used FT-IR to measure systematically the degree of isocyanate consumption with the addition of nanofillers to PURFs. In addition, authors also investigated the density, closed-cellular morphology, thermal conductivity and mechanical properties of PURNCFs consisting of nanosilica and nanoclay particles with the experimental data that was obtained from the reaction kinetics study of the system [4].

Figure 5.21 shows thermal conductivity values of pure PURF and PURNCFs containing various types of nanoclay and nanosilica particles with 5% concentrations. Based on the results, it was shown that the conductivity values of PURNCFs consisting of two types of clays are much lower compared to that of pure PURF [4]. The

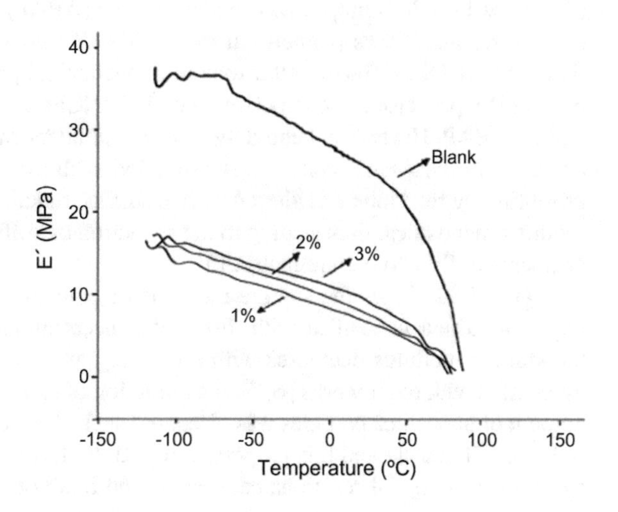

Fig. 5.20 Storage modulus of pure PURF and PURNCFs consisting of modified nanosilica with different contents. Reproduced with permission from Ref. [6]. Copyright The Author(s), 2010

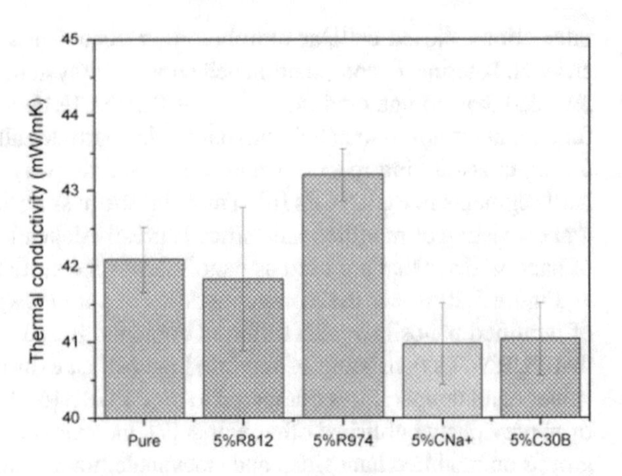

Fig. 5.21 Thermal conductivity values of pure PURF and PURNCFs consisting of different nanoparticles with 5% concentrations. Reproduced with permission from Ref. [4]. Copyright 2018 Elsevier Ltd.

reason behind this finding was explained such that the decrease of cell size in nanoclay filled PURNCFs leads to the reduction of heat transfer by radiation and also the open cell content which increases thermal conductivity does not change that much in the clay filled PURNCFs [4, 39]. However, on the other hand, it was revealed that PURNCFs consisting of nanosilicas displayed similar or higher thermal conductivity values compared to that of pure PURF. This observation was also described such that these two PURNCFs are composed of high ratios of open cell contents which lead to substantial amounts of heat transfer by radiation despite the fact that they have much smaller cell sizes compared to pure PURF [4].

Earlier on, 3-Aminopropyltriethoxysilane (APTS) was used as a surface modifying molecule and this modified nanosilica was used as a nanofiller in PURNCFs [5]. Furthermore, the effects of using APTS on morphology, mechanical, thermomechanical, physical and thermal properties of PURNCFs were compared with those of PURNCFs filled with nanosilica particles which were surface modified with *n*-(2-aminoethyl)-3-aminopropyltrimethoxysilane (AEAP). In accordance with experimental results, it was pointed out that PURNCFs consisting of nanosilica modified with APTS exhibited better dynamic mechanical properties but inferior static mechanical properties in comparison with PURNCFs consisting of nanosilica modified by AEAP. The reason behind the poor dynamic mechanical properties that were observed in the AEAP system was explained such that the stoichiometric ratio of cross-linking reactions was altered by two surface functional groups associated with modified nanosilica, thus leading to the reduction of stiffness or the number of hard segments in PURNCF bulk matrix [5].

Figure 5.22 shows the storage modulus of pure PURF and PURNCFs consisting of modified nanosilica with different concentrations. As shown in Fig. 5.22, the storage modulus decreases with increasing amount of modified nanosilica. In agreement with other works [6, 7], the reduction of storage modulus with increasing amount of nanosilica particles was discussed such that cross-linking density reduced in PURNCF matrix and this property reduced the formation of hard domains in the system [5]. Also, the reactions between amino functional groups on nanosilica and

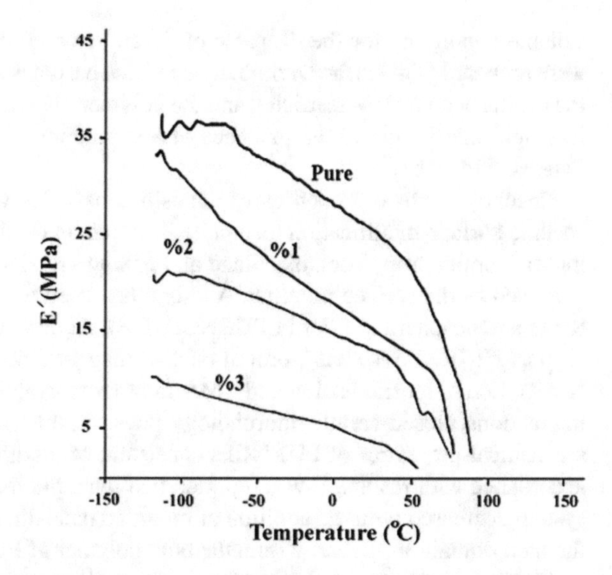

Fig. 5.22 Storage modulus of pure PURF and PURNCFs consisting of modified nanosilica with different concentrations. Reproduced with permission from Ref. [5]. Copyright Koninklijke Brill NV, Leiden, 2011

isocyanate functional groups in PURNCF matrix paved the way for disturbing the stoichiometric ratio during sample preparation [5–7]. It was previously shown that in PURNCFs consisting of modified nanosilicas, static mechanical properties exhibited improvements but their dynamic mechanical properties decreased much more in comparison with those of pure PURF [5–7]. The underlying mechanism behind these results was explained such that amounts of crosslinking density and hard domains in bulk matrix reduced but their values increased in interfacial regions, and thus stiffness in interfacial areas elevated while its value diminished in bulk matrix. Consequently, based on this mechanism, it was pointed out that the bulk matrix controls dynamic mechanical properties whereas the interfacial area is responsible for static mechanical properties [5].

In accordance with previously published major PURCNF works containing different surface properties of nanosilica particles [5–7], using surface modified nanosilica particles basically improves mechanical properties however decreases dynamic mechanical properties of PURNCFs. Especially, using surface modified nanoparticles having more than one amino functional group in PURNCFs decreases dynamic mechanical properties much more compared to system containing nanosilica having one amino functional group [5]. Since the crosslinking density is reduced in bulk PURNCFs filled with surface modified nanoparticles, the rigidity, stiffness or storage modulus and glass transition temperature values drop with incorporation of surface modified nanosilica particles. The reason behind the decrease of storage modulus and glass transition temperature in PURNCFs consisting of surface modified nanosilica particles in comparison with pure PURF can be explained such that chemical crosslinks between urethane linkages of PU chains are mainly disturbed via chemical bonding interactions between surface functional groups of nanofillers and urethane linkages of PU chains [5–7]. Thus, this physical mechanism provides suitable conditions for the decrease of T_g and storage modulus in PURNCFs that were prepared with surface modified nanosilica particles since molecular motions of PU chains become less restricted and the polymer matrix is transformed into a much less rigid matrix due to the presence of less amount of crosslinking points within PURNCF matrix.

Recently, surface properties of nanosilica particles were altered by utilizing a specific surface modification technique in which an original dipodal silane consisting of 3-aminopropyltriethoxysilane and *gamma*-glycidoxypropyltrimethoxysilane was used as the surface modifier. Authors then used this surface modified nanosilica as a nanospherical filler in PURNCFs [28]. Authors used characterization tools such as FT-IR, TGA, SEM, proton nuclear magnetic resonance spectroscopy (1H-NMR), DMA, tensile testing and TMA in order to evaluate types of intermolecular interactions, closed-cellular morphology, physical, thermal, mechanical and thermo-mechanical properties of PURNCFs consisting of modified nanosilica particles. In accordance with results, it was reported that dynamic mechanical properties of the system decreased with the addition of modified nanosilica due to the suppression of the hard domain formation within the bulk polymer of PURNCF system [28].

Figure 5.23 shows alternation of linear thermal expansion coefficient of PURNCFs with respect to amount of modified nanosilica. According to experimental data, it was shown that inclusion of modified nanosilica particles to PURFs increases linear thermal expansion coefficient. However, it was also revealed that PURNCFs containing higher contents of modified nanosilica particles have much lower linear thermal expansion coefficients [28]. These experimental findings pointed out that adding small amounts of modified nanosilica particles (up to 1%) diminished dimensional firmness of PURNCFs because of elastic character reduction of cell walls which led to higher thermal expansion coefficients. But, at higher additions of modified nanosilica (more than 1%) to PURFs, dimensional firmness of foams was enhanced but their linear thermal expansion coefficient values were diminished [28]. Moreover, based on TMA results, it was pointed out that incorporation of 2% modified nanosilica to PURFs reduced glass transition temperature values due to reduction of hard phase amount and increase of polymer chain mobility [28].

Previously, halogen-free flame-retardant PURNCFs were fabricated by incorporating nanohybrid particles of graphene oxide and nanospherical silica and a flame

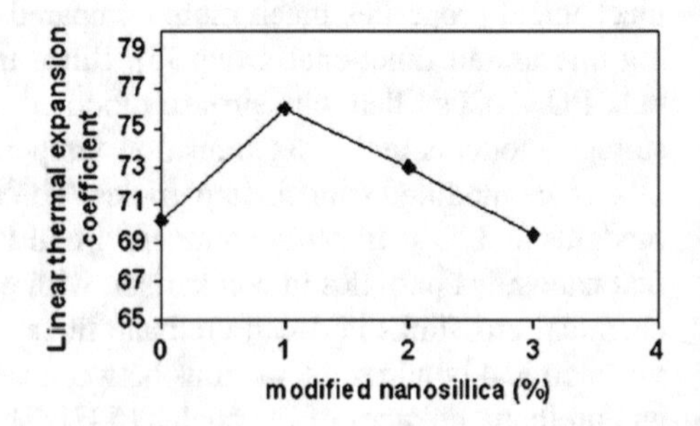

Fig. 5.23 Alternation of linear thermal expansion coefficient of PURNCFs with respect to amount of modified nanosilica. Reproduced with permission from Ref. [28]. Copyright Koninklijke Brill NV, Leiden, 2010

retardant (dimethyl methylphosphonate) in various contents [30]. Authors investigated mechanical, flame retardant and thermal properties of PURNCFs containing various contents of dimethyl methylphosphonate and nanosilica/graphene oxide hybrid. Based on the results, it was shown that flame retardant and mechanical properties of PURNCFs extensively improved by incorporating both flame retardant and nanohybrid particle compared to pure PURF and PURFs containing only dimethyl methylphosphonate [30].

Table 5.3 shows foam density, thermal conductivity and mechanical properties of pure PURF and PURNCFs consisting of nanoparticles with different compositions. In accordance with thermal conductivity data of PURNCFs in Table 5.3, it is clearly seen that incorporation of flame retardant or both nanohybrid particle and flame retardant to PURFs does not really change thermal conductivity values [30]. As shown in Table 5.3, density values of PURFs increase with incorporation of flame retardant compared to that of pure PURF. In addition, it was shown that density values of PURNCFs consisting of both flame retardant and nanohybrid particle are much higher compared to PURNCFs containing only flame retardant. The reason behind this finding was explained such that the presence of nanohybrid particles act as heterogeneous nucleation media in terms of reducing median cell diameter via more uniform distributions within the PURNCF matrix [30].

Formerly, superparamagnetic Fe_3O_4/silica nanoparticles were incorporated into PURFs in order to prepare new magnetic PURNCFs by using a one-step synthesis route [32]. Authors prepared PURNCFs consisting of maximum 3% of magnetic nanoparticles, and they also used various characterization tools such as FT-IR, DMA and TGA on PURNCFs in order to systematically evaluate closed-cellular structure and intermolecular interactions, thermomechanical, and thermal decomposition

Table 5.3 Thermal conductivity, density and mechanical strengths of pure PURF and PURNCFs consisting of nanoparticles with different compositions

Sample code	Density (kg m^{-3})	Compressive strength (MPa)	Specific compressive strength (MPa g^{-1} cm^{-3})	Thermal conductivity (W m^{-1} K^{-1})
Pure RPU	49.3	0.27	5.47	0.0241
16DMMP/RPU	49.4	0.26	5.26	0.0242
20DMMP/RPU	53.7	0.25	4.65	0.0240
24DMMP/RPU	54.5	0.21	3.85	0.0241
1SNGO/19DMMP/RPU	58.8	0.30	5.10	0.0241
2SNGO/18DMMP/RPU	58.5	0.33	5.64	0.0239
3SNGO/17DMMP/RPU	56.2	0.28	4.98	0.0244
4SNGO/16DMMP/RPU	49.8	0.25	5.02	0.0244
2GO/18DMMP/RPU	48.8	0.26	5.33	0.0243

properties of nanocomposite foams. In accordance with results, it was revealed that thermal, mechanical, and magnetic properties of PURNCFs were improved compared to pure PURF [32].

Figure 5.24 shows storage modulus values of pure PURF and PURNCF consisting of 3% magnetic nanoparticles with respect to temperature. As shown in Fig. 5.24, it was pointed out that storage modulus increases but glass transition temperature decreases with the addition of 3% magnetic nanoparticles to PURFs [32]. The reason behind the reduction of glass transition temperature with the addition of magnetic particles was due to the reduction in the formation of cross-linking points between the polymer chains of PURNCF system [32, 40, 41]. As a consequence of this finding, it is well-noted that the decrease of crosslinking density in PURNCFs filled with magnetic particles directly increases the movement of polymer chains. Then, glass transition temperature diminishes due to higher mobility of PU chains in PURNCFs consisting of superparamagnetic Fe_3O_4/silica nanoparticles. Thus, this result clearly shows that glass transition temperature is mainly affected by the value of cross-linking density in bulk foam matrix. However, on the other hand, the storage modulus increase with respect to magnetic nanoparticle addition can be explained such that the presence of 3% magnetic nanoparticles enhances the rigidity of PURNCFs, thus leading to higher values of storage modulus compared to pure PURF.

Previously, synthesis of PURNCFs containing silica aerogel nanoparticles with efficient thermal insulation properties was reported thoroughly [33]. In the study, PURNCFs were prepared in two methods in which silica aerogel was added to MDI component in the first method, and in the second procedure silica aerogel was added to the polyol component by including various concentrations of silica aerogel nanoparticles in the range of 1–5 wt%. Authors used many characterization techniques to explore influences of incorporating these nanoparticles on thermal, mechanical, morphological and surface properties of PURNCFs. In accordance with water contact angle and thermal conductivity data, it was shown that thermal conductivity values diminished and also values of hydrophobicity elevated in cyclopentane blown

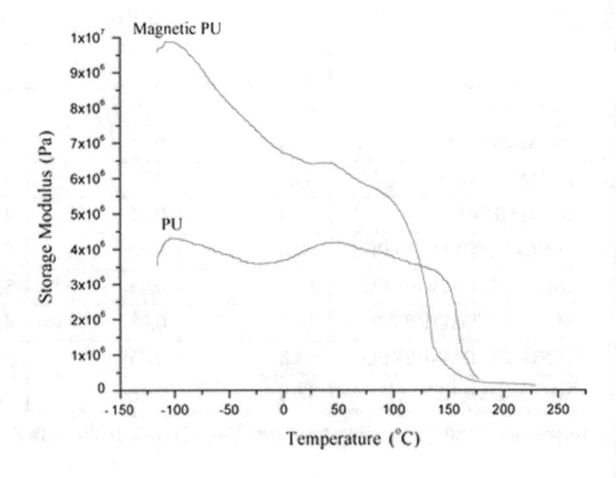

Fig. 5.24 Storage modulus of pure PURF and PURNCF consisting of 3% magnetic nanoparticles. Reproduced with permission from Ref. [32]. Copyright Springer-Verlag 2012

PURNCFs with incorporation of 5 wt% silica aerogel nanoparticles compared to those of pure PURF [33].

Table 5.4 displays thermal conductivity values of pure PURF and PURNCFs containing various concentrations of silica aerogel nanoparticles which were prepared from MDI dispersions. Generally, it is very well-known that various important factors such as foam density, closed-cellular morphology, the blowing agent, sample preparation conditions and also additive types can affect the thermal conductivity behavior of PURFs. According to experimental data in Table 5.4, it is clearly seen that PURNCFs consisting of CFC-11 as the blowing agent displayed the lowest thermal conductivity values [33]. Furthermore, for all blowing agents, the thermal conductivity values of PURNCFs which were prepared in MDI or polyol dispersions are much lower compared to pure PURF. In addition, it was also revealed that the thermal conductivity decreases with increasing amount of silica aerogels in comparison with pure PURF, thus leading to better thermal insulation properties in PURNCFs [33].

Influences of utilizing commercial nanoparticles on closed-cellular morphology, thermal, mechanical and thermomechanical properties of PURFs were studied in detail [34]. Authors used various sorts of nanoparticles such as titanium dioxide nanoparticles with spherical geometries and halloysite clay nanotubes with rod-like geometries as nanofillers to produce PURNCFs consisting of titania nanocrystals and nanohalloysite with improved properties. According to experimental findings of the study, it was pointed out that PURNCFs consisting of 10% of these commercial nanoparticles had improved thermal and mechanical properties. In addition, it was pointed out that the thermal properties improved by adding titania and halloysite nanoparticles in the range of 4–8% [34].

Table 5.5 shows thermal conductivity values of pure PURF and PURNCFs consisting of HNTs and TiO_2 with different compositions. Based on the results, it was shown that PURNCFs consisting of HNTs with up to 8% concentration exhibit almost the same thermal conductivity values 3 and 10 days after their preparation

Table 5.4 Thermal conductivity values of pure PURF and PURNCFs containing different contents of silica aerogels which were prepared from MDI dispersions

Silica aerogel (wt%)	Blowing agent			
	CFC-11	Cyclopentane	Normal pentane	Normal hexane
0%	0.0209 ± 0.0003	0.0314 ± 0.0001	0.0301 ± 0.0001	0.0345 ± 0.0002
1%	0.0189 ± 0.0002	0.0295 ± 0.0003	0.2879 ± 0.0002	0.0328 ± 0.0001
3%	0.0178 ± 0.0001	0.0277 ± 0.0002	0.0266 ± 0.0001	0.0308 ± 0.0002
5%	0.0171 ± 0.0002	0.0268 ± 0.0001	0.0257 ± 0.0001	0.0299 ± 0.0001

Table 5.5 Thermal conductivity values of pure PURF and PURNCFs consisting of HNTs and TiO_2 with different compositions

Type	Thermal conductivity (W/m K) 3 days after (mean value ± SD)	Thermal conductivity (W/m K) 10 days after (mean value ± SD)
Neat PU foam	0.0254 ± 0.0006	0.0262 ± 0.0003
TiO_2 2%	0.0246 ± 0.0003	0.0249 ± 0.0003
TiO_2 4%	0.0244 ± 0.0007	0.0261 ± 0.0006
TiO_2 6%	0.0292 ± 0.0015	0.0342 ± 0.0003
TiO_2 8%	0.0254 ± 0.0001	0.0262 ± 0.0005
TiO_2 10%	0.0329 ± 0.0007	0.0343 ± 0.0003
HNTs 2%	0.0263 ± 0.0018	0.0263 ± 0.0003
HNTs 4%	0.0242 ± 0.0004	0.0309 ± 0.0008
HNTs 6%	0.0239 ± 0.0001	0.0262 ± 0.0002
HNTs 8%	0.0236 ± 0.0001	0.0262 ± 0.0005
HNTs 10%	0.0245 ± 0.0009	0.0343 ± 0.0003

HNTs: Halloysite clay nanotubes; TiO_2: Titania

[34]. However, with the addition of 10% HNT to PURF, thermal conductivity value of PURNCF increased substantially 10 days after their fabrication. Furthermore, most of PURNCFs consisting of TiO_2 with up to 8% concentration exhibited almost the same thermal conductivity values 3 and 10 days after their preparation. In addition, it was also pointed out that thermal conductivity value of PURNCF containing 10% titanium dioxide nanoparticles increased in 3 and 10 days after their production. The reason behind this finding was explained such that the presence of 10% nanoparticles increases the viscosity, and thus this property leads to break apart of many cells and also decreases the well-dispersion of nanoparticles due to their aggregation behavior. Thus, because of this mechanism, PURNCFs consisting of higher contents of titanium dioxide nanoparticles (about 10%) end up having higher thermal conductivity values [34].

The synthesis of PURNCFs filled with disilanolisobutyl polyhedral oligomeric silsesquioxane nanoparticles was formerly reported [36]. During synthesis, authors chemically incorporated disilanolisobutyl POSS nanoparticles with two hydroxyl groups directly on PU backbone to produce hybrid POSS-based PURNCFs. Authors examined effects of using POSS nanoparticles on closed cellular morphology such as amount of closed cells, specific geometry of cells, foam density and thermal conductivity and compressive mechanical properties of PURNCFs. According to results, it was shown that population of cells diminished and average area of cells increased by incorporating 1.5 and 2.0 wt% disilanolisobutyl POSS compared to pure PURF. On the contrary, it was also shown that incorporation of 0.5 wt% disilanolisobutyl POSS to PURFs led to elevation in amount of cells and reduction in size of cells compared to pure PURF [36].

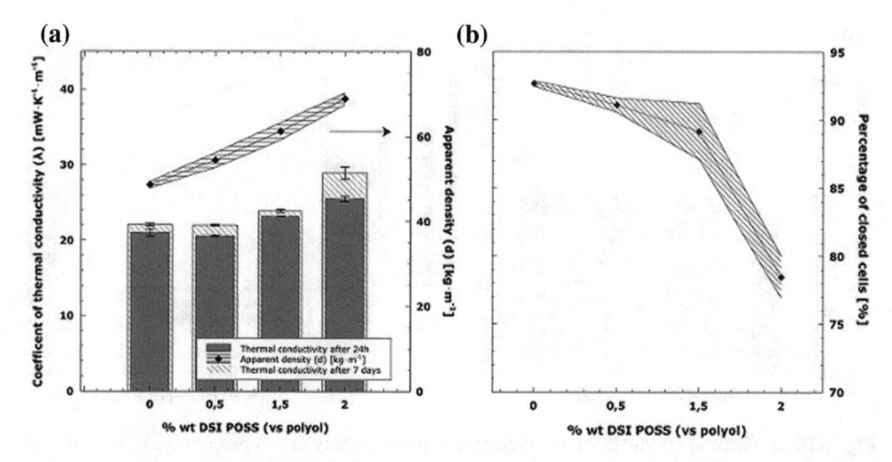

Fig. 5.25 **a** Thermal conductivity coefficient and foam density and **b** percentage of closed cells in pure PURF and PURNCFs containing different amounts of disilanolisobutyl POSS nanoparticles. Reproduced with permission from Ref. [36]. Copyright 2018 John Wiley & Sons, Ltd.

Figure 5.25 shows (a) thermal conductivity coefficient and foam density and (b) percentage of closed cells in pure PURF and PURNCFs containing different amounts of disilanolisobutyl POSS nanoparticles. Based on the results, it was pointed out that thermal conductivity coefficient values increased with increasing amount of disilanolisobutyl POSS nanoparticles. In addition, it was reported that PURNCF consisting of 2.0 wt% disilanolisobutyl POSS nanoparticles exhibited the largest increase in thermal conductivity coefficient since this foam consists of mainly open cells within its cellular structure [36]. In addition, based on apparent density results, it was also shown that the density of PURNCFs increased with increasing amount of disilanolisobutyl POSS nanoparticles. The increase of density with increasing amount of POSS particles was explained such that during the reactive foaming process, the cell growth was slowed down by higher values of viscosity as a result of increasing the gelling reaction speed, and thus this leads to bubble collapse and formation of foams with higher density [36].

Recently, modified POSS nanoparticles as suspended groups and POSS nanoparticles as chemical crosslinks were used in the fabrication of PURNCFs [24]. Thermal conductivity and water absorption behaviours, closed-cellular morphology and mechanical properties of PU rigid nanocomposite foams consisting of 0–15 wt% POSS were characterized thoroughly. Based on FT-IR results, it was shown that there is a good amount of interaction between POSS nanoparticles and PURNCF matrix. It was also revealed that cell size elevated and cell density diminished with incorporation of chemically crosslinked POSS nanoparticles to pure PURF. However, on the other hand, it was reported that the addition of suspended POSS nanoparticles decreased cell sizes and elevated cell density compared to those of pure PURF [24].

Figure 5.26 shows (a) thermal conductivity coefficient and foam density and (b) percentage of closed cells in pure PURF and PURNCFs containing different amounts

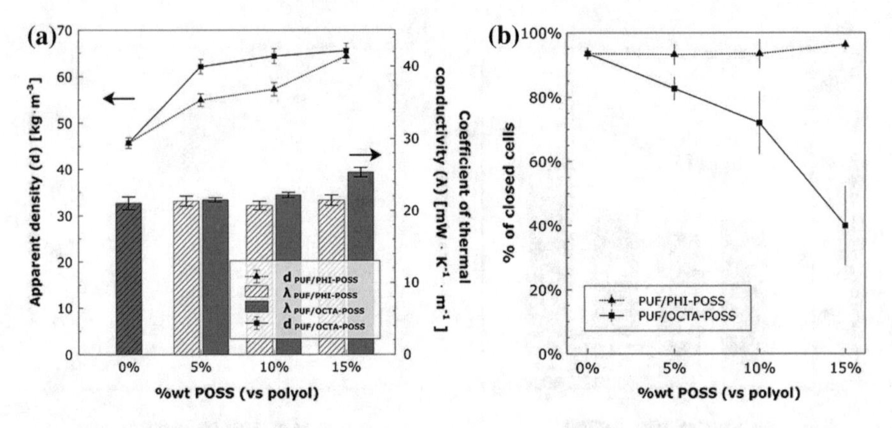

Fig. 5.26 **a** Thermal conductivity coefficient and foam density and **b** percentage of closed cells in pure PURF and PURNCFs containing different amounts of POSS nanoparticles as chemical crosslinks or suspended POSS nanoparticles. Reproduced with permission from Ref. [24]. Copyright 2015 John Wiley & Sons, Ltd.

of chemically crosslinked or suspended POSS nanoparticles. Based on the results, it was revealed that thermal conductivity coefficient elevated with incorporation of 15 wt% POSS nanoparticles as chemical crosslinks, while it did not change with the addition of same amount of suspended POSS nanoparticles compared to pure PURF [24]. The reason behind this observation was explained such that the close content of cells was moderately increased in the presence of highest amount of suspended POSS nanoparticles while the addition of POSS nanoparticles as chemical crosslinks substantially reduced the ratio of closed cells in PURNCFs. In addition, it was also revealed that incomplete crystallization of POSS nanoparticles as chemical crosslinks within PURNCF matrix and formation of these POSS crystals disturbed the closed-cellular morphology by opening up some of the closed cells and by this way diminished closed cell amount within PURNCF matrix [24].

5.5 Thermal Degradation and Flammability

Previously, a surface alteration method of fumed nanosilica was performed by exploiting *N*-(2-aminoethyl)-3-aminopropyltrimethoxysilane molecule as a surface refining instrument, and these surface adjusted nanosilica particles were incorporated into PURFs [7]. Authors used several characterization techniques such as FT-IR, TGA, SEM, tensile testing, DMA and TMA to explore physical, thermal, mechanical and thermomechanical properties of PURNCFs consisting of modified nanosilica. In accordance with experimental results, it was found out that increasing amount of modified nanosilica in PURNCFs paved the way for enhancing both thermal stability and mechanical properties [7].

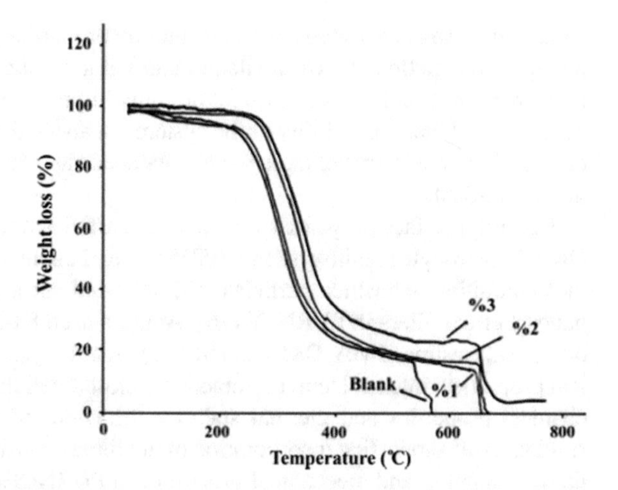

Fig. 5.27 TGA data of pure PURF and PURNCFs consisting of modified nanosilica with different compositions. Reproduced with permission from Ref. [7]. Copyright 2009 Society of Plastics Engineers

Figure 5.27 shows TGA data of pure PURF and PURNCFs containing different amounts of modified nanosilica particles. Based on the results, it was shown that increasing amount of modified nanosilica particles enhanced thermal stability of PURNCFs since modified nanosilica particles function as thermal insulators within the PURNCF matrix [7]. In addition, it was reported that weight loss associated with pure nanosilica particles consisting of no surface modification was found to be 4% at 700 °C. The reason for this finding was stated such that this weight loss corresponds to adsorbed water molecules on the surface of pure nanosilica. In contrast, the weight loss associated with pure modified nanosilica particles was found as 19% in the plateau region of TGA curve. Thus, based on these results, the organic content on the modified nanosilica was calculated as 4% [7].

As shown in Fig. 5.27, the thermal stability increased linearly with the content of modified nanosilica. This result also shows that the thermal stability was not affected negatively with the addition of higher modified nanosilica contents since the surface functional groups of modified nanosilica particles prevented any kind of aggregation in the system. Thus, this type of mechanism in PURNCFs consisting of higher amounts of modified nanoparticles enhanced thermal degradation temperatures in PURNCFs compared to those of pure PURF. The well-dispersion of modified nanosilica leads to the reduction of cell size and increase in cell density, and these factors also contributed to the positive thermal stability improvement of PURNCFs since favorable interactions between the polymer chains and modified nanosilica particles occur during the reactive foaming process.

Also, using nanosilica particles that are surface modified with amino silanes as nanospherical fillers in PURFs provides the basis for enhanced thermal stability and thermal decomposition temperatures compared to those of nanosilica that are surface modified with other types of silanes [5–7, 28]. Furthermore, using both silane modified spherical nanosilica particles and chemically compatible flame retardant agents in PURNCFs is also another alternative method in terms of using these types of surface modified spherical nanofillers in the fabrication of PURNCFs with extra enhanced thermal decomposition and stability properties [30]. The establishment of favorable interactions between silane modified nanosilica particles and compatible flame retardants in PURNCFs might not only improve thermal decomposition temperatures and thermal stability of the system but also reduce flammability properties of PURNCFs consisting of these hybrid silanized nanosilica particles and compatible flame retardants.

Recently, surface properties of nanosilica particles were altered by using gamma-Glycidoxypropyltrimethoxysilane (GDS) and diethanol amine, successively and these modified nanosilica particles with adjusted surface properties were used as nanospherical fillers in PURNCFs [6]. Authors used FT-IR, thermal analysis methods, tensile testing, TMA, DMA and SEM in order to systematically analyze mechanical properties, intermolecular interactions, closed-cellular morphology, thermomechanical properties and thermal stability behaviour of the system. According to results, it was shown that incorporation of modified nanosilica enhanced the values of thermal stability and mechanical properties in PURNCFs. Thus, thermal decomposition temperatures and thermal stability of PURNCF system consisting of modified nanosilica are greatly dependent on the surface properties of nanoparticles that are used as nanofillers in the system.

Figure 5.28 displays TGA data of pure PURF and PURNCFs containing modified nanosilica particles with different contents. In accordance with experimental data, it was shown that inclusion of two times surface modified nanosilica particles into PURFs enhanced thermal stability values of PURNCFs [6]. The reason behind thermal stability improvement was explained such that the heat transport to PURNCF matrix was reduced due to higher specific thermal capacity values of silica nanoparticles. As a consequence of this mechanism, the mass loss of PURNCFs consisting of modified nanosilica particles decreased noticeably. In addition, it was reported that the highest thermal stability was obtained in PURNCF containing 3% modified nanosilica by increasing the GDS concentration [6]. Furthermore, based on TGA results of pure nanosilica and two times surface modified nanosilica particles, the

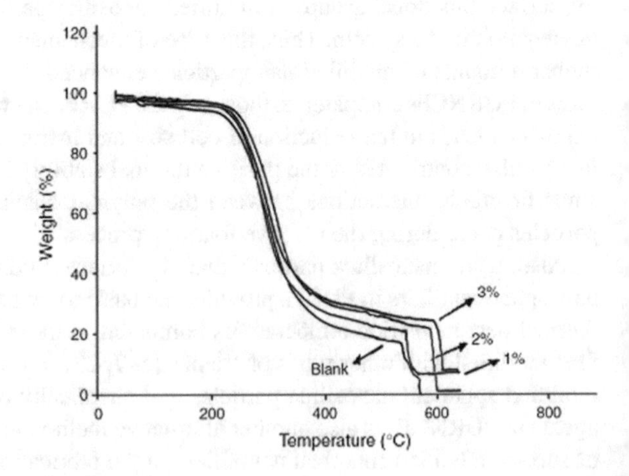

Fig. 5.28 Thermograms of pure PURF and PURNCFs consisting of modified nanosilica with different contents. Reproduced with permission from Ref. [6]. Copyright The Author(s), 2010

weight loss at 600 °C associated with pure nanosilica particles was reported as 4%. This result was associated with adsorption of water molecules on the surface of pure nanosilica particles. In addition, it was also pointed out that weight loss associated with modified nanosilica particles in the plateau region of TGA data was observed as 22% [6].

Formerly, 3-Aminopropyltriethoxysilane (APTS) was used as a surface modifying molecule and this modified nanosilica was used as a nanofiller in PURNCFs [5]. Furthermore, the effects of using APTS on morphology, mechanical, thermomechanical, physical and thermal decomposition properties of PURNCFs were compared with those of PURNCFs filled with nanosilica particles which were surface modified with *n*-(2-aminoethyl)-3-aminopropyltrimethoxysilane (AEAP). Authors used characterization tools such as FT-IR, TGA, SEM, DMA, tensile testing and TMA in order to comprehend physical, thermal, mechanical and thermomechanical properties of PURNCFs consisting of modified nanosilica particles. In accordance with TGA experimental results, it was pointed out that PURNCFs consisting of modified nanosilica exhibited enhanced thermal degradation temperatures up to optimum modified nanosilica content. Moreover, it was also revealed that further addition of modified nanosilica particles greater than the optimum content diminished the values of thermal degradation temperatures and thermal stability in PURNCFs [5].

Figure 5.29 presents TGA data of pure PURF and PURNCFs consisting of modified nanosilica with different compositions. Based on the results, it was shown that 4% weight loss at 700 °C in pure nanosilica particles consisting of no surface modification was associated with adsorbed water molecules on the surface of pure nanosilica particles. In addition, it was revealed that concentration of organic functional groups on modified nanosilica surface was reported as 14% according to TGA results [5]. Also, based on TGA results of PURNCFs consisting of modified nanosilica, it was pointed out that thermal stability increases with inclusion of

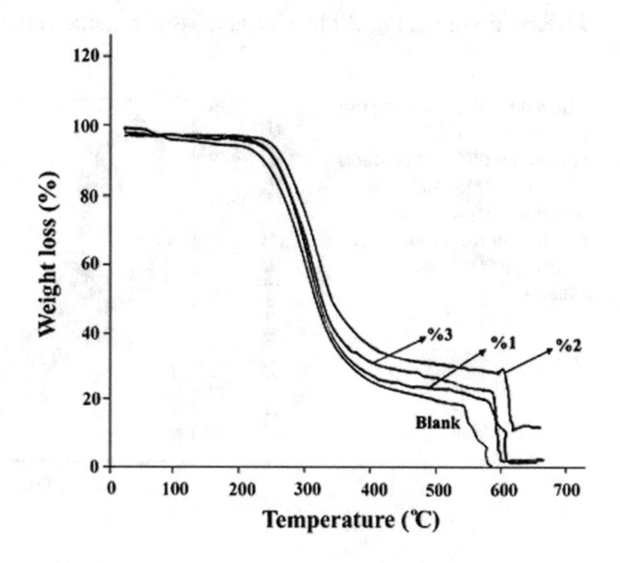

Fig. 5.29 TGA curves of pure PURF and PURNCFs consisting of modified nanosilica with different compositions. Reproduced with permission from Ref. [5]. Copyright Koninklijke Brill NV, Leiden, 2011

2% functionalized nanosilica. However, on the other hand, it was also shown that the presence of 3% modified nanosilica in PURNCFs clearly decreases the thermal stability. The reason behind this observation was explained such that PURF system could not be completely crosslinked due to unfavorable interactions between surface functional groups of modified nanosilica particles and polymer matrix, thus reducing cross-linking density in PURNCF system containing 3% modified nanosilica [5].

Earlier on, halogen-free flame-retardant PURNCFs were fabricated by incorporating nanohybrid particles of graphene oxide and nanospherical silica and a flame retardant (dimethyl methylphosphonate) in various contents [30]. Authors investigated mechanical, flame retardant and thermal properties of PURNCFs containing various contents of dimethyl methylphosphonate and nanosilica/graphene oxide nanohybrid particle. Based on the results, it was shown that flame retardant and mechanical properties of PURNCFs extensively improved by incorporating both flame retardant and nanohybrid particle compared to pure PURF and PURFs containing only dimethyl methylphosphonate [30].

Figure 5.30 shows heat release rate (HRR) data of pure PURF and PURNCFs consisting of nanoparticles with different contents. Based on the results, it was shown that heat release rate data of PURF consisting of 20 phr flame retardant, PURNCF consisting of 2 phr GO and 18 phr flame retardant and PURNCF filled with 2 phr nanohybrid particle and 18 phr flame retardant diminished in comparison with those of pure PURF. Specifically, it was reported that PHRR and mHRR values of PURNCF filled with 18 phr DMMP and 2 phr nanohybrid particle reduced by 44.6 and 45.4%, respectively in comparison with those of pure PURF [30].

As depicted in Fig. 5.30, the second HRR peak which has much smaller intensity compared to the first HRR peak was suppressed with the addition of nanoparticles and flame retardant to PURFs. Specifically, the second HRR peak of PURNCF consisting of 2 phr nanohybrid particle and 18 phr flame retardant was substantially suppressed in comparison with those of PURF consisting of 20 phr flame retardant, PURNCF containing 2 phr GO and 18 phr flame retardant and also pure PURF.

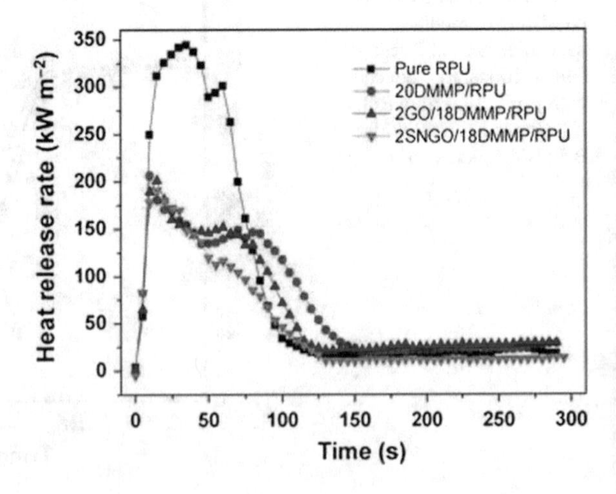

Fig. 5.30 HRR data of pure PURF and PURNCFs containing different amounts of nanoparticles and flame retardant. Reproduced with permission from Ref. [30]. Copyright 2018 Taylor & Francis

Based on these results, the reason behind reduction of PHRR and mHRR values of PURNCFs with incorporation of both flame retardant and nanohybrid particle was discussed such that flame retardant produced PO_2 free radicals which caused deactivation of these free radicals during the burning event of PURNCF matrix [30, 42]. In addition, the presence of nanosilica particles on the surface of nanohybrid particle prevented thermal decomposition of GO sheets and also strengthened the physical barrier property of PURNCF against transport of heat and volatile gases and hence improved flame retardant properties of PURCNF system [30].

These results clearly showed that using a nanohybrid particle which combines positive properties of two different types of nanoparticles within its structure besides the presence of flame retardant enhances thermal stability and flammability properties of PURCNFs. In polymer nanocomposites, using hybrid nanoparticles instead only one type of nanoparticle could be used as an efficient method to improve thermal stability and thermal decomposition temperatures. However, in this regard, choosing different nanoparticles which are physically and chemically compatible with each other is the most important criterion since uncompatible nanoparticles could not form the required synergy during synthesis procedure of hybrid nanoparticles. As a consequence of this, if this kind of nanohybrid particle which does not have compatible nanoparticle components is used as a nanofiller in polymer nanocomposites, then thermal stability and thermal decomposition temperatures could not be enhanced compared to the case of using only one type of nanoparticle in polymeric materials. Thus, for this reason, different types of nanoparticle components which can be used for the synthesis of hybrid nanoparticles should be chosen carefully such that they meet the above criterion for the fabrication of advanced polymer nanocomposites with enhanced thermal stability values.

Formerly, surface properties of nanosilica particles were adjusted by utilizing a specific surface modification technique in which an original dipodal silane consisting of 3-aminopropyltriethoxysilane and *gamma*-glycidoxypropyltrimethoxysilane was used as the surface modifier. Authors then used this surface modified nanosilica as a nanospherical filler in PURNCFs [28]. Basically, the polar surface functional groups such as hydroxyl and amine groups on the modified nanosilica were responsible for the well-dispersion of nanosilica particles within the PURNCF matrix. Authors used characterization tools such as FT-IR, TGA, SEM, proton nuclear magnetic resonance spectroscopy (1H-NMR), DMA, tensile testing and TMA in order to evaluate types of intermolecular interactions, closed-cellular morphology, physical, thermal, mechanical and thermomechanical properties of PURNCFs consisting of modified nanosilica particles. In accordance with results, it was pointed out that thermal properties such as thermal stability and thermal decomposition temperatures of PURNCFs consisting of modified nanosilica particles enhanced with incorporation of nanosilica particles [28].

Figure 5.31 shows TGA data of pure PURF and PURNCFs containing different amounts of modified nanosilica. Based on the results, it was shown that weight loss at 600 °C associated with pure nanosilica particles consisting of no surface modification was found out as 4% [28]. This amount of weight loss in pure nanosilica particles without any surface modification was associated with adsorbed water molecules onto

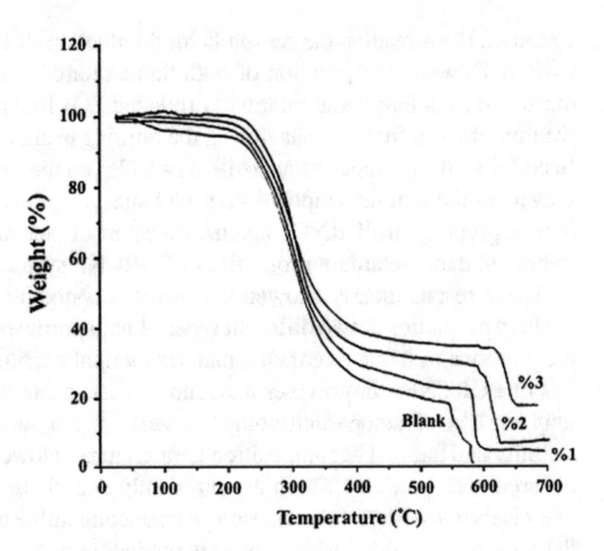

Fig. 5.31 TGA data of pure PURF and PURNCFs containing different amounts of modified nanosilica particles. Reproduced with permission from Ref. [28]. Copyright Koninklijke Brill NV, Leiden, 2010

nanosilica particles. In addition, it was also revealed that concentration of organic functional groups on modified nanosilica particles was reported as 14% [28]. In contrast, based on TGA results of PURNCFs consisting of modified nanosilica particles, it was shown that the thermal stability was improved with higher amounts of modified nanosilica incorporation due to thermal insulation property of nanosilica particles [28].

As shown in Fig. 5.31, the thermal stability increased linearly with the content of modified nanosilica. This result also shows that thermal stability was not affected negatively with incorporation of higher amounts of modified nanosilica since surface functional groups of modified nanosilica particles prevented any kind of aggregation in the system, thus enhanced thermal stability and thermal degradation temperatures in PURNCFs compared to those of pure PURF. The well-dispersion of modified nanosilica particles leads to the reduction of cell size and increase in cell density, and these important factors also contribute to positive thermal stability enhancement of PURNCFs since favorable interactions between polymer chains and modified nanosilica particles occur during reactive foaming process.

Previously, superparamagnetic Fe_3O_4/silica nanoparticles were incorporated into PURFs in order to prepare new magnetic PURNCFs by using a one-step synthesis procedure [32]. The process of producing nanoparticles on substrates from a colloidal suspension or gel state was used to synthesize nanoparticles consisting of both core and shell layers, and the synthesized nanoparticles were characterized by using TEM, XRD, FT-IR in order to understand structure and morphology of nanoparticles. Moreover, authors prepared PURNCFs consisting of maximum 3% of magnetic nanoparticles, and they also used various characterization tools such as FT-IR, DMA and TGA on PURNCFs in order to systematically evaluate closed-cellular morphology and intermolecular interactions, thermomechanical, and thermal decomposition properties of nanocomposite foams. In accordance with TGA results, it was revealed

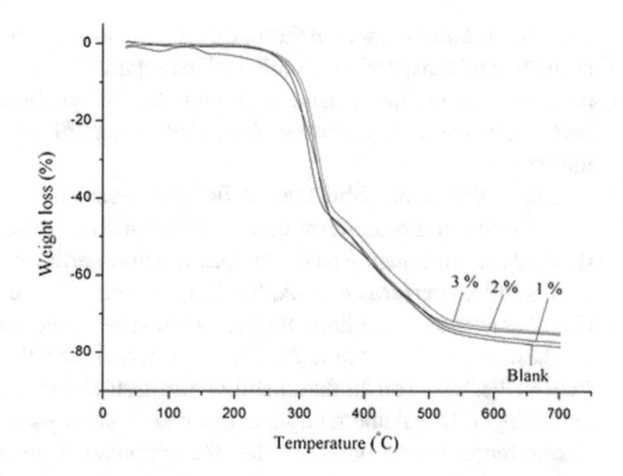

Fig. 5.32 TGA data of pure PURF and nano Fe_3O_4/SiO_2-filled PURNCFs consisting of different compositions. Reproduced with permission from Ref. [32]. Copyright Springer-Verlag 2012

that thermal stability and thermal decomposition temperatures of PURNCFs were improved compared to those of pure PURF [32].

Figure 5.32 shows the TGA curves of pure PURF and nano Fe_3O_4/SiO_2-filled PURNCFs consisting of different compositions. In accordance with experimental data, it was revealed that thermal stability and remaining parts of carbon in PURNCFs were enhanced with higher amounts of Fe_3O_4/SiO_2 incorporation due to the high thermal capacity of Fe_3O_4/SiO_2 core-shell nanoparticles [32]. Moreover, the reason behind increase of thermal stability was explained such that covalent bonding interactions are formed between polyol component of PURNCF system and hydroxyl functional groups of nanosilica coating. In addition, it was also revealed that the char formation in PURNCFs consisting of Fe_3O_4/SiO_2 nanoparticles was moderately improved in comparison with pure PURF [32].

Using more than one type of spherical nanoparticle with different properties in PURNCFs definitely improves thermal stability and thermal decomposition temperatures much more than using only one type of spherical nanoparticle in PURNCFs. Furthermore, introducing compatible surface functional groups to these nanoparticles enhances intermolecular interactions between surface functional groups of these binary nanoparticles mixtures. Moreover, besides using one type of spherical nanofiller or surface modified spherical nanofiller in PURNCFs, binary or ternary spherical nanoparticle mixtures consisting of at least two types of nanoparticles with different properties and geometries can be used in order to fabricate PURNCFs with more superior properties. Alternatively, using core-shell nanoparticles of Fe_3O_4 and SiO_2 in PURNCFs resulted in more enhancement of thermal stability and char formation due to favorable interactions among nanosilica and Fe_3O_4 particles and PU chains [32]. Basically, covalent bonding interactions between hydroxyl functional groups on silica nanoparticles and polyol component of PURNCFs are responsible for the enhancement of thermal stability and thermal decomposition temperatures. Thus, consequently, cell sizes of foams reduced and their cell density values increased with the incorporation of core-shell nanoparticles of Fe_3O_4 and SiO_2 into PURFs. As a result, the enhancement in thermal decomposition properties was observed because of the fact that dispersion of core-shell nanoparticles in PURNCFs was facilitated by the covalent bonding of nanosilica particles and positive synergy was provided by favourable interactions between core-shell nanoparticles and PU chains in PURNCF matrix [32].

Salasinska et al. fabricated PURNCFs consisting of flame retardants having no halogens in the presence of nanofillers such as multi-walled carbon nanotubes (MWCNTs) and nano-sized titanium dioxide particles by using one-step method [43]. Authors performed TGA, limiting oxygen index, cone calorimeter and smoke chamber tests and gas chromatography with mass spectrometry experiments in order to learn more about flame retardant properties and thermal stability behavior of PURNCFs. Based on the flammability results, it was shown that using both nanofillers and halogen-free flame retardants resulted in synergistic effects in PURNCFs [43]. Furthermore, it was observed that the combustion process was obstructed by the carbonized layer whose formation was facilitated positively in the presence of nanofillers. Moreover, it was revealed that the amount of thermal degradation products was reduced with the incorporation of nanofillers [43].

Figure 5.33 shows effects of using nanoparticles and flame retardants consisting of no halogens on the heat release performance of PURNCFs. Based on the heat release curves, it was shown that maximum rate of heat release diminished in most of PURFs. However, it was also reported that the addition of small amount of nanoparticles as a partial exchange for small content of flame retardants led to reduction in the intensity of heat release due to the formation of positive synergistic interactions between nanofillers and flame retardants containing no halogens [43]. Previously, the presence of synergistic effects between nanofillers and flame retardants containing no halogens was reported in previously published works [43–46]. In addition, it was also revealed in Fig. 5.33 that the lowest pHRR values were obtained for PURNCF consisting of 15% zinc borate and 0.5% MWCNTs and PURNCF consisting of 15%

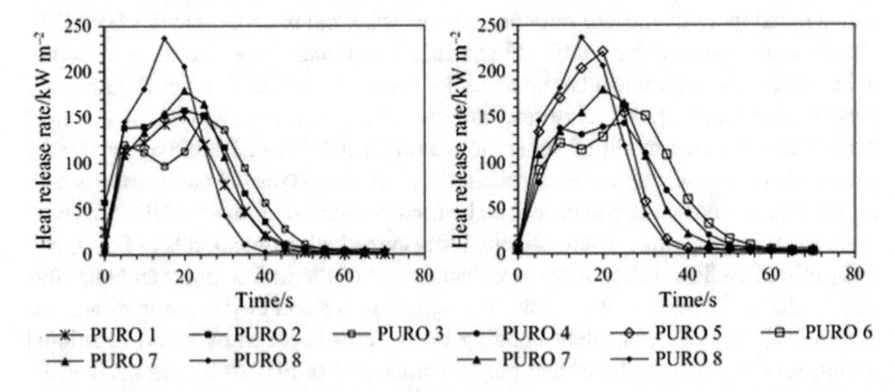

Fig. 5.33 Heat release rate data of pure PURF and PURNCFs containing different amounts of flame retardants without any halogens and nanofillers such as MWCNTs and nano-sized titanium dioxide particles. Reproduced with permission from Ref. [43]. Copyright The Author(s) 2017

flame retardant and 3% nano-sized titanium dioxide and their pHRR values were reduced by 37 and 30%, respectively compared to pure PURF [43].

Previously, effects of using various types of nanoparticles such as titanium dioxide nanoparticles with spherical geometries, nanoclay particles with plate-like geometries and carbon nanofibers (CNFs) with rod-like shapes on closed-cellular morphology, thermal and mechanical properties and thermal stability of PURFs were explored in depth [35]. Authors used the well-known sonication method for homogenously distribution of nanoparticles in PMDI component of PURF which was then immediately mixed with the polyol component by using a mechanical stirrer. In accordance with experimental data, it was revealed that mechanical and thermal properties of PURNCFs consisting of only 1 wt% nanoparticles exhibited significant enhancement compared to those of pure PURF. In addition, it was also reported that maximum improvement in terms of thermal stability and thermal decomposition temperatures was found for PURNCFs consisting of CNFs whereas TiO_2 filled PURNCFs displayed the minimum performance [35].

Figure 5.34 shows the TGA curves of (a) pure PURF, and PURNCFs consisting of (b) TiO_2, (c) nanoclay and (d) CNF. Based on the results, it was shown that the decomposition profiles in all the samples occurs via two step degradation process. In addition, it was reported that that the decomposition temperatures at 50% mass loss for PURNCFs consisting of nanoparticles are higher than pure PURF. More specifically, it was pointed out that PURNCFs consisting of CNFs exhibited the maximum elevation of 18 °C in terms of decomposition temperatures at 50% mass

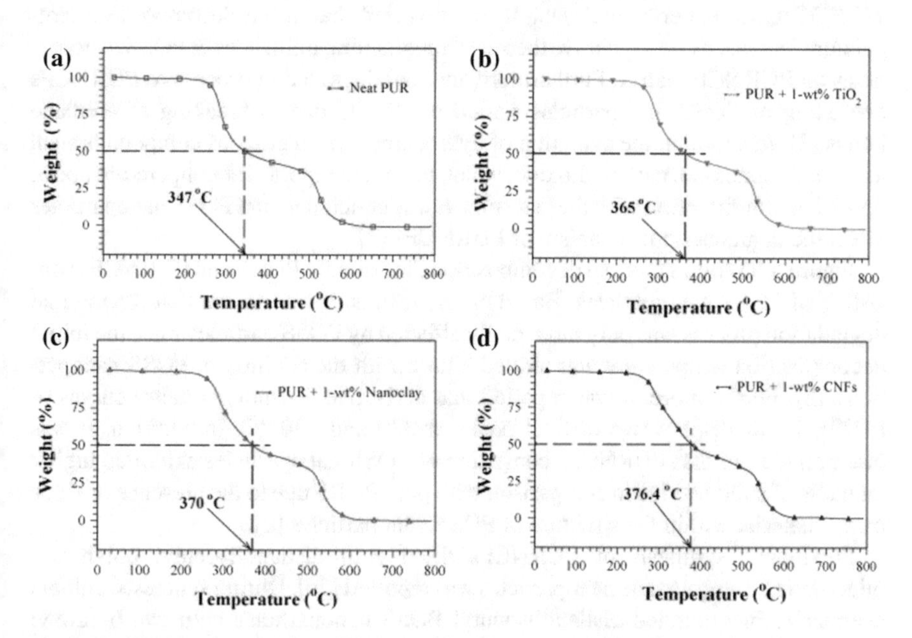

Fig. 5.34 TGA curves of **a** pure PURF, and PURNCFs consisting of **b** TiO_2, **c** nanoclay and **d** CNF. Reproduced with permission from Ref. [35]. Copyright 2007 Elsevier B.V.

loss while PURNCFs containing TiO_2 nanoparticles displayed minimum elevation of about 7 °C in comparison with pure PURF [35].

Thus, in accordance with experimental data, it is well noted that inclusion of small amounts (about 1%) of nanoparticles into PURFs definitely increases the thermal stability and thermal degradation temperatures due to the reinforcing ability and uniform distribution of nanoparticles throughout PURNCF matrix. Enhancement of thermal stability with incorporation of 1% nanoparticles occurs basically as a result of better and uniform distribution of rigid nanoparticles within polymer matrix compared to the case of using higher concentrations of nanoparticles as nano-sized fillers in PURNCFs. The well-dispersion of nanoparticles at smaller concentrations leads to higher degrees of favorable interactions between polymer chains and surface of nanoparticles due to higher surface areas of nanoparticles consisting of very limited amount of aggregation. Thus, nanoparticles with smaller concentrations, limited amount of aggregation behavior and uniformly dispersion state within polymer matrix can be able to function as thermal blockades against transport of heat and evaporative gases more efficiently during the combustion process of PURNCFs.

Earlier on, Pagacz et al. explored thermal degradation behaviour of PURNCFs consisting of two functionalized POSS nanoparticles by using FTIR spectroscopy attached thermogravimetric (TG) analysis which was also combined with quadrupole mass spectroscopy (QMS) [2]. Authors basically used POSS nanoparticles as suspended groups to the main PU chain and also other type of POSS nanoparticles as chemical crosslinks because of their eight hydroxyl side chains. Based on TG/FTIR/QMS experimental data, it was revealed that initial decomposition temperature increased due to positive thermal strengthening influences of POSS nanoparticles on PURNCF matrix. Furthermore, thermal degradation process of PURNCFs consisting of POSS nanoparticles started at 150 °C due to breaking of urethane bonds [2]. In addition, the evolution of hydrocarbons, and gaseous compounds such as carbon monoxide, carbon dioxide, water, etc. were reported at temperature above 350 °C. In confirmation with these results, it was concluded that POSS nanoparticles affect the degradation mechanism of PURNCFs [2].

Figure 5.35 shows thermal decomposition data of pure PURF and PURNCFs consisting of POSS nanoparticles. Based on the results, it was shown that the thermal degradation process was only moderately affected by POSS particles since the initial decomposition temperature was shifted a little with the addition of POSS nanoparticles [2]. Furthermore, it was reported that derivative thermogravimetry curves of PURFs in air display two distinct peaks at 320 and 530 °C. In addition, it was also pointed out that PURNCFs consisting of POSS nanoparticles exhibited higher amounts of solid residue in comparison with pure PURF due to the presence of silica based material within the structure of POSS nanoparticles [2].

Previously, synthesis of PURNCFs filled with disilanolisobutyl polyhedral oligomeric silsesquioxane nanoparticles was reported [36]. During synthesis, authors chemically incorporated disilanolisobutyl POSS nanoparticles with two hydroxyl groups directly on PU backbone to produce hybrid POSS-based PURNCFs. Authors examined effects of using POSS nanoparticles on closed cellular morphology such as

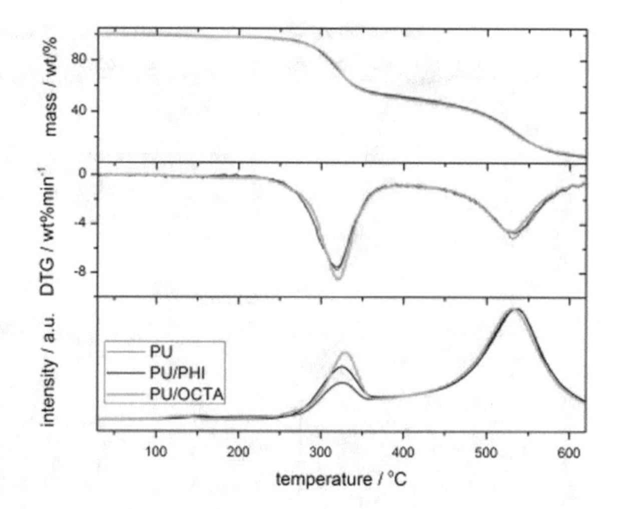

Fig. 5.35 Thermal degradation profiles of pure PURF and PURNCFs consisting of POSS nanoparticles. Reproduced with permission from Ref. [2]. Copyright 2016 Elsevier B.V.

amount of closed cells, specific geometry of cells, foam density and thermal conductivity and compressive mechanical properties of PURNCFs. According to results, it was shown that population of cells diminished and average area of cells increased by incorporating 1.5 and 2.0 wt% disilanolisobutyl POSS compared to pure PURF. On the contrary, it was also shown that incorporation of 0.5 wt% disilanolisobutyl POSS to PURFs led to elevation in amount of cells and reduction in size of cells compared to pure PURF. In accordance with TGA data, the degradation process of PURFs did not change that much with inclusion of POSS nanoparticles into PURFs [36].

Figure 5.36 displays TGA and derivative thermogravimetry (DTG) profiles of pure PURF and PURNCFs consisting of disilanolisobutyl POSS nanoparticles in (a, c) argon and (b, d) air environments. In accordance with experimental results, it was shown that thermal degradation mechanism of PURFs is composed of a two-stage process in which the intensity of the second stage in air is much higher compared to that of argon environment [2, 36, 47]. In addition, as observed from Fig. 5.36, thermal degradation behaviour does not change that much with the addition of POSS nanoparticles to PURFs. Furthermore, it was reported that in argon environment, the solid residue that was left after thermal degradation was found to be moderately higher in the PURNCF sample consisting of 2.0 wt% disilanolisobutyl POSS nanoparticles in comparison with pure PURF. The reason behind this finding was explained such that thermally stable siliceous residue in PURNCFs acts as thermal barrier layers against the diffusion of heat and volatile gases during the combustion process. However, higher amounts of char residues were not observed in TGA profiles of PURNCFs consisting of POSS nanoparticles in the air environment [36].

In accordance with previously published works [2, 5–7, 28, 36], performing surface modification of nanosilica or POSS nanoparticles is a specific and useful method for the creation of compatible nanospherical surfaces with PU matrix, and also this method was shown to be very effective concerning the improvement of thermal

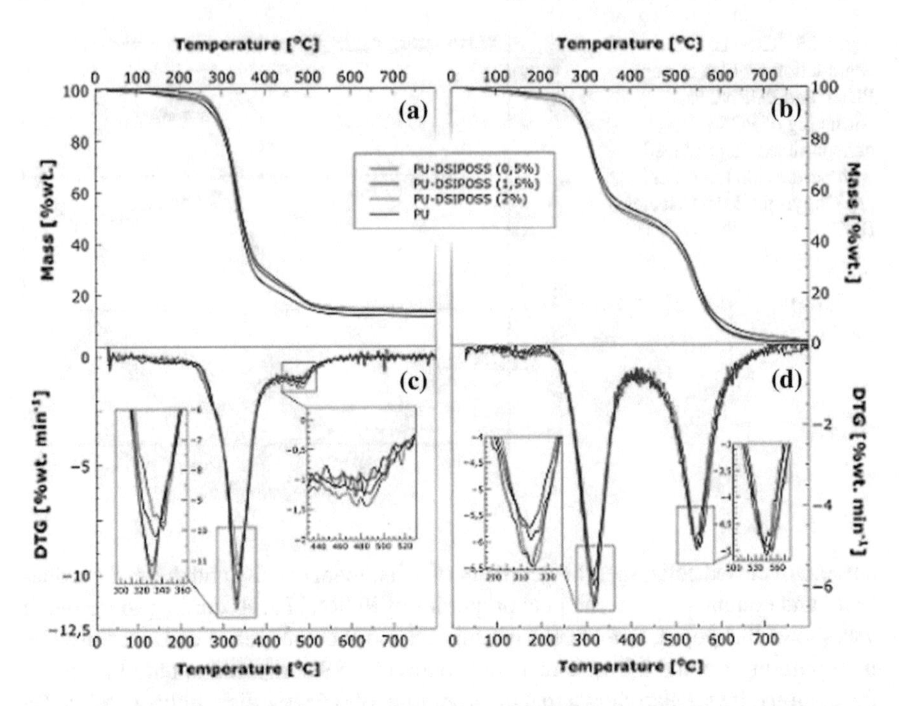

Fig. 5.36 TGA and DTG profiles of pure PURF and PURNCFs consisting of disilanolisobutyl POSS nanoparticles in different gas environments. Reproduced with permission from Ref. [36]. Copyright 2018 John Wiley & Sons, Ltd.

stability and thermal decomposition temperatures in PURNCFs. More precisely, if surface functional groups on POSS or silica nanoparticles have amino or silanol groups within their chemical structures, then using these nanofillers that are surface modified with these specific functional groups as nanospherical fillers in PURFs provides the basis for enhanced thermal stability and thermal decomposition temperatures compared to those of silica or POSS nanoparticles that are surface modified with other types of surface modifying agents [2, 5–7, 28, 36]. On the other hand, thermal stability and thermal decomposition behaviour of PURNCFs were improved with the addition of nanohybrid particles of nanosilica/graphene oxide [30] and superparamagnetic Fe_3O_4/silica nanoparticles [32]. Thus, using a hybrid nanoparticle instead of one type of nanoparticle in PURNCFs was shown to be a very efficient method in which synergetic effects of combining diverse characteristics of two different types of nanoparticles paved the way for increasing degradation rates [30, 32]. In addition, PURNCF containing nanosilica particles modified with amino silane molecules displayed enhanced thermal stability since amine groups of silane molecules on modified nanosilica particles could form more powerful and increased number of intermolecular interactions with urethane groups [5–7, 28]. Thus, in analogous with these results, using hydrophilic nanosilica particles with higher number of silanol functional groups in the presence of other types of modified nanoparticles with optimum contents in PURFs enhanced thermal stability and thermal decomposition temperatures since favorable Hydrogen bonding interactions could take place among silanols of nanosilica particles, urethane linkages of PU molecules and surface functionalities of other types of modified nanoparticles [48]. All of these experimental data clearly show that adjusting surface properties of spherical nanoparticles and also incorporation of second type of additive such as nanoparticles or flame retardants are very important in the production of PURNCFs with elevated thermal stability and thermal decomposition temperatures.

References

1. Akkoyun, M., Suvaci, E.: Effects of TiO_2, ZnO, and Fe_3O_4 nanofillers on rheological behavior, microstructure, and reaction kinetics of rigid polyurethane foams. J. Appl. Polym. Sci. **133**(28) (2016)
2. Pagacz, J., Hebda, E., Michalowski, S., Ozirnek, J., Sternik, D., Pielichowski, K.: Polyurethane foams chemically reinforced with POSS-thermal degradation studies. Thermochim. Acta **642**, 95–104 (2016)
3. Javni, I., Zhang, W., Karajkov, V., Petrovic, Z.S., Divjakovic, V.: Effect of nano- and micro-silica fillers on polyurethane foam properties. J. Cell. Plast. **38**(3), 229–239 (2002)
4. Santiago-Calvo, M., Tirado-Mediavilla, J., Ruiz-Herrero, J.L., Rodriguez-Perez, M.A., Villafane, F.: The effects of functional nanofillers on the reaction kinetics, microstructure, thermal and mechanical properties of water blown rigid polyurethane foams. Polymer **150**, 138–149 (2018)
5. Nikje, M.M.A., Tehrani, Z.M.: The effects of functionality of the organifier on the physical properties of polyurethane rigid foam/organified nanosilica. Des. Monomers Polym. **14**(3), 263–272 (2011)
6. Nikje, M.M.A., Tehrani, Z.M.: Polyurethane rigid foams reinforced by doubly modified nanosilica. J. Cell. Plast. **46**(2), 159–172 (2010)
7. Nikje, M.M.A., Tehrani, Z.M.: Thermal and mechanical properties of polyurethane rigid foam/modified nanosilica composite. Polym. Eng. Sci. **50**(3), 468–473 (2010)
8. Liu, W.S., Kong, J.H., Toh, W.E., Zhou, R., Ding, G.Q., Huang, S., Dong, Y.L., Lu, X.H.: Toughening of epoxies by covalently anchoring triazole-functionalized stacked-cup carbon nanofibers. Compos. Sci. Technol. **85**, 1–9 (2013)
9. Ding, J., Maitra, P., Wunder, S.L.: Characterization of the interaction of poly(ethylene oxide) with nanosize fumed silica: surface effects on crystallization. J. Polym. Sci., Part B: Polym. Phys. **41**(17), 1978–1993 (2003)
10. Kweon, J.O., Noh, S.T.: Thermal, thermomechanical, and electrochemical characterization of the organic-inorganic hybrids poly(ethylene oxide) (PEO)-silica and PEO-silica-$LiClO_4$. J. Appl. Polym. Sci. **81**(10), 2471–2479 (2001)
11. Sun, Y.Y., Zhang, Z.Q., Moon, K.S., Wong, C.P.: Glass transition and relaxation behavior of epoxy nanocomposites. J. Polym. Sci., Part B: Polym. Phys. **42**(21), 3849–3858 (2004)
12. Burgaz, E.: Thermomechanical analysis of polymer nanocomposites. In: Huang, X., Zhi, C. (eds.) Polymer Nanocomposites. Springer, Cham (2016)
13. Raghavan, S.R., Walls, H.J., Khan, S.A.: Rheology of silica dispersions in organic liquids: new evidence for solvation forces dictated by hydrogen bonding. Langmuir **16**(21), 7920–7930 (2000)
14. Barthel, H., Rosch, L., Weis, J.: In: Auner, N., Weis, J. (eds.) Organosilicon Chemistry II: From Molecules to Materials, pp 761–777. VCH Publishers, Weinheim (1996)
15. Khan, S.A., Zoeller, N.J.: Dynamic rheological behavior of flocculated fumed silica suspensions. J. Rheol. **37**(6), 1225–1235 (1993)

16. Cassagnau, P.: Melt rheology of organoclay and fumed silica nanocomposites. Polymer **49**(9), 2183–2196 (2008)
17. Wu, J., Mather, P.T.: POSS polymers: physical properties and biomaterials applications. Polym. Rev. **49**(1), 25–63 (2009)
18. Matejka, L., Amici Kroutilova, I., Lichtenhan, J.D., Haddad, T.S.: Structure ordering and reinforcement in POSS containing hybrids. Eur. Polym. J. **52**, 117–126 (2014)
19. Logakis, E., Pandis, C., Pissis, P., Pionteck, J., Potschke, P.: Highly conducting poly(methyl methacrylate)/carbon nanotubes composites: investigation on their thermal, dynamic-mechanical, electrical and dielectric properties. Compos. Sci. Technol. **71**(6), 854–862 (2011)
20. Wu, Y., Li, L.G., Feng, S.Y., Liu, H.Z.: Hybrid nanocomposites based on novolac resin and octa(phenethyl) polyhedral oligomeric silsesquioxanes (POSS): miscibility, specific interactions and thermomechanical properties. Polym. Bull. **70**(12), 3261–3277 (2013)
21. Blanco, I., Abate, L., Bottino, F.A., Bottino, P., Chiacchio, M.A.: Thermal degradation of differently substituted cyclopentyl polyhedral oligomeric silsesquioxane (CP-POSS) nanoparticles. J. Therm. Anal. Calorim. **107**(3), 1083–1091 (2012)
22. Lee, A., Lichtenhan, J.D.: Viscoelastic responses of polyhedral oligosilsesquioxane reinforced epoxy systems. Macromolecules **31**(15), 4970–4974 (1998)
23. Fina, A., Monticelli, O., Camino, G.: POSS-based hybrids by melt/reactive blending. J. Mater. Chem. **20**(42), 9297–9305 (2010)
24. Hebda, E., Ozimek, J., Raftopoulos, K.N., Michalowski, S., Pielichowski, J., Jancia, M., Pielichowski, K.: Synthesis and morphology of rigid polyurethane foams with POSS as pendant groups or chemical crosslinks. Polym. Adv. Technol. **26**(8), 932–940 (2015)
25. Ray, S.S., Bousmina, M.: Biodegradable polymers and their layered silicate nano composites: in greening the 21st century materials world. Prog. Mater Sci. **50**(8), 962–1079 (2005)
26. Cao, X., Lee, L.J., Widya, T., Macosko, C.: Polyurethane/clay nanocomposites foams: processing, structure and properties. Polymer **46**(3), 775–783 (2005)
27. Han, X.M., Zeng, C.C., Lee, L.J., Koelling, K.W., Tomasko, D.L.: Extrusion of polystyrene nanocomposite foams with supercritical CO_2. Polym. Eng. Sci. **43**(6), 1261–1275 (2003)
28. Nikje, M.M.A., Tehrani, Z.M.: Novel modified nanosilica-based on synthesized dipodal silane and ıts effects on the physical properties of rigid polyurethane foams. Des. Monomers Polym. **13**(3), 249–260 (2010)
29. Modesti, M., Lorenzetti, A., Besco, S.: Influence of nanofillers on thermal insulating properties of polyurethane nanocomposites foams. Polym. Eng. Sci. **47**(9), 1351–1358 (2007)
30. Wang, Z.Z., Li, X.Y.: Mechanical properties and flame retardancy of rigid polyurethane foams containing SiO_2 nanospheres/graphene oxide hybrid and dimethyl methylphosphonate. Polym. Plast. Technol. **57**(9), 884–892 (2018)
31. Bernal, M.M., Pardo-Alonso, S., Solorzano, E., Lopez-Manchado, M.A., Verdejo, R., Rodriguez-Perez, M.A.: Effect of carbon nanofillers on flexible polyurethane foaming from a chemical and physical perspective. RSC Adv. **4**(40), 20761–20768 (2014)
32. Nikje, M.M.A., Nejad, M.A.F., Shabani, K., Haghshenas, M.: Preparation of magnetic polyurethane rigid foam nanocomposites. Colloid Polym. Sci. **291**(4), 903–909 (2013)
33. Nazeran, N., Moghaddas, J.: Synthesis and characterization of silica aerogel reinforced rigid polyurethane foam for thermal insulation application. J. Non-Cryst. Solids **461**, 1–11 (2017)
34. Lorusso, C., Vergaro, V., Conciauro, F., Ciccarella, G., Congedo, P.M.: Thermal and mechanical performance of rigid polyurethane foam added with commercial nanoparticles. Nanomater. Nanotechnol. **7**, 1–9 (2017)
35. Saha, M.C., Kabir, M.E., Jeelani, S.: Enhancement in thermal and mechanical properties of polyurethane foam infused with nanoparticles. Mater. Sci. Eng. A: Struct. **479**(1–2), 213–222 (2008)
36. Hebda, E., Bukowczan, A., Ozimek, J., Raftopoulos, K.N., Wronski, S., Tarasiuk, J., Pielichowski, J., Leszczynska, A., Pielichowski, K.: Rigid polyurethane foams reinforced with disilanolisobutyl POSS: synthesis and properties. Polym. Adv. Technol. **29**(7), 1879–1888 (2018)

37. Seo, W.J., Jung, H.C., Hyun, J.C., Kim, W.N., Lee, Y.B., Choe, K.H., Kim, S.B.: Mechanical, morphological, and thermal properties of rigid polyurethane foams blown by distilled water. J. Appl. Polym. Sci. **90**(1), 12–21 (2003)
38. Saha, M.C., Kabir, M.E., Jeelani, S.: Effect of nanoparticles on mode-I fracture toughness of polyurethane foams. Polym. Compos. **30**(8), 1058–1064 (2009)
39. Estravis, S., Tirado-Mediavilla, J., Santiago-Calvo, M., Ruiz-Herrero, J.L., Villafane, F., Rodriguez-Perez, M.A.: Rigid polyurethane foams with infused nanoclays: relationship between cellular structure and thermal conductivity. Eur. Polym. J. **80**, 1–15 (2016)
40. Ferry, J.D.: Viscoelastic Properties of Polymers, 3rd edn. Wiley, New York (1980)
41. Macosko, C.W.: Rheology: Principles, Measurements, and Applications (Advances in Interfacial Engineering). Wiley, New York (1993)
42. Feng, F.F., Qian, L.J.: The flame retardant behaviors and synergistic effect of expandable graphite and dimethyl methylphosphonate in rigid polyurethane foams. Polym. Compos. **35**(2), 301–309 (2014)
43. Salasinska, K., Borucka, M., Leszczynska, M., Zatorski, W., Celinski, M., Gajek, A., Ryszkowska, J.: Analysis of flammability and smoke emission of rigid polyurethane foams modified with nanoparticles and halogen-free fire retardants. J. Therm. Anal. Calorim. **130**(1), 131–141 (2017)
44. Wang, X., Hu, Y., Song, L., Yang, H., Yu, B., Kandola, B., Deli, D.: Comparative study on the synergistic effect of POSS and graphene with melamine phosphate on the flame retardance of poly(butylene succinate). Termochim. Acta **543**, 56–164 (2012)
45. Yang, F., Nelson, G.L.: Combination effect of nanoparticles with flame retardants on the flammability of nanocomposites. Polym. Degrad. Stab. **96**(3), 270–276 (2011)
46. Isitman, N.A., Kaynak, C.: Nanoclay and carbon nanotubes as potential synergists of an organophosphorus flame-retardant in poly(methyl methacrylate). Polym. Degrad. Stab. **95**(9), 1523–1532 (2010)
47. Chattopadhyay, D.K., Webster, D.C.: Thermal stability and flame retardancy of polyurethanes. Prog. Polym. Sci. **34**(10), 1068–1133 (2009)
48. Burgaz, E., Kendirlioglu, C.: Thermomechanical behavior and thermal stability of polyurethane rigid nanocomposite foams containing binary nanoparticle mixtures. Polym. Test. **77**, 105930 (2019). https://doi.org/10.1016/j.polymertesting.2019.105930